River Biota

DIVERSITY AND DYNAMICS

SELECTED EXTRACTS FROM

THE RIVERS HANDBOOK

EDITED BY

GEOFFREY PETTS
BSc, PhD
Environmental Research and Management
The University of Birmingham

AND

PETER CALOW
DSc, PhD, CBiol, FIBiol
Department of Animal and Plant Sciences
The University of Sheffield

b
Blackwell
Science

Editorial Offices:
Osney Mead, Oxford OX2 0EL
25 John Street, London WC1N 2BL
23 Ainslie Place, Edinburgh EH3 6AJ
238 Main Street, Cambridge
Massachusetts 02142, USA
54 University Street, Carlton
Victoria 3053, Australia

Other Editorial Offices:
Arnette Blackwell SA
224, Boulevard Saint Germain
75007 Paris, France

Blackwell Wissenschafts-Verlag GmbH
Kurfürstendamm 57
10707 Berlin, Germany

Zehetnergasse 6
A-1140 Wien
Austria

First published 1996

Set by Setrite Typesetters, Hong Kong
Printed and bound in Great Britain
at the Alden Press Ltd,
Oxford and Northampton

DISTRIBUTORS

Marston Book Services Ltd
PO Box 269
Abingdon
Oxon OX14 4YN
(*Orders*: Tel: 01235 465500
Fax: 01235 465555)

USA
Blackwell Science, Inc.
238 Main Street
Cambridge, MA 02142
(*Orders*: Tel: 800 215-1000
617 876-7000
Fax: 617 492-5263)

Canada
Copp Clark, Ltd
2775 Matheson Blvd East
Mississauga, Ontario
Canada, L4W 4P7
(*Orders*: Tel: 800 263-4374
905 238-6074)

Australia
Blackwell Science Pty Ltd
54 University Street
Carlton, Victoria 3053
(*Orders*: Tel: 03 9347 0300
Fax: 03 9349 3016)

A catalogue record for this title
is available from the British Library

ISBN 0-86542-716-X

Library of Congress
Cataloging-in-Publication Data

Rivers handbook. Selections.
River biota: selected extracts
from the Rivers handbook/edited by
Geoffrey Petts and Peter Calow.
p. cm.
Includes bibliographical references
and index.
ISBN 0-86542-716-X
1. Stream animals. 2. Stream plants.
3. Stream ecology.
I. Petts, Geoffrey E. II. Calow, Peter.
III. Title.
QL145.R582 1996
574.5′26323—dc20 96-6929
CIP

Contents

List of Contributors

P. D. ARMITAGE *Institute of Freshwater Ecology, The River Laboratory, East Stoke, Wareham, Dorset BH20 6BB, UK*

P. B. BAYLEY *Department of Fisheries and Wildlife, Oregon State University, Nash Hall 104, Corvallis, Oregon 97331-3803, USA*

P. CALOW *Department of Animal and Plant Sciences, University of Sheffield, PO Box 601, Sheffield S10 2UQ, UK*

K. W. CUMMINS *Pymatuning Laboratory of Ecology and Department of Biological Sciences, University of Pittsburgh, PA 15260, USA*

A. M. FOX *Center for Aquatic Plants, University of Florida, FL 32606, USA*

J. A. GORE *Center for Environmental Research and Service, The Troy State University System, Troy, Alabama 36082, USA*

A. G. HILDREW *School of Biological Sciences, Queen Mary and Westfield College, Mile End Road, University of London, London E1 4NS, UK*

H. W. LI *Department of Fisheries and Wildlife, Oregon State University, Nash Hall 104, Corvallis, Oregon 97331-3803, USA*

E. P. McELRAVY *Department of Entomological Sciences, University of California, Berkeley, CA 94720, USA*

L. MALTBY *Department of Animal and Plant Sciences, University of Sheffield, PO Box 601, Sheffield S10 2UQ, UK*

R. H. NORRIS *Water Research Centre, Canberra University of Canberra, PO Box 1, Belconnen ACT, Australia 2616*

G. E. PETTS *School of Geography, University of Birmingham, Edgbaston, Birmingham B15 2TT, UK*

V. H. RESH *Department of Entomological Sciences, University of California, Berkeley, CA 94720, USA*

C. S. REYNOLDS *Institute of Freshwater Ecology, Windermere Laboratory, Cumbria, LA22 0LP, UK*

A. K. WARD *Department of Biology, University of Alabama Tuscaloosa, AL 35487-0344, USA*

R. G. WETZEL *Department of Biology, University of Alabama, Tuscaloosa, AL 35487-0344, USA*

Preface

If the extent to which books are 'borrowed' from libraries is a measure of their success, then personal experience with the extent to which *The Rivers Handbook* volumes have gone missing from our own shelves has to be encouraging! Another factor is, of course, price; and we have been very much aware that the *Handbook* in its original form was not affordable for most students. So following some encouragement we have decided to produce less expensive versions. In doing this we could have simply published softback editions of the originals. But, again following encouragement, we opted instead to take a selection of chapters from the *Handbooks* and reorganize these into three groupings intended to be especially helpful for supporting course work (for undergraduate and postgraduate programmes) in river ecology. The result has been the development of three books: *River Biota*; *River Flows and Channel Forms* and *River Restoration* of which this is one.

Each book opens with a completely new chapter, presenting general principles and indicating how the rest of the book is structured around these. All the other chapters are taken, with some updating, from the parent *Handbooks*. Our hope, therefore, is that the repackaging brings benefits in terms of both availability and convenience for a broader readership and we would welcome feedback on this. As ever, we are grateful to authors of individual chapters for their co-operation in compiling the reorganized versions and especially to the Publishers for their encouragement and support in this venture.

Geoffrey E. Petts
Peter Calow

1: Introduction

P. CALOW AND G. E. PETTS

1.1 INTRODUCTION

This book is about river biota. The word *river* is used broadly to mean all freshwater systems, of whatever size, with unidirectional flow. We do this as a matter of convenience, not because we do not recognize differences in ecology between small fast-flowing streams and large slow-flowing rivers, but because in the real world it is often difficult to recognise sharp distinctions between the two, and the transitions may not even be very regular. Our aim is rather to describe the kinds of organisms that live throughout these extended ecosystems, and to discuss some of the key ecological processes in which they take part.

River ecosystems are important in a number of ways: they provide corridors through the landscape and interact with the catchments that surround them; they provide aesthetic enjoyment and recreational opportunities; they act as important navigational routes for people and goods; they provide an important source of potable water and sometimes food, and; they provide convenient sinks for some of our human wastes. What distinguishes them from other aquatic systems is that they flow *in one direction* and it is this one feature that makes the ecological systems they contain distinctive in both structure and process.

As with all ecosystems, river systems involve a complex interaction of biota with their physical and chemical environment. The biota represent a rich diversity of microorganisms, plants and animals. All are influenced in how they behave and/or where they live within the system and/or in their own morphology, to a greater or lesser extent by the unidirectional currents to which they are exposed. Very generally, we shall see that these influence mobility and orientation behaviour of animals, microhabitat selection (often to reduce exposure), and favour streamlined morphologies. High flow rates, that often occur erratically after heavy rainfall, can lead to natural disturbances that wash biota away and that not only introduces unpredictability into community structures and processes but also favours the evolution of recolonization abilities.

Similarly, ecosystem processes, involving energy fluxes and cycles of matter, are importantly influenced by these flow conditions. Particular stretches of rivers are open to both energy and matter. Thus processes operating within a particular stretch are influenced by the input of biomass produced upstream of it, and also to the side as runoff and fall-in from the catchment; and in turn produce biomass that is exported into downstream stretches and, with flooding, into the surrounding catchment. The flow conditions have important consequences for nutrient dynamics; for the cycles of carbon, nitrogen and phosphorus, etc., that are typical of most ecosystems, are extended into continuous spirals in river systems.

In the following chapters we deal first of all with the taxa, before describing some of the important processes in which they are involved. The Chapters on taxa (2–6) are organized in a trophic hierarchy addressing primary producers and microbial decomposers first and then invertebrate consumers and fishes later. This leads naturally to an explicit consideration of trophic interactions and food webs (Chapter 7), and to the processes that drive these interactions at the base of the food chains, namely decomposition (Chapter 8) and primary production (Chapter 9). Our understanding of community composition and ecosystem processes depend importantly on an ability to obtain reliable information through sampling programmes and this is dealt with in Chapter 10. Finally, because we make considerable use of rivers, they are subject to much human disturbance in both their physical and chemical conditions. How the biota are influenced by these disturbances, and in turn can be used to provide information on them, is addressed in the two final chapters (11 and 12) that deal respectively with hydrological and water quality issues.

The next two sections of this Chapter are intended to provide general orientations to these other Chapters by highlighting a few key aspects of the hydrology and process ecology of rivers. For more detailed information on the hydrology, the reader is referred to the companion volume *River Flows and Channel Forms* (Petts & Calow 1996a). Another companion volume *River Restoration* (Petts & Calow 1996b), addresses issues concerned with the disturbance of river systems by human activities, its management and remediation and provides a detailed background to Chapters 11 and 12. Finally, for a more detailed treatment of river ecology and management the reader should refer to the parent volumes *The Rivers Handbooks* (Calow & Petts 1992 & 1994).

1.2 PHYSICAL BASIS OF RIVER ECOLOGY

Section 1.1 introduced the point that the all-important feature of river ecosystems for the biota they contain and the ecosystem processes that occur within them is that they flow in one direction, by gravity, from source to sea. A detailed treatment of these physical properties and their ecological consequences is given in Petts & Calow (1996a). It is sufficient here to note that this unidirectional flow leads to a longitudinal gradient in channel slope (from steep to flat), hydraulic characteristics (especially from shallow to deep), river morphology (in general from thin and shallow to broad and deep) and sediment size (from coarse to fine) from source to sink (Fig. 1.1). These determine local conditions which in turn influence local community structures and ecosystem processes.

Also of importance, though, are local interactions between the flowing water and the solid surfaces that constitute bed and banks. These generate a wide variety of habitats reflecting the planform and cross-sectional variation of hydraulic conditions. Three general velocity zones can be recognized: riffles (shallows), pools (deeps) and channel margins. Under high-flow conditions velocity through riffles and pools may be similar; but under lower flow conditions relatively high velocities are maintained in the downstream face of riffles as compared with pools, sustaining favourable diffusion gradients in nutrients and gases for biota. Low-flow velocities in pools and lateral margins may support different biota from riffles, not only because of lower shear stresses but also because of less favourable diffusion conditions. Again the implications of these local conditions for the distribution and abundance of biota will be obvious in all the chapters on all biota but are especially emphasized in the chapter on algae (Chapter 2).

1.3 ECOSYSTEM PROCESSES

A basic diagram of energy flow through a general ecosystem is given in Fig. 1.2 and this is illustrated for the River Thames in Fig. 1.3. Energy can flow into ecosystems from sunlight through photosynthesis, but whereas this must be the sole input for the biosphere as a whole, particular ecosystems might derive energy from other sources as already elaborated organic material. River systems are open to the input of matter and hence energy from upstream and the surrounds as either dissolved or particulate material. Thus, for most ecosystems the predominant source of energy income is from primary production and in rivers the sources of this are algae (Chapter 2) and macrophytes (Chapter 3).

The present understanding of the relationship between the supply of energy from primary production and the demand from herbivores in rivers is that the former is several times greater than the latter, and this is particularly the case for macrophytes (Chapter 3). Hence, most of this primary production is yielded to the detritus. Moreover, with some variation between and within systems (as discussed below), biomass produced outside the system, and entering as particulate or dissolved organic material, is the major source of energy income. Microbial decomposers thus play

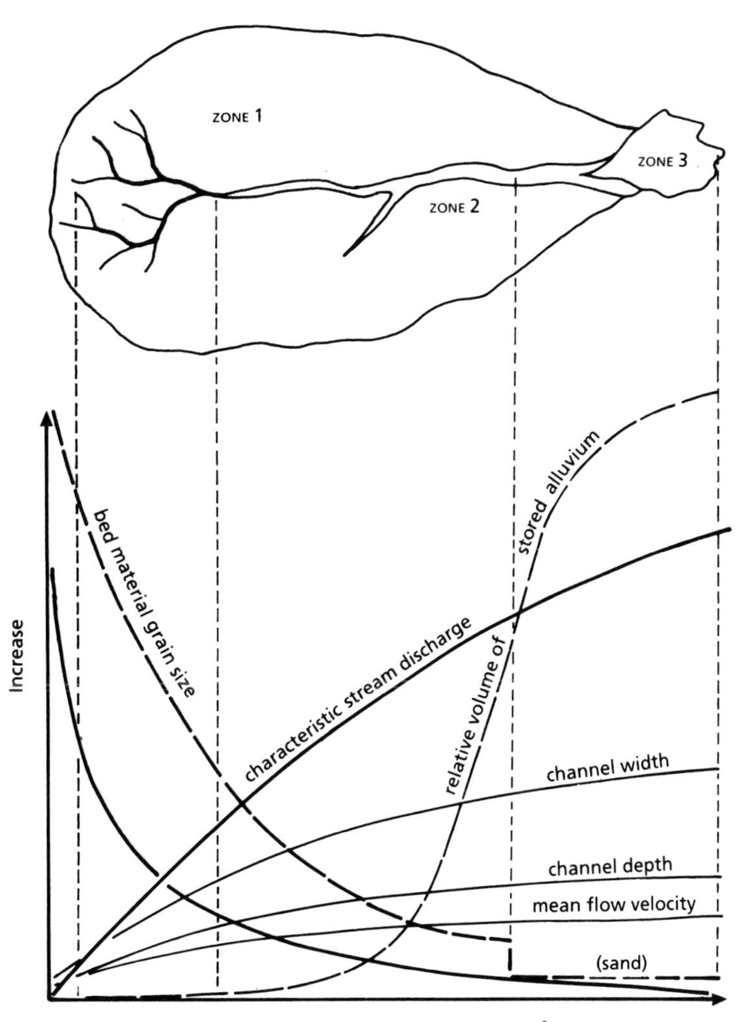

Fig. 1.1 Schematic representation of the variation in channel properties through a drainage basin (based on a concept of Schumm 1977).

a key role in the yield of this dead organic material to consumers, either simply by conditioning the detritus or by converting it into microbial biomass that is more easily utilized by the metazoan consumers.

The incorporation of dissolved organic matter, from the water column via a microbial/protozoan loop, is discussed in Chapter 8. Such loops appear to be of considerable importance in the economy of marine and 'still' freshwater systems and involve water-column microbes. Likely to be of more importance in the utilization of dissolved organic matter in flowing waters are the 'microbial slimes' that adhere to all submerged surfaces.

Whether the action of microbial decomposers is treated as primary or secondary production in lotic systems has been the subject of some debate (Wiegert & Owen 1971), but this is probably largely semantic.

A most important feature of lotic systems, is that they change in systematic fashion from source to sink. Because of the likelihood of shading from surrounding vegetation in upland stretches, heterotrophic processes based on decomposition rather than primary production tend to dominate in upper reaches; primary production becomes

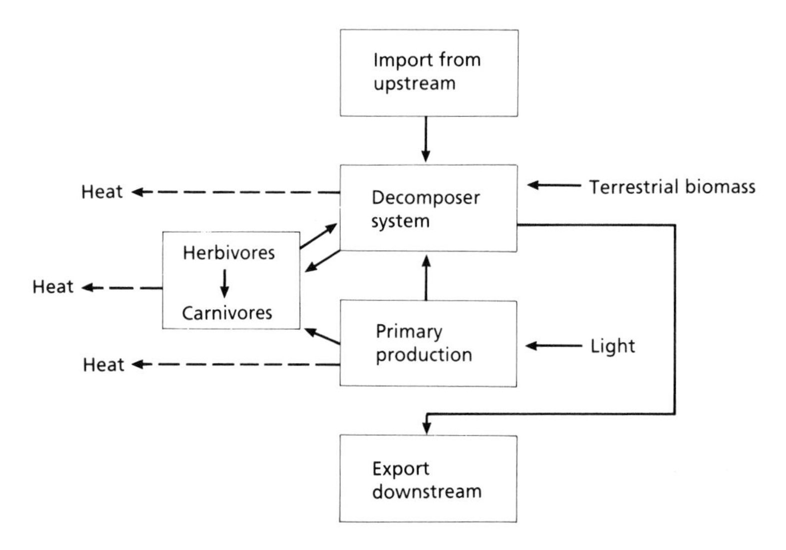

Fig. 1.2 Energy flow through a stretch of river (very general representation).

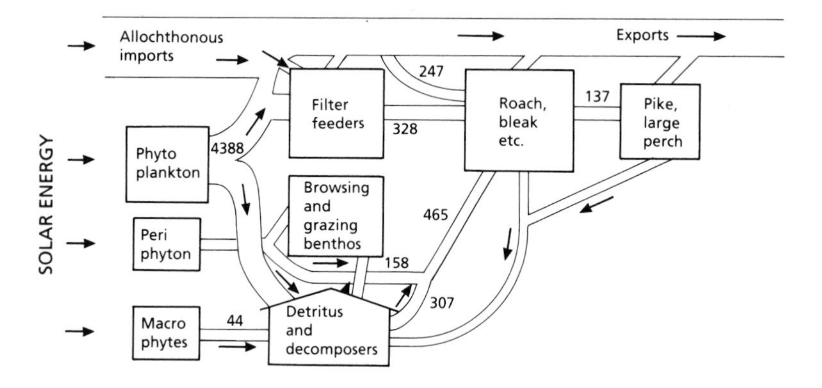

Fig. 1.3 Energy flow through the River Thames, UK (units are kcal m^{-2} $year^{-1}$) (after Mann 1980).

more important in middle reaches; because of sediment loads, causing shading, heterotrophic processes are likely to become dominant again in lower courses.

These shifts in major sources of energy income along the lengths of rivers, together with the way the energetics of the ecosystems in one stretch are influenced by the energetics of upstream stretches (by import of production and washed out detritus) and in turn influence the energetics of downstream stretches (by export), will lead to a longitudinal continuum of ecosystem processes. It is arguable that there are predictable changes to these processes, along the continuum and that these drive trophodynamics which in turn determine trophic structure (Chapter 5). This is the basis of the river continuum concept.

That there is a continuum of energetic processes in systems that are longitudinally extended is almost beyond doubt — apart from isolated and semi-isolated pockets that might persist in these continua (e.g. dead-zones see Chapter 2). But how general and predictable are the continua in both functional and structural terms through space and time?

A first general comment is that though function determines structure (e.g. food webs and community structure) it would appear that, ecologically, many variants of structure are compatible with a particular type of functioning dependent upon other conditions (Odum 1962). What this means is that the longitudinal variation in processes need not be related very soundly to predictable changes in biota.

Second, these general longitudinal patterns within systems can be disturbed by pollution, land management (e.g. forestation or deforestation), and also potentially by local discontinuities in natural features such as geology and gradients. Between systems, changes in the latter and also water chemistry are likely to cause variation in the longitudinal patterns in function and structure (Winterbourn 1986).

Finally, an obvious feature of flowing-water systems is their temporal variability — driven by variability in physical conditions. These, together with physical variability along the length of a river, are likely to lead to a patchiness in susceptibility to scouring and washing away of communities and hence to a patchiness, at any one time, in the development of community and ecosystem characteristics in particular places (Statzner & Higler 1986; Townsend 1989). The developmental stage in recolonization (succession) is known to have an effect on energetic function and community structure (Williams & Hynes 1976). The extent to which this patchiness, dependent on temporal variability, dominates the longitudinal pattern may vary from one system to another.

1.4 STRUCTURE OF THE BOOK

In summary, then, this book considers river biota in four interrelated ways.

1 Chapters 2–6 describe taxa, largely grouped trophically, and describe major adaptations and features of their ecologies.

2 Chapters 7–9 describe the bases of ecosystem processes in terms of the pathways of trophic interactions and the rules governing them, and the key production processes related to both photosynthesis *and* decomposition.

3 Chapter 10 considers the principles, practices and problems associated with making reliable observations on river organisms. These are crucial in our description and understanding of river biota and the part they play in ecosystem processes.

4 Chapters 11–12 turn to the applied issues of how river biota are impacted by human activities and how in turn they can be used as indicators of these effects in river management programmes.

REFERENCES

Calow P, Petts GE. (1992) *The Rivers Handbook*, vol 1. Blackwell Scientific Publications, Oxford. [1.1]

Calow P, Petts GE. (1994) *The Rivers Handbook*, vol 2. Blackwell Scientific Publications, Oxford. [1.1]

Mann KH. (1980) The total aquatic system. In: Barnes RSK, Mann KH (eds) *Fundamentals of Aquatic Ecosystems*, pp 185–220. Blackwell Scientific Publications, Oxford. [1.3]

Odum EP. (1962) Relationships between structure and function in the ecosystem. *Japanese Journal of Ecology* **12**: 108–18. [1.3]

Petts GE, Calow P. (1996a) *River Flows and Channel Forms*. Blackwell Science, Oxford. [1.1, 1.2]

Petts GE, Callow P. (1996b) *River Restoration*. Blackwell Science, Oxford. [1.1]

Schumm SA. (1977) *The Fluvial System*. John Wiley and Sons, New York. [1.2]

Statzner B, Higler B. (1986) Stream hydraulics as a major determinant of invertebrate zonation patterns. *Freshwater Biology* **16**: 127–37. [1.3]

Townsend CR. (1989) The patch dynamics of stream community ecology. *Journal of the North American Benthological Society* **8**: 36–50. [1.3]

Wiegert RG, Owen DF. (1971) Trophic structure, available resources and population density in terrestrial vs. aquatic ecosystems. *Journal of Theoretical Biology* **30**: 69–81. [1.3]

Williams DD, Hynes BN. (1976) The recolonisation mechanisms of stream benthos. *Oikos* **27**: 265–77. [1.3]

Winterbourn MT. (1986) Recent advances in our understanding of stream ecosystems. In: Polunin N (ed) *Ecosystem Theory and Applications*, pp 240–68. John Wiley & Sons, London.

2: Algae

C.S.REYNOLDS

2.1 INTRODUCTION

Although they often comprise a conspicuous component of fluvial plant life and are sometimes the dominant primary producers, algae scarcely have an easy time in rivers. Most of the problems relate to the supposedly persistent and unidirectional passage of water (Fig. 2.1). It is not simply a matter of susceptibility to irreversible removal and downstream transport, for variability in the flow may lead alternatively to movement or suspension of substratum material, with a concomitant increase in turbidity and deprivation of light, to siltation and burial by disentrained sediment and, perhaps, to desiccation. Even when the physical environment is apparently less exacting, such biotic factors as comparative rates of arrival, growth or selective removal (as food) by animals have profound effects upon the composition, abundance and temporal change in the algal assemblages present and, potentially, to the functional organization of the fluvial ecosystem.

The purpose of this chapter is to explore these selective mechanisms in greater detail and to make some general deductions about their impact upon the structure and maintenance of the algal assemblages of flowing waters.

Notwithstanding the recent rapid growth of ideas about ecosystem function in running waters (Boudou & Ribeyre 1989) or the long history of diligent phycological investigations of the algal species constituting the fluvial communities (e.g. Round 1981), the factors that regulate their spatial and temporal distributions remain incompletely understood. Neither is there yet a generally recognized paradigm to accommodate past findings or one that will identify priorities for future study. There are good reasons why this should be the case (see Blum 1960; Whitton 1975); not least of these are the complexities of the interactions and the logistical difficulties of sampling rivers adequately, either in space or in time. So, while it is timely to propose some sort of synthesis, it is necessary to emphasize that the opinions expressed are not unequivocally established and some may well prove to be controversial.

The chapter does not set out to provide a comprehensive list of algal species in rivers, neither is it an intention necessarily to review the key studies in detail. Several major works (Blum 1956; Hynes 1970; Whitton 1975; Round 1981) already fulfil these roles and should be consulted in preference. Here, the approach is to identify the interactions and limitations governing algal production in the principal habitats of the river, only incidentally introducing the important algal genera and species, and then to emphasize those features impinging upon the contribution of algae to fluvial ecosystems.

2.2 HABITATS

In much the same way as it is possible to distinguish among the fragments of fluvial environments—open channel flow, the containing substratum, weedy backwaters etc.—and the inherent variability attaching to each, so there are clearly delimited microhabitats in rivers, each supporting its own distinctive association of biota. A recurrent theme of this chapter is that, because rivers are able to maintain a spatially—and temporally—diverse array of microhabitats, they collectively offer an almost infinitely 'patchy' environment. It is therefore important to

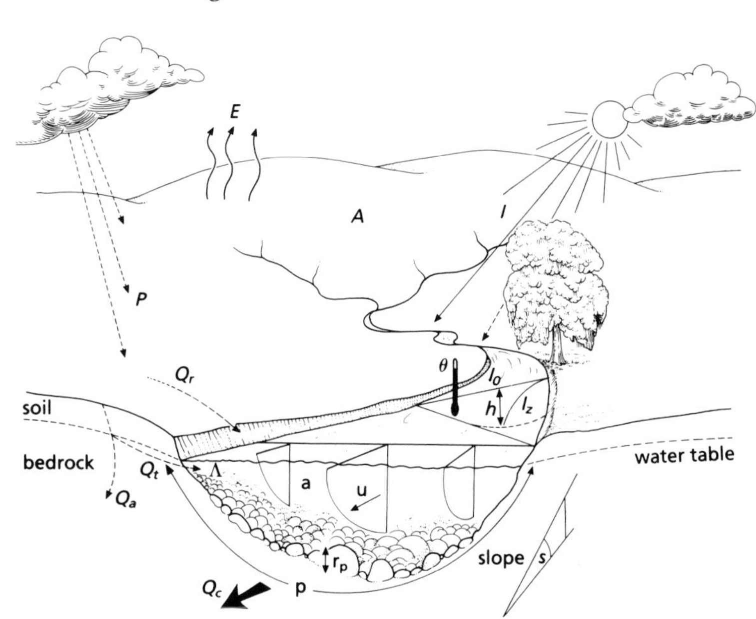

Fig. 2.1 Environmental characters of rivers and their catchments influencing the ecology of fluvial algae. The terms are identified and their quantitative ranges are interrelated in the text.

recognize the various categories of attached, benthic algae—from the larger 'macrophytic' forms to the variety of microscopic 'algal turfs' and films growing on a variety of available surfaces—as well as those planktonic species suspended in the water.

'Macrophytic algae'

To survive in a given location of flowing water, benthic algae are assumed to require, obviously, the means of physical attachment to the solid bounds of channel but, less obviously perhaps, a resistance to physical damage by the flow itself. River algae seem to have adopted two quite different approaches to these problems. The 'macrophytic algae', so styled because their sizes are sometimes quite comparable to those of aquatic mosses, liverworts and small flowering plants, are the most conspicuous when present in any quantity. These are, morphologically, relatively complex, either essentially tubular (e.g. *Lemanea*, *Enteromorpha*) or coenocytic threads (e.g. *Cladophora*), dense tufts of uniseriate filaments (e.g. *Oedogonium*, *Ulothrix*, *Stigeoclonium*) or conspicuously branched structures reinforced with biomineral deposits of carbonate (e.g. *Chara*, *Nitella*).

Generally, they become firmly attached to stones and rocks as well as other available solid surfaces but they are sufficiently flexible to allow water to flow through and around their distal ends. In this way, filaments of *Cladophora*, for example, can grow to lengths sometimes exceeding 4 m. Most macrophytic species are presumed to have upper limits of tolerance to velocity and turbulence. Although these are not necessarily quantified, there are differing views about whether these species are selectively favoured by their tolerance or even their reliance upon fast flows (Whitton 1975) or whether the ability to survive in, and rapidly recolonize from, the adjacent slow-moving boundary layers is crucial.

Epiliths and endoliths

The second approach to damage-resistant, self-maintenance is to exploit continuously the microphytic boundary-layer habit. Whereas the crustose or felt-like turfs of (usually) unicellular algae attain a thickness of a few tens of micrometres, the 'bed-layer' adjacent to the solid surface on which they grow may sometimes extend several millimetres beyond (Fig. 2.2). Within this microzone, the flow is retarded and becomes lami-

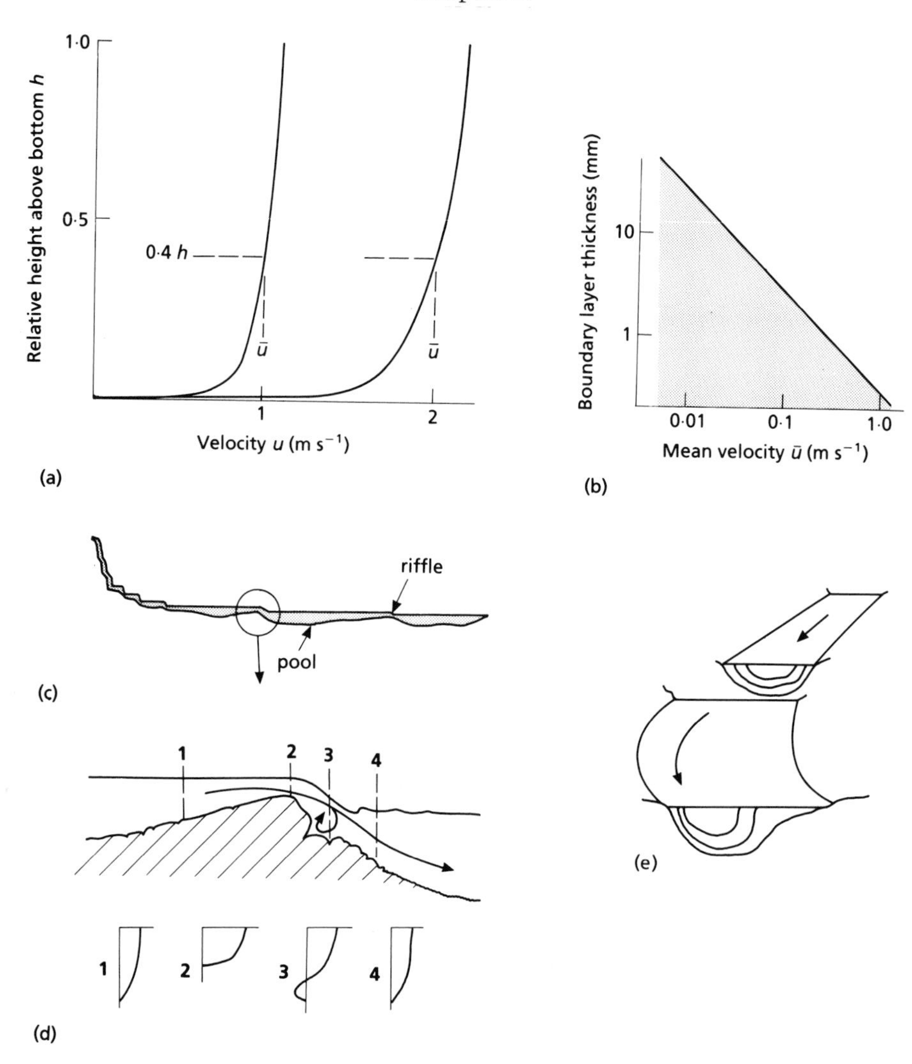

Fig. 2.2 Velocity profiles in rivers. (a) Distribution of velocity (u) against relative water depth of flows over smooth surfaces averaging $\bar{u} = 1$ or $2\ \mathrm{m\ s^{-1}}$; note that u at $0.4\ \mathrm{h} \simeq \bar{u}$. (b) The thickness of the viscous bed layer as a function of mean velocity (after Smith 1975). (c) Long profile of a river, to show changes in gradient and depth. (d) A portion is enlarged to represent a pool–riffle–pool sequence, through which the velocity profiles (1–4) are distorted and, in places, reversed. (e) Sagittal sections of a river channel, with velocity contours to show eccentric displacement towards the outside bend of the channel. Based on various sources.

nar so that, even at main channel velocities approaching $2\ \mathrm{m\ s^{-1}}$, it is possible for these microphytic algae to maintain their positions. Nevertheless, the epilithic growths on hard surfaces forming the 'wetted perimeter' of the channel have to be firmly attached, either by sticky, gelatinous secretions, as in the diatom *Cocconeis* and the cyanophyte *Chamaesiphon*, or by means of a stalked 'hold-fast', exemplified by the pennate diatoms, *Achnanthes, Cymbella* and *Gomphonema*. In the encrusting Rhodophytes (e.g. *Hildenbrandia, Lithoderma*), a flattened thallus is closely applied to the stone surface, while low-growing cushions of filamentous cyanophytes, such as *Phormidium* and *Rivularia*, exploit and, if sufficiently compact, extend the boundary layer.

In some softer rocks, especially limestones, algae may be partly (e.g. *Gongrosira*) or wholly endolithic (e.g. *Schizothrix*), living just beneath the stone surface. In some calcareous headwater streams, calcium carbonate is precipitated as travertine (also called 'tufa' or 'marl') as dissolved carbon dioxide in the spring water equilibrates with the atmosphere. The process is enhanced by the photosynthetic consumption of carbon dioxide by algae, such as *Oocardium* and *Rivularia*, and mosses, which then deposit delicate and elaborate 'casts' about themselves on the surfaces of riffles and waterfalls.

Epiphytes

In addition to furnishing further surfaces colonizable by algae, the blades and stems of submerged macrophytes will, if sufficiently dense, create a local microenvironment of reduced velocity and turbulence (Hydraulics Research Laboratory 1985; Dawson 1989). Thus, while it is possible to recognize the analogous associations of epipelic diatoms (*Cocconeis, Navicula* and *Gomphonema* spp. on younger leaves; *Achnanthes, Meridion* and others are said to be more often associated with older or dying leaves; Margalef 1960), macrophyte beds in flowing water may provide (albeit temporary) refuges for larger unicells (e.g. *Closterium*), for the less robust filamentous and colonial forms (e.g. *Aulacoseira, Oncobyrsa*) and, indeed, to non-attached algae (including *Cryptomonas*) and free-swimming animals such as daphniids. The presence of these more distinctive 'epiphytes' in the main flow depends greatly upon the extent of seasonal dieback and regrowth of the macrophyte beds and of the frequency of flushing events.

Epipsamms and epipels

Owing either to spatial variability in the velocity structure in the channel or to temporal fluctuation in the bulk discharge, many rivers will provide zones of low flow wherein water displacement is sufficiently weak for deposits of gravels, sands and silts to acquire temporary stability. These locations furnish opportunities for the growth of algae which, while not apparently well-adapted to vigorous flow, are nevertheless sufficiently invasive to exploit them. Epipsammic algae attach to sandgrains. Epipelic algae are generally associated with fine sediments. The species concerned form loose mats or films made up of numerous gliding (*Oscillatoria, Phormidium*) or aggregating filaments (*Microcoleus, Mastigocladus*), nets (*Hydrodictyon*) or clusters of unicells (*Nitzschia, Caloneis, Surirella* and *Cymatopleura*). While suitable microhabitats persist, these algae collectively can represent a significant and expanding component of the fluvial microflora. At the same time, however, they remain vulnerable to increases in the discharge and the consequently enhanced shear forces and flushing rates.

Suspended algae

The live algae to be found in suspension in the open water may, to a greater or lesser extent, reflect the components of the various benthic associations, in proportion related to species-specific rates of dislodgement and flushing out by the flow. However, this benthic component is augmented, sometimes overwhelmingly, by habitually free-living species common in the plankton of lakes and ponds. Several long-standing tenets about this planktonic element, or 'potamoplankton' (Zacharias 1898), for instance that it originated primarily from standing waters in the catchment or that its substantial development could be manifest only in either very long or extremely sluggish rivers, are currently undergoing revision. Certainly, growth and replication of unicellular and simple coenobial species occur whilst in fluvial suspension; if these criteria are used to characterize the potamoplankton (*cf.* Margalef 1960), then downstream development of populations may indeed be traced back to inocula from lentic waters in the catchment (Reynolds & Glaister 1992). Yet, equally, others may well originate in the benthos: Swale (1969) showed the plankton of the middle Severn, western Britain, to be frequently dominated by a species of *Navicula*.

The most prolific growths of planktonic algae in larger rivers appear not to be clearly derived from either source, whereas their distribution, in space as well as in time, would indicate that populations are effectively native to rivers. The most frequently encountered of these are the smaller species of centric (*Cyclotella* and, especially, *Stephanodiscus*) and pennate diatoms (*Nitzschia*, *Asterionella*), as well as a variety of green algae, such as *Chlorella*, *Ankistrodesmus*, *Scenedesmus* and *Pediastrum*, and euglenoids (*Trachelomonas*, *Phacus*); the rarer reports of dominance by other limnetic species, including *Dinobryon*, *Cryptomonas*, *Microcystis* and *Oscillatoria*, may be confined to just those long, sluggish pool-like rivers or to occasions of severe drought, once thought to apply to all potamoplankton.

Although there is often an evident trend for planktonic algae to be more abundant downstream (Greenberg 1964; Hynes 1970), neither the scale of downstream increase (Reynolds & Glaister 1992) nor the synchrony of maxima at different locations along the length of the river (Lack *et al* 1978) can be explained simply by reference to the downstream passage of a growing inoculum of cells. Reynolds' (1988) preliminary discussion of this paradox deduced that the presence of non-flowing water—whether in slow-flowing, intermediate reaches (e.g. the Sudanese White Nile; Prowse & Talling 1958), in impounded stretches of the water course (Talling & Rzóska 1967; Décamps *et al* 1984), in side arms (Wawrik 1962) and cuts (Moss & Balls 1989) or in a series of retentive in-channel 'dead-zones', as envisaged and defined by Valentine and Wood (1977; see also Carling 1994)—was crucial to the maintenance and dynamics of plankton in rivers. The results of subsequent investigations (Reynolds & Glaister 1989; Reynolds *et al* 1991) are beginning to amplify these contentions: potamoplankton is confined neither to particularly large nor particularly long rivers.

2.3 CONTROLLING FACTORS

As in lakes, where the understanding of the ecological factors regulating the dynamics of algal production has substantially evolved, the growth of autotrophs in running water is essentially dependent upon an adequate supply of light energy and dissolved inorganic nutrients, while species selection, abundance and dominance are influenced by competitive interactions and herbivorous consumers. In addition, river plankton is reliant upon maintenance in suspension if growth is to offset its seemingly inevitable consignment to the sea, while for the algae attached to the surface of gravel, stones and higher plants, size stability and orientation of the substratum, together with the frequency of its movement, contribute to the complex of microhabitats available. Many of the critical variables in streams and rivers are therefore related to, or are determined by, the physical properties of channel flow. It is expedient, therefore, to determine the ambient capacity factors—those 'regional features' of localities that govern the attachable quantities and composition of flowing-water algal communities—and then to consider the regulatory roles of seasonal and increasingly stochastic variability in channels.

Persistent 'regional' factors

The basic morphology of channels is ultimately governed at the catchment scale: the location with respect to latitude; the horizontal distance and vertical descent to the sea; the geology of bedrock and its tectonic history; its geochemical lability and resistance to erosion; and its climate, with particular reference to the amount of rainfall and the seasonality of its distribution. Scaling affects the characteristics of entire drainage basins (e.g. their areas, configurations, aspects, generated flows), through those of particular reaches (channel-form substrata, turbidities) to those distinguishing the diversity of microhabitats available for biological exploitation by algae (Fig. 2.1).

Discharge

The linking property is the net discharge (Q), itself related to the balance between the precipitation input (P) less evapotranspirational loss (E) per unit of topographical catchment (A) upstream of the point of measurement:

$$Q = A(P - E) \quad (1)$$

and where

$$Q = Q_r - Q_t + Q_a \quad (2)$$

Q_r is the flow across the ground surface ('run-off'), Q_t is the fraction percolating laterally to the channel as groundwater ('throughflow') and Q_a is the fraction of percolating water penetrating to deeper aquifers and which does not necessarily re-enter the local surface drainage. The yield to channel flow (Q_c) is thus equivalent to $Q_r + Q_t$.

Fluvial chemistry

The capacity of the river to support algal production is initially related to the general ionic environment and, in particular, to the supply of nutrient resources to the water. When annual rainfall is distributed evenly through the seasons and throughflow dominates the drainage to the channels, the prolonged, intimate contact between the bedrock-derived mineral particles and the interstitial water ensures a steady, nutrient-enriched water flow. In contrast, where there is scant soil cover or the ground is already waterlogged, the supply to channels is dominated by scarcely modified run-off and is very responsive to the periodicity of precipitation events.

In this way, headwater streams rising in hard-rock, mountainous regions of high precipitation yield weakly ionic, nutrient-poor waters in which the algal growth capacity may be severely limited. Such waters are particularly susceptible to acidification, carrying numerically sparse algal populations, perhaps characterized by the diatoms *Eunotia* and *Pinnularia*. Streams draining catchments based on soft or porous rocks (e.g. shales, limestone chalk, evaporites) or on depositional lowlands are generally of high ionic strength, alkaline and potentially rich in nutrients: at times, they may support heavy epipelic growths featuring (*inter alia*) *Navicula* spp., *Caloneis*, *Gyrosigma*. This potential may be further enhanced by agricultural activities in the catchment and by the products of sewage arising from human settlement.

Within these extremes, the total load of a solute (Λ) is generally a function of discharge, of the form:

$$\Lambda = k\, Q_c^{\,f} \quad (3)$$

where k is a local constant depending upon the availability of a given chemical and its solubility in water; f is generally less than 1, the concentration being diluted by higher discharges (curve 1 in Fig. 2.3). Although this applies to most major ions in well-mineralized waters, the pattern is not unique. More vigorous displacement of deep soil-water constituents by throughflow flushing can raise the concentration of (e.g.) N-species or organic matter in streams (curve 2 in Fig. 2.3), or be ultimately diluted by sustained run-off (curve 3), as in the elution of H^+ ions from peat soils.

Channel form

The depth of water in the river, its gradient and the 'roughness' of the channel play a leading role in delimiting algal microhabitats. From its first definable rills to its coastal outfall, the typical drainage basin is based upon the progressive confluence of streams of increasing rank (order) and increasing discharge capacity. Following the direction of slope, the shape and pattern are in-

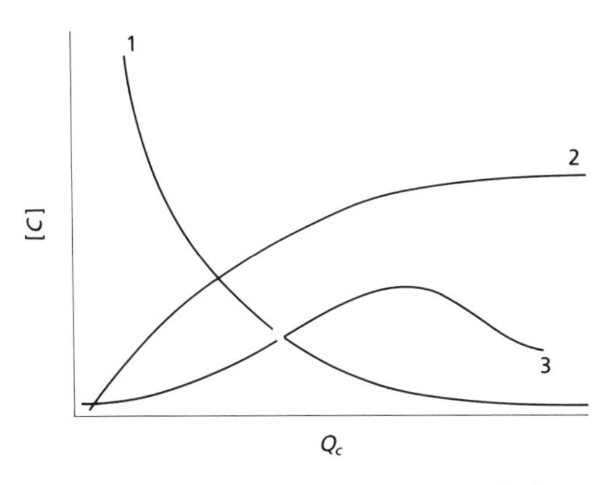

Fig. 2.3 Idealized curves of concentrations of solute concentration ([C]) as a function of fluctuations in channel flow (Q_c). Most catchment-derived determinants conform to curve 1 but the displacement of deep soil-water solutes conforms more closely to curve 2, while curve 3 describes the elution and ultimate dilution of H^+ ions from acid soils. Redrawn from Meybeck *et al* (1989).

fluenced by the gradient (Leopold *et al* 1964). Ideally, the long profile of a river tends toward a catena; much of the work of downcutting, removal of eroded material and downstream deposition can be interpreted as the dynamic progression towards the ideal. On a smaller scale, natural channels follow a series of steps, manifest as cascades in rocky mountain streams and as alternating deeper pools and faster flowing shallow areas ('riffles') on gentler slopes. Lengthwise variations in the profile and sinuosity of the channel, as well as the implied sequence of erosion and deposition, contribute to a downstream alteration in the regional blend of local microenvironmental characters, associated with the progressive increase in ambient discharge. The algal habitats generated in channels are as varied as the geographical distribution in annual run-off generated by rainfall (P), between virtually zero (Atacama Desert, Chile and the eastern Sahara) and the 11.5 m frequently experienced at Kauai, Hawaii ($Q_c/A = 0.35$ m^3 km^2 s^{-1}), and as the range in drainage basin areas. The largest, the Amazon Basin (7.05×10^6 km^2) delivers to the sea some 6.7×10^{12} m^3 annually, at an average rate of 212×10^3 m^3 s^{-1} ($Q_c/A \sim 0.03$ m^3 km^{-2} s^{-1}). Algal communities have been studied in headwater streams where the reported discharges were eight to nine orders of magnitude less (Reynolds 1988). Throughout this range, the form and dimensions of the channel vary continuously to contain the flow of water, so the predominant habitat structure and, hence, the composition of the species assemblages present, respond to related variations in width (b), depth (h) and mean velocity ($\bar{u}$) in terms of the discharge,

$$b = k_1 (Q_c)^{\chi}$$

$$h = k_2 (Q_c)^{\psi}$$

$$\bar{u} = k_3 (Q_c)^{\omega}$$

Moreover, because at any point Q_c approximates to the product $bh\bar{u}$, the coefficients and exponents are cumulatively related:

$$k_1 k_2 k_3 = 1$$

$$\chi + \psi + \omega = 1$$

In other words, width, depth and velocity increase downstream, although less rapidly than does Q_c. Neither are the proportions constant; the ratio of $b:h$ is greatest in low-gradient channels through unconsolidated materials (e.g. fluvoglacial deposits), while in adjacent pool and riffle sections carrying instantaneously similar discharge, changes in cross-sectional area, bh, are reciprocated by changes in mean velocity therethrough:

$$\bar{u} = Q_c/bh \quad (4)$$

Velocity

One further generalization concerning mean velocities through a channel section is that expressed by the familiar Manning equation (5):

$$\bar{u} = 1/n \, (a/p)^{2/3} \, s^{1/2} \quad (5)$$

where a is the cross-sectional area of the channel and p is its wetted perimeter (see Fig. 2.1; the quotient a/p is also known as the *hydraulic radius* and, in wide, smooth channels, approaches a value close to $a/b = h$); s is the slope in the sense $1/x$, where x is the horizontal distance per unit vertical descent; $1/n$ is a factor correcting for frictional resistance to flow, due, for example, to a rough or shallow bottom, or to abundant macrophytes in the channel. Ven Te Chow (1981) has recently updated estimates of n between ~0.03 in large, open channels, 0.04–0.06 in smaller ones and up to ~0.15 in channels heavily beset with macrophyte growth. Except in cascading reaches confined within rocky gorges, mean velocity rarely exceeds the ~2 m s^{-1} at which most rivers accommodate the flow by enlarging their channels.

Velocity structure

Nevertheless, the range $0 < u < 2$ m s^{-1} embraces a spectrum of environments from weedy, silted backwaters to boulder-strewn rapids and deep, pool-like reaches. This variability characterizes rivers from source to outfall and by no means in a regular progression along their long profile.

It is equally important to emphasize that much the same range of velocities is encountered within individual reaches, at scales in the order of a few metres to a few millimetres. The classic vertical profile (i.e. in the z direction) of instantaneous velocities (u_z) in the downstream (x direction) reveals flow to be more rapid away from the frictionable bounding surfaces, adjacent to which the velocity approaches zero (see Fig. 2.2(a)). As the channel section alters through adjacent pools and riffles, so the velocity profile is altered downstream (see Figs. 2.2(c) and (d)). Moreover, considerable lateral differences in velocity (i.e. in the y direction) are evident on bends where flow is centrifugally displaced (see Fig. 2.2(e)).

This variability in velocity structure is of profound significance to the ecology of fluvial algae. The presence of a near-bed boundary layer, perhaps a few millimetres in thickness, at the surfaces of stone and macrophytes furnishes a microhabitat suitable for the settlement and development of epipelic algae of lesser dimensions (generally <0.1 mm), yet which is sufficiently close to the main flow and its supply of dissolved gases and nutrients for favourable diffusion gradients to be maintained. Pool-like reaches may be better tolerated than shallow riffles by macrophytic algae, provided suitable sites for attachment are available; yet again, they benefit from the advective replenishment of mainflow solutes. Moreover, it is the lateral 'dead-zones' of reduced flow that retain water—progressively renewed by fluid exchange with the mainflow (Young & Wallis 1987)—that provide the enhanced opportunities for the growth and otherwise paradoxical downstream increase of planktonic algae.

Turbulence and shear

The condition for the persistence of planktonic cells in the open channel, as elsewhere, is that the motion of the water is sufficiently turbulent (Margalef 1960; Carling 1994) constantly to redisperse suspended particles throughout the water depth. In flowing waters, turbulence is generated between adjacent water layers travelling at different velocities. At low velocities, viscosity overcomes the turbulence: the layers slide over one another in laminar flows. The ratio of the inertial to the viscous forces is expressed by the dimensionless Reynold's number, Re:

$$Re = \rho_w ul/\nu \qquad (6)$$

where ρ_w is the density of the water, ν is its viscosity and l is the appropriate length dimension; in the present consideration, it is equated with the depth of water, h. As Re increases from 500 towards 2000, laminar flow breaks rapidly into turbulence (e.g. Smith 1975). This transition is inserted in the plot of h versus u in Fig. 2.4(a) to illustrate the conditions for the onset of turbulence in channels.

The intensity of the turbulence is quantifiable from velocity fluctuations in each plane, $\pm u'$, $\pm v'$ and $\pm w'$ about the mean downstream velocity, u, in the x, y and z directions respectively. To overcome the rapid tendency to zero of these series, it is customary to refer each quantity to its non-vanishing root mean square value. For example:

$$u' = [(\pm u')^2]^{1/2} \qquad (7)$$

The scale of turbulent intensity is the time-averaged product of the change of momentum imposed upon the main motion in the x direction; this is also known as the friction velocity, u^*. Considering the vertical component:

$$u^* = (u' \ w')^{1/2} \qquad (8)$$

The turbulence associated with flow applies a horizontal stress, called shear (τ), which is transmitted through adjacent vertical layers:

$$\tau_{xz} = \rho_w \ (u' \ w') \qquad (9)$$

whence:

$$u^* = (\tau_{xz}/\rho_w)^{1/2} \qquad (10)$$

While turbulence is generally proportional to the velocity and the vertical distance above the bed (Carling 1994), the stress on the bottom can be derived from its transmission through the velocity profile (du/dz); assuming the mean velocity, $\bar{u}$, to be located at $0.4h$ above the bed (see Fig. 2.2(a)),

$$u^* = (\tau_0/\rho_w)^{1/2} = 0.4h \ (du/dz) \qquad (11)$$

Rearranging:

$$du/dz = u^*/0.4h \qquad (12)$$

and integrating:

$$u = 2.5 \ u^* \ \ln(h/c) \qquad (13)$$

c is an integration constant with a finite value related to the roughness of the bed: r_p corresponds to the height an object projects from the bed and:

$$c \sim r_p/30 \qquad (14)$$

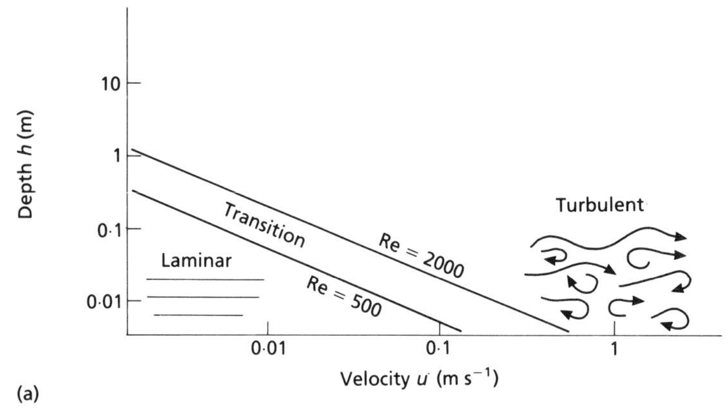

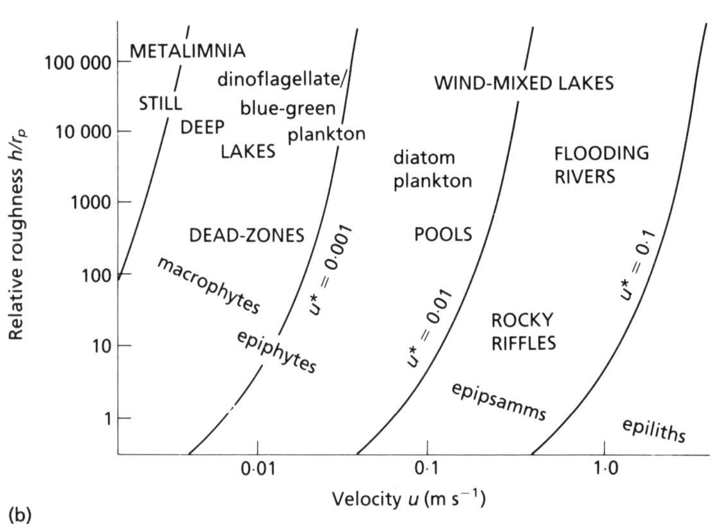

Fig. 2.4 (a) Organization of flow near solid surfaces, as a function of velocity and water depth; and expressed in terms of the Reynolds Number (Re, dimensionless); flow is nearly always turbulent ($Re > 200$) except at low current speeds in shallow water, where $Re < 500$. Ordered structure breaks down to full turbulence in the transitional range, $500 < Re < 2000$. Redrawn from Smith (1975). (b) The depth divided by the height of surface projections—'relative roughness'—is substituted on the vertical axis in order to represent the distributions of freshwater habitats and of their associated dominant algae. The inserted contours of shear velocity (u^*) assume unidirectional gravitational or wind-driven currents.

Then:

$$u_z = 2.5\ u^*\ \ln\ (30h/r_p) \qquad (15)$$

so at $0.4h$,

$$u = 2.5\ u^*\ \ln\ (12h/r_p) \qquad (16)$$

Depending on the ratio h/r_p, from approaching 1 for very stony streams to >5000 for deeper flows over fine silts, the value of u^* can range between 1/6 and 1/30 u.

The relationships are ecologically instructive. For a given change in u, the stress applied to the bed increases geometrically and the thickness of the non-turbulent boundary layers adjacent to solid surfaces is correspondingly compressed (see Fig. 2.2(b)). If the bottom is 'rough', there will be many areas, chiefly on the downstream side of large boulders, where flow is weaker (see Fig. 2.2(d)). Local differences in shear will often influence the microscale distribution and abundance of attached algae and, equally, render some areas more susceptible to sediment deposition than others. This principle is extended in the construction of Fig. 2.4(b), which depicts the general range of aquatic algal habitats in terms of their relative roughness (h/r_p) and mean velocity (u). Isopleths of friction velocity (u^*) are superimposed. The plot distinguishes among the approximate characteristics of lakes, rivers, pools and riffles and suggests the distributions of the major algal life forms.

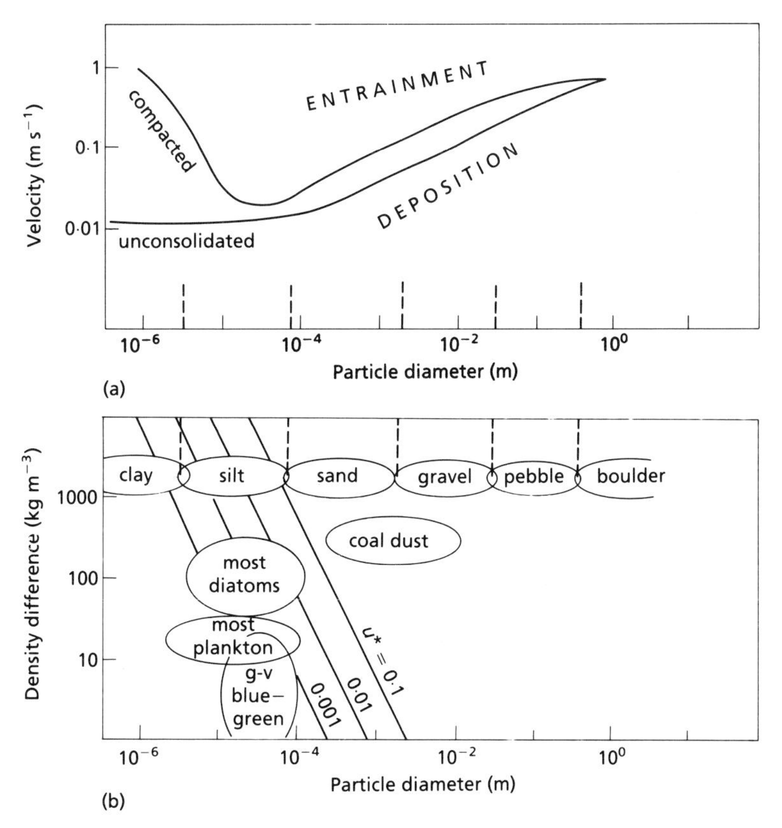

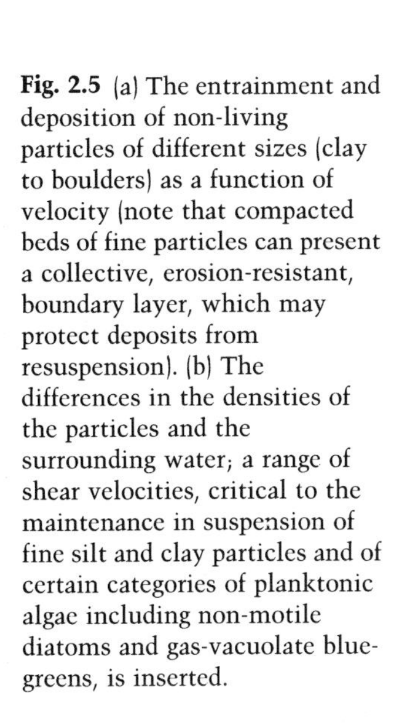

Fig. 2.5 (a) The entrainment and deposition of non-living particles of different sizes (clay to boulders) as a function of velocity (note that compacted beds of fine particles can present a collective, erosion-resistant, boundary layer, which may protect deposits from resuspension). (b) The differences in the densities of the particles and the surrounding water; a range of shear velocities, critical to the maintenance in suspension of fine silt and clay particles and of certain categories of planktonic algae including non-motile diatoms and gas-vacuolate blue-greens, is inserted.

Entrainment and settlement

Generally, the greater the velocity, the greater is the size of the largest particles moved and of the smallest unmoved (Fig. 2.5(a); for further discussion, see Carling 1994). The uniformly low local roughness coefficients offered by compacted fine sediments in large, gently graded, lowland rivers protects them from entrainment: $r_p <$ boundary layer thickness. The threshold for resuspension may depend upon the mean velocity accelerating to the point of channel enlargement! So far as benthic algae are concerned, the less frequently disturbed surfaces – large stones and boulders – should be more suitable habitats for large or slower-growing species. On the other hand, the significant presence of diatom growths on fine sediment surfaces will occur only in slow-flowing periods when the boundary layer is least compressed.

The criterion for entrainment of particles, including planktonic algae, whether introduced into the river from external sources or resuspended from the bottom by turbulent shear, is that turbulent velocity fluctuations substantially exceed the intrinsic settling rate (w_s) of the particle. Humphries and Imberger (1982) proposed:

$$[(w')]^{1/2} > 15w_s \quad (17)$$

where w_s is the intrinsic settling rate of the particle which, subject to the criterion that the flow of water over the moving particle remains laminar, is described by the Stokes equation (18):

$$-\ w_s = d^2\ g(\rho_a - \rho_w)\cdot(18_v\ \phi) \quad (18)$$

d being the diameter of the particle (or the diameter of a sphere of identical volume), ρ_a is its density, ρ_w is the density of the water, and η is its viscosity; ϕ is a 'form-resistance factor' compensating for distortions from the spherical form; g is gravitational acceleration. In the plot of particles in terms of their test-density difference and diameter (Fig. 2.5(b)), contours of critical u^*

values are inserted, assuming that $2u^*$ is adequate to meet the entrainment criterion (equation 17). It is evident that, under the turbulence conditions encountered in the majority of larger streams and rivers, clay and fine silt particles, as well as the cells of most phytoplankton, will be substantially entrained in the flow. Moreover, while the entrainment criterion continues to be satisfied but a basal boundary layer is maintained, the rate of population change due to loss from suspension (σ) by settlement is largely a function of water depth, not of velocity as might be supposed (Reynolds *et al* 1990):

$$N_t/N_0 = e^{-w_s t/h} \qquad (19)$$

where N_0 and N_t are the populations of particles separated by an interval t. Nevertheless, the faster is the rate of flow, the further downstream material will be transported in the same unit time before loss from suspension is complete.

Turbidity

Fine, non-living particles in suspension interfere with the penetration of underwater light and so constitute an important environmental factor in the growth and distribution of both attached and planktonic algae. Put at its simplest, the incident light penetrates beyond the reflective, uneven surface of the water (I_0) is reduced exponentially by absorption and scatter so that at depth z, the remaining light (I_z) is approximately given by:

$$I_z = I_0 \, e^{-\varepsilon z} \qquad (20)$$

where ε is the coefficient of monochromatic light extinction. This coefficient comprises two separate effects (absorbance in particular wavebands and backscatter) of several components, including the pure water itself, together with dissolved colour (ε_w), the suspended algae (ε_a, which is a product of the algae present and their specific areal absorbance; $\varepsilon_a = n.\varepsilon_s$) and the load of non-living clay and silt particles in suspension (ε_p). In rivers, where turbidity is often mainly due to non-living suspensoids, it is useful to express the suspended load as a light-extinction coefficient comparable with that for the algae. From a series of measurements of suspended solids in the Bristol Channel (Hydraulics Research Laboratory 1981) and the depth–time distribution of planktonic photosynthesis (Joint & Pomroy 1981) it has been possible to approximate ε_p as 20 m^{-1} (kg $m^{-3})^{-1}$. The impact on the penetration of a given light income, say up to 2 mmol photon m^{-2} s^{-1}, of suspended-solid loads in the observed range (1 g to 2 kg m^{-3}) is represented in Fig. 2.6(a). In many rivers, loads >0.1 kg m^{-3} may begin seriously to limit the amount of light reaching the bottom. For many bottom-dwelling attached algae which are unable to adapt to extreme shade (e.g. by raising cell-specific chlorophyll sufficiently to enhance their photosynthetic efficiencies), turbidity reduces the opportunities for survivorship of the population. For planktonic cells in a turbid river and exposed to extreme fluctuations in light, from $I_0 \rightarrow I_z$ and back within intervals of minutes, it is important to exploit the available energy as effectively as possible. One way to achieve this is to increase the light-harvesting capacity of the cells by increasing pigment content/unit biomass. As a result, low-light adapted cells have shortened response times in which to raise the photosynthetic rate, but have a widened absorption spectrum and an increased risk of photoinhibition near the surface.

As both the amount and longevity of particles in suspension as well as the distance that they are transported downstream are all likely to be properties of the flow through the individual reach, these quantities will vary with discharge. Figure 2.6(b) is redrawn from a derivation in Reynolds (1988) of ε_p ($+\varepsilon_w$) based on plankton chlorophyll, photosynthesis and inferred photic depth, at differing discharges in a section on the River Thames near Reading, as measured by Lack (1971) and Kowalczewski and Lack (1971); its inclusion here is to illustrate the sensitivity of the phytoplankton-carrying capacity of a lowland river to discharge-led variability in the suspended-solid load.

Seasonal factors

The principal factors regulating algal production in lakes are day length, solar energy income and, in turn, water temperature, especially in the higher latitudes (Lund 1965). This may be expected to hold for rivers, although seasonal

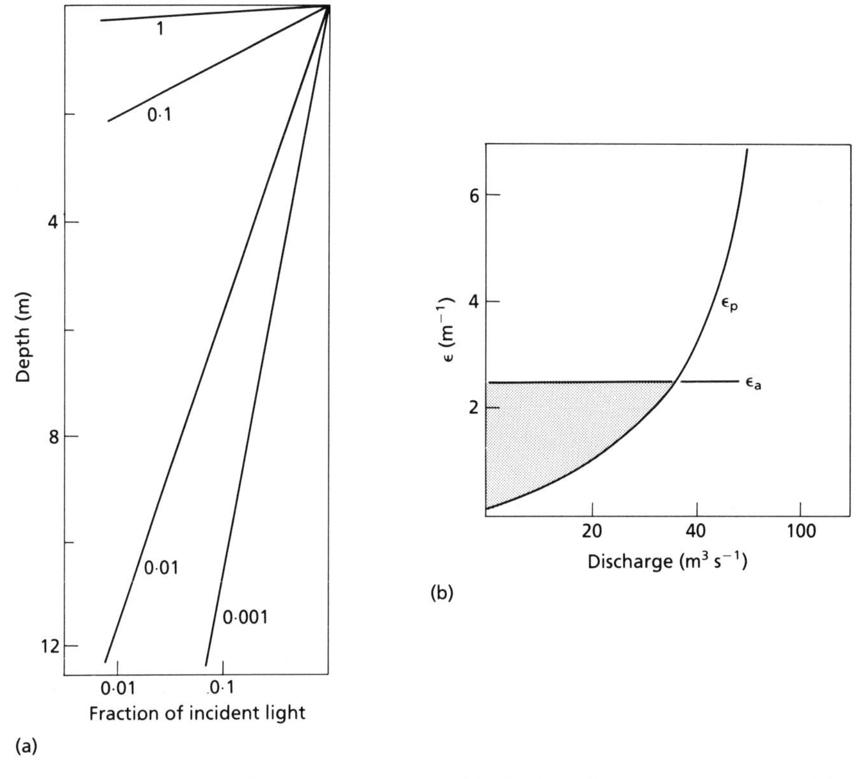

Fig. 2.6 Water clarity as a function of flow. (a) The fraction of the incident light penetrating the surface reaching into the water column at the selected concentrations of clay particles (in kg per m^3 of suspending water). (b) The turbidity response in the River Thames, due to non-living material (mostly clay), as a function of discharge (ε_p) and the extinction (ε_a) of a self-shaded concentration of planktonic diatoms in the River Thames at Reading; the stippled area therefore denotes capacity for net growth increment of fluvial phytoplankton. Revised representation of data derived by Reynolds (1988) from data presented in Lack (1971).

variations in discharge may have a more immediate impact upon the survival and growth of algae in given reaches. As with other aspects of fluvial phycology, there is a dearth of studies in which the effects of the various factors have been satisfactorily separated.

Light

In clear headwater streams, the incident light penetrating to the bed of the stream is barely modified either in intensity, spectral composition or duration. Many of the benthic chlorophytes are known, or are supposed, to tolerate high light intensities or, alternatively, are less adaptable to low average light intensities than representatives of algal groups with a greater capacity for chromatic adaptation. The filamentous green algae, *Ulothrix* and *Cladophora*, are among the genera said to be influenced by increasing day length; *Hildenbrandia* and *Batrachospermum* are genera of red algae said to be tolerant of a shading by overhanging trees and herbs (Whitton 1975) and which, under temperate deciduous forests, will be of increasing relative importance through the summer. Algae may be excluded from deeper surfaces during episodes of high turbidity which may well regulate the growth of phytoplankton independently of other factors. Williams (1964) attributed the winter growth of planktonic diatoms in central North American rivers, despite sustained flows and near-freezing temperatures, to the fact that little silt was washed in while the ground itself remained frozen.

Temperature

In rivers, temperature is assumed to respond more rapidly to increased solar radiation fluxes and, conversely, to lose heat to a cooler atmosphere more quickly than do lakes, by virtue of their relative shallowness and more vigorous, more complete vertical mixing. On the other hand, the general weakness of density stratification, save where it is supported by salinity gradients, means that the higher and lower temperature extremes at the surfaces of lakes (θ35°C, <0°C) are rarely reached in rivers. Mean temperatures are generally supposed to increase from source to mouth. On a smaller scale, there is also provisional evidence that horizontal temperature differences can develop between the more retentive dead-zones in lowland river reaches and the main channel flow, under appropriate warming or cooling episodes (Reynolds & Glaister 1989).

So far as the responses of algae are concerned, few data are available to verify the supposition that algae grow faster at higher temperatures. If river algae conform to patterns of growth of their limnetic counterparts (Reynolds 1989; Nielsen & Sand-Jensen 1990), then it would be expected that the algae of larger unit surface area:volume ratio ($>10^6$ m^2 m^{-3}) are the more rapidly growing at 20°C and only about half as fast at 10°C ($Q_{10} \sim 2$); species with relatively lower unit area:volume ratio ($<10^5$ m^2 m^{-3}) are both slower growing and relatively more sensitive to lowered temperatures ($Q_{10} \geq 3$). Most rivers appear to support attached and suspended algae belonging to the first category, so it is not easy to attribute distributions of species, either in time or in space, to water temperature alone. However, the optimal seasonal production of planktonic algae (e.g. Lack 1971) and of benthic assemblages (e.g. Phinney & McIntyre 1965) is clearly responsive to the interactions of lengthening days and rising temperatures (Fig. 2.7).

Grazing

Various gastropods, certain mayflies, hydroptilid caddisflies, mites and specialized cyprinid fish feed on epilithic and epiphytic algae. Grazing tends to vary quasi-seasonally, reflecting the proliferation of algae, to some extent, the life-

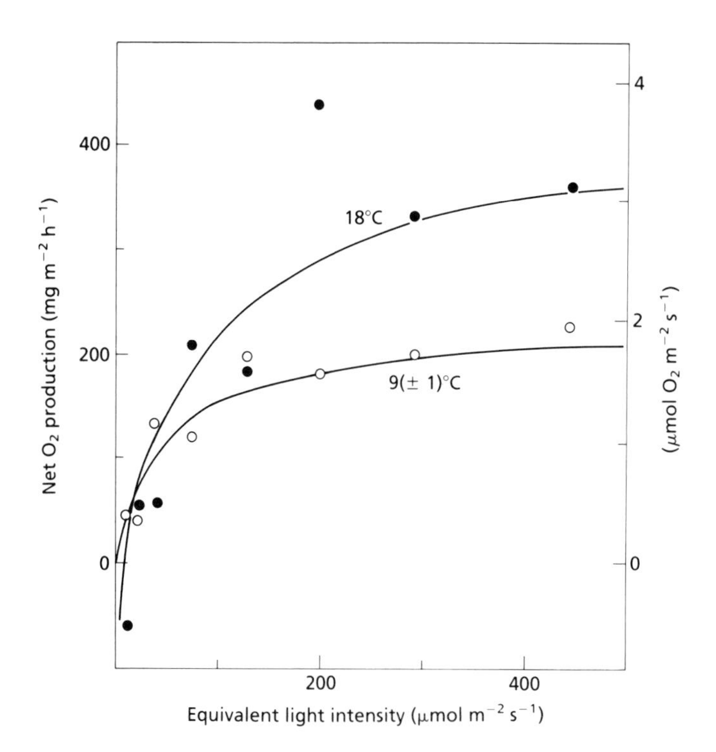

Fig. 2.7 Curves of net oxygen production by photosynthesizing benthic epiphytic algal communities, within two temperature ranges, as a function of light availability. Such data are relatively scarce, these examples being derived from the data of Phinney and McIntyre (1965).

cycles of the consumers they sustain and, perhaps, the ability of the algal community to 'recover' after episodes of heavy grazing. Algal production may also sustain, in part, a variety of benthic filter feeders, from simuliid larvae to bivalve molluscs, but their presence will rarely exert a direct effect upon the algal production. There have been many studies on the food requirements, feeding rates of a selection of common invertebrates and, in general terms, on the impact of grazing on the food resource (for reviews see Gregory 1983; Lamberti & Moore 1984). Gregory (1983) has also provided one of the few critical evaluations of the plant–herbivore relationship in lotic communities, wherein the influence of grazing on the structure of the algal assemblage was investigated. Attached diatoms are a frequent food source for many grazing animals, whereas filamentous forms and blue-green algal species are often supposed to be rejected. Nevertheless, their faster growth rates and their higher productivity (*sensu* production per unit biomass) at relatively low biomass can still enable films of diatoms to dominate heavily grazed benthic communities. On the other hand, the availability of suitable algae may regulate the growth rates, abundance and distribution of direct herbivores and, in turn, the extent to which the lotic ecosystem is supported by autochthonous-based production as opposed to catchment-derived, allochthonous components.

Stochastic factors

Both plants and animals in running waters are nevertheless governed to a large extent by external disturbances, notably the occurrence of high water flows. In headwater streams, episodes of enhanced surface run-off generally have the effect of reducing friction to permit faster velocities of streamflow; as discussed above, the principal effects on algae are expressed through the movement and, perhaps, overturning of larger stones and through the compression of the boundary 'logarithmic' (see Carling 1994; Fig. 2.2(b)) layer which exposes more prominent algae to mechanical stress. In deeper, low-gradient lowland reaches, where channel flow is already well equilibrated to fluvial geometry, the response to increased discharge may be relatively minor in terms of the channel section but, nevertheless, of a magnitude to effect the resuspension and transport of fine materials, with all the consequential impacts on turbidity, light penetration and redeposition rates. Such effects are generically similar from river to river but their characteristics will be unique to each system, according to river size, channel section, nature of the bed and seasonality of floods. Neither will there be, necessarily, in any given river, a direct relationship between turbidity load and discharge; indeed, a hysteretic effect of turbidity would be anticipated, first increasing abruptly and then falling back, following an abrupt change from one velocity state to another.

Except in regions where rainfall is itself markedly seasonal in distribution, it seems reasonable to suppose that whereas the annual alternation of the seasons remains a primary driving factor in the ecology of stream algae, the further the distance downstream, the more the seasonality is suppressed by the responses to frequent discharge variations. In both cases, the community structure is liable to rapid reversal and re-initiation. In the headwaters, the 'reset mechanism' (Gregory 1983) operates every autumn; in lowland rivers, the incidence of elevated discharges may be prompted several times per year. The impact is particularly evident in the comparison (Fig. 2.8) of the responses of phytoplankton in the lower River Thames during a dry and a much wetter summer, as observed and subsequently simulated by Whitehead and Hornberger (1984).

2.4 ECOLOGICAL PROCESSES

It was implied at the outset that survival prospects of algae in rivers are confronted by numerous problems. The subsequent sections have shown, however, that fluvial algae are apparently well adapted to exploit particular niches within the vast spectrum of potential microhabitats consequential upon the interactions of water chemistry, hydrology, channel-form and gradient, substratum, suspended-load and, ultimately, velocity parameters. Moreover, variability among these properties often occurs over short distances

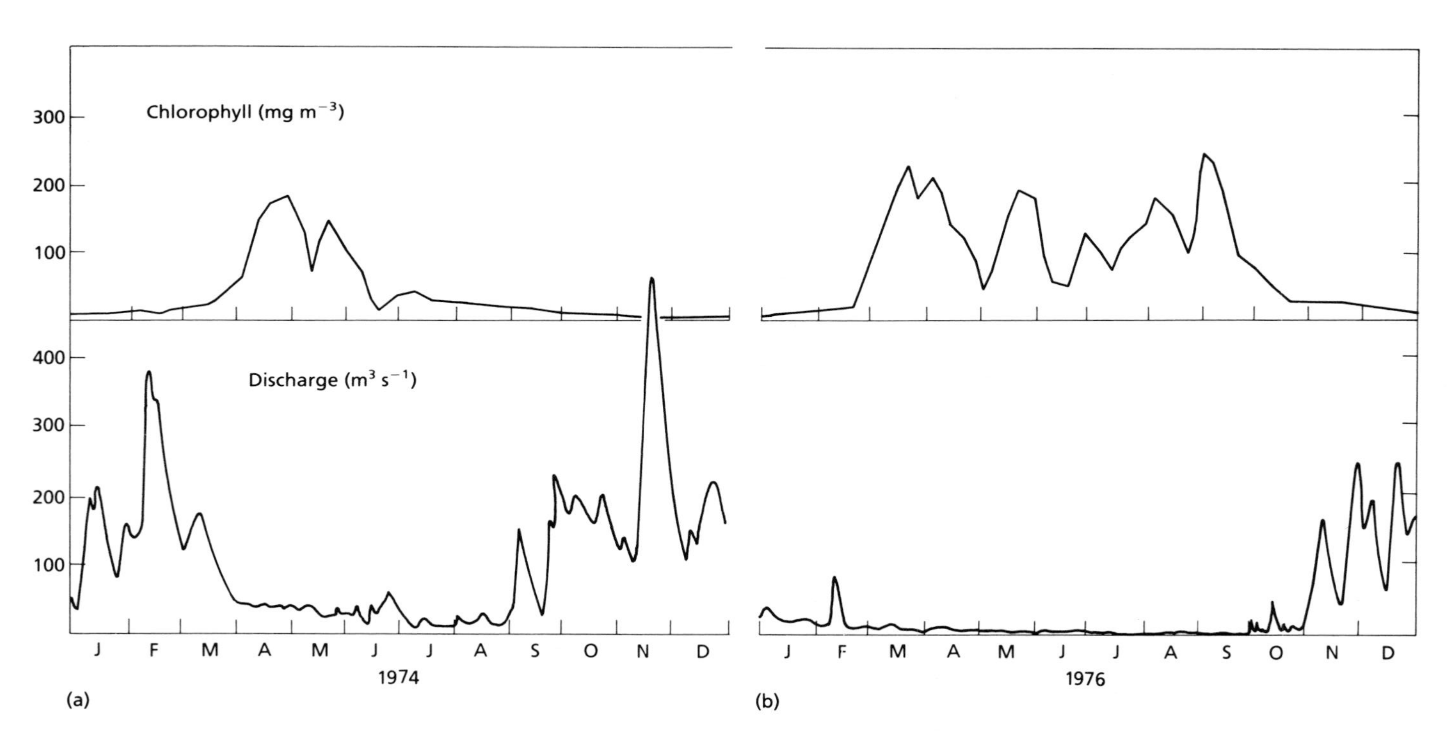

Fig. 2.8 Comparisons of chlorophyll *a* concentrations in (a) the River Thames between Staines and Teddington in relation to contemporaneous discharge fluctuations during 1974 (having a cool wet summer) and (b) 1976 (an exceptional drought year). Redrawn fom data plotted in Whitehead and Hornberger (1984).

in space and over short intervals of time. Velocity and turbulence relate the spatial and temporal dimensions.

As with the understanding of other ecosystems (Wiens 1989), recognition of the importance of scaling effects is essential to defining key processes regulating ecological structure and function. Thus, although the longevity of a given water system is likely to be comparable to the timescale of the geomorphological evolution of its drainage basin, perhaps measurable in millions of years ($>10^{14}$ s), the mean residence time of flowing water may be a matter of a few hours to little more than a year or two (10^4-10^8 s). At the same time, the flowing water is constantly mixed, by turbulence, in the lateral and vertical directions (at scales of 10^1-10^3 s; Reynolds 1988), bringing positive benefits to both attached and suspended algae in enhanced dispersal, nutrient renewal and distribution, the removal of wastes and the exchange of respiratory gases. Other, negative, aspects include added turbidity and wash-out of suspended algal populations, although for plankton and, indeed, other species, survival is assisted by the presence of refuges ('dead-zones') along the channel.

In Fig. 2.9, these scales of 'environmental variability' are compared with the timescales of the various algal responses. Whereas the organisms' physiological responses at scales $<10^4$ s will determine their fitness to function successfully in flowing water, the rate of cell replication (or generation time) is the final expression of the ability of a species to survive and recruit new cells to, and perhaps dominate, particular communities. Because, in many natural rivers, cell division times generally exceed 10^5 s, the key scales of variability are those operating at comparable frequency. This may bias the selective advantage among potentially competing planktonic algae towards species tolerant of high-frequency disturbance ('ruderal' *R*-species in the terminology of Grime 1979) or, briefly, towards those fast-growing opportunists (*C*-species) that quickly take advantage of intervals of more favourable growth (declining velocity, reduced turbidity, elevated nutrient concentration) associated with low discharge. Selective pressures are

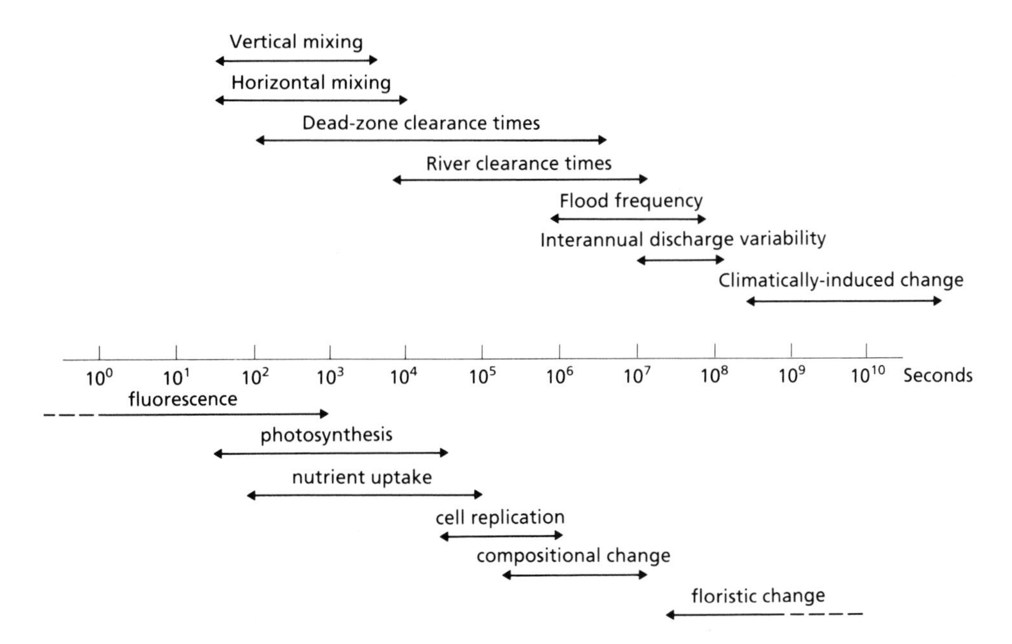

Fig. 2.9 Temporal dimensions of environmental fluctuations (against a common logarithmic scale, in seconds) and the hierarchy of algal responses, from the biochemistry and physiology of individual cells to their relative growth rates and the consequent adjustments in relative abundance and dominance of the community.

generally not in favour of the predominance of the relatively slow-growing but more persistent dinoflagellates and colonial cyanobacteria, save in those long or seasonally sluggish rivers or in extensively regulated catchments dominated by standing waters (Prowse & Talling 1958; Décamps *et al* 1984; Krogmann *et al* 1986; Mohammed *et al* 1989). Given a sufficient period of downstream travel, it is not unusual to find such limnetic forms as *Microcystis* developing in, and sometimes dominating, the lower reaches of slow-moving rivers. Thus, where disturbance is substantially damped out, species describable as *S*-strategists in Grime's (1979) terminology may be selected.

While not subject to the same continuous tendencies towards downstream elimination of the reproductive stock, attached algae are nevertheless sensitive to discharge variability. This is consequential not only upon the increased turbidity during flood, but also upon the compression of the boundary layers, and upon direct mechanical stress on protruding or pedunculate epiliths. Encrusting growths may be more resistant and possibly benefit from frequent disturbances to improve periodically their access to light and nutrient resources. The ability of accelerating velocities to move and turn over increasingly larger (heavier) stones constitutes a disturbance-frequency scale on which, at one extreme, algal colonization is effectively prevented and, at the other, it can proceed unchecked by movement of the substratum.

This counterposition of opportunities and destructive perturbations plays a major part in regulating the spatial and temporal distributions of the main species. It also helps to explain sharp seasonal and interannual contrasts in the abundance of certain species which are sometimes, although usually erroneously, attributed to pollution or enrichment of the system concerned.

The dynamic approach to the structure of lotic algal communities was prominently advanced by Margalef (1960). He envisaged a series of comparable ecological successions, each moving towards a climactic stage but each susceptible to the abrupt intervention of destructive mechanical forcing, 'resetting' (Gregory 1983) the system to an earlier successional condition: in short, what is now referred to as Connell's (1978) Intermediate Disturbance Hypothesis. Under this scheme, the sequence in which species dominate is not necessarily a true ecological succession in the sense of Odum (1969). Moreover, different successional stages can be maintained simultaneously in adjacent zones.

Epilithic successions may be initiated on appropriate 'young' surfaces, such as stones freshly exposed to the light during a recent spate. Depending upon the water flow and the chemical nature (acidity, nutrient content), the pioneer community may be dominated by encrusting diatoms, in which *Ceratoneis* species are frequently represented or, as recently shown experimentally by Elber and Schanz (1990), by the chrysophyte *Hydrurus foetidus*. Communities often featuring species of *Achnanthes* and *Meridion* maintain the succession, yielding place to larger diatom species of *Synedra* and *Aulacoseira*, or to red or blue-green algae in particular streams, before the eventual overgrowth by macrophytic species, including mosses, if the river bed is not disturbed or turned over once again. It is of interest that, for instance, Arnold and Macan (1969) observed dense growths of moss (*Amblystegium eurhynchioides*) to be associated only with large boulders and bands of solid rock exposed in the bed of a cascading hill stream, whereas the smallest stones carried the least growth of any kind; intermediate-sized stones and small boulders sometimes supported dense growths of epilithic diatoms, notably *Meridion* spp. and, later in spring, of *Lemanea*, and, in summer, of *Ulothrix* or *Nostoc*. On each occasion that the total biomass of attached algae increased, it is probable that the number of species represented diminished or an increasing fraction of the total biomass was invested in a diminishing number of species (Hynes 1970).

The precise species representation and dominance through this successional stage is highly variable form river to river, even from location to location within a given river or at the same point from year to year. Again, stochastic events and circumstances contribute to this variability but certain species-associations may occur preferentially in certain kinds of rivers, with identifiable 'regional' characteristics. For instance, according to Margalef's (1960) scheme, an *Aulacoseira–Navicula–Rhoicosphenia–Cymbella–Diploneis* association is frequently encountered in limestone mountains. *Phormidium* and *Lyngbya* are often well represented, as an encrusting growth habit, while *Amphipleura*, and other genera, e.g. *Chaetophora*, *Chara* and *Zygnema*, may be prominent in the quieter pools and backwaters. Mosses such as *Fontinalis* and *Fissidens* may quickly tend to dominance. In streams running over rather less alkaline granites, sandstones and slates, the analogous associations feature *Eunotia*, *Fragilaria*, *Melosira*, *Nitzschia* and *Pinnularia* spp., with *Oedogonium* and *Tribonema* often dominating in pool-like conditions, where the succession proceeds to liverworts and mosses and such flowering plants as *Callitriche* and *Ranunculus*. In nutrient-enriched, hard-water streams, the prevalent *Cocconeis–Melosira–Navicula–Surirella* assemblages often become overgrown by *Cladophora*, *Stigeoclonium* or, in some instances, by *Enteromorpha*. Similarly, humic and saline flows each support distinctive and successive algal assemblages. In every case, however, the deeper, more turbid, pools of downstream reaches become increasingly hostile to the growth of benthic algae, usually because of light deficiency, although a lack of suitably firm substrata, save for such constructions as piers or piling, may often be as important. In such downstream reaches, however, the principal survival opportunities for autotrophs pass increasingly to the phytoplankton.

The principal assemblages represented in the potamoplankton (namely the *Stephanodiscus–Nitzschia–Asterionella* group, prominent in moderate-to-large rivers, and the *Chlorella–Ankistrodesmus–Scenedesmus*–euglenoid group of more general occurrence in smaller, nutrient-rich streams), together with those more typically limnetic associations of Reynolds (1987) recognized to survive in slow-flowing reaches of rivers (notably the mesotrophic *Dinobryon–Sphaerocystis* grouping, the *Oscillatoria agardhii–Oscillatoria redekei* of shallow, well-mixed basins and the eutrophic summer grouping that includes *Microcystis*), appear to be selected by locally obtaining conditions of water depth, light availability and fluid displacement, rather than in accord with a strict successional progression. The most familiar transition, that between the diatoms and green-algal/euglenoid associations, cannot be regarded as being autogenic; it is most commonly mediated in time by a change in discharge (and especially in depth), whereas spatial transitions from greens (upstream) to diatoms (downstream) may move down or up the river with seasonal changes in discharge (Sabater & Muñoz 1990).

In both the benthos and the plankton, successional sequences are generally brief, being subject to frequent truncation or reversal by abrupt changes in the ambient environment. Indeed, successions in flowing water can scarcely culminate in a climactic, internally regulated equilibrium, analogous to (say) the forest climax or the pond hydrosere (Symoens *et al* 1988); while water continues to flow through a defined channel and remains subject to externally driven fluctuations in discharge, the succession cannot move to its climax without the intervention of large-scale factors, such as climate change-mediated differences in hydrology or geomorphological alterations to the drainage basin. Rather, successions are initiated as pioneer species develop after a recent flood and they progress with the establishment of other plants, yet they fail to evolve beyond the flow-imposed plagioclimax condition, dominated by macrophytic algae, aquatic bryophytes or flowering plants.

Subclimactic communities are noted for their species diversity, which, in combination with the stochastic nature of the initial colonization and the subsequent development, contributes to the maintenance of the variety of lotic habitats necessary to the survival of a wide range of species.

It is the frequency of the environmental fluctuations and the periodicity and intensity of the major events which govern the ecology of many elements of the fluvial biota (see for example Hildrew & Townsend 1987) and of the fluvial algae in particular. In the representation of aquatic habitats against axes of resource availability and frequency of disturbance (Fig. 2.10(a)), most rivers will tend towards the upper right-hand corner. Following the logic of Grime (1979), exclusion of the biologically untenable contingency of low

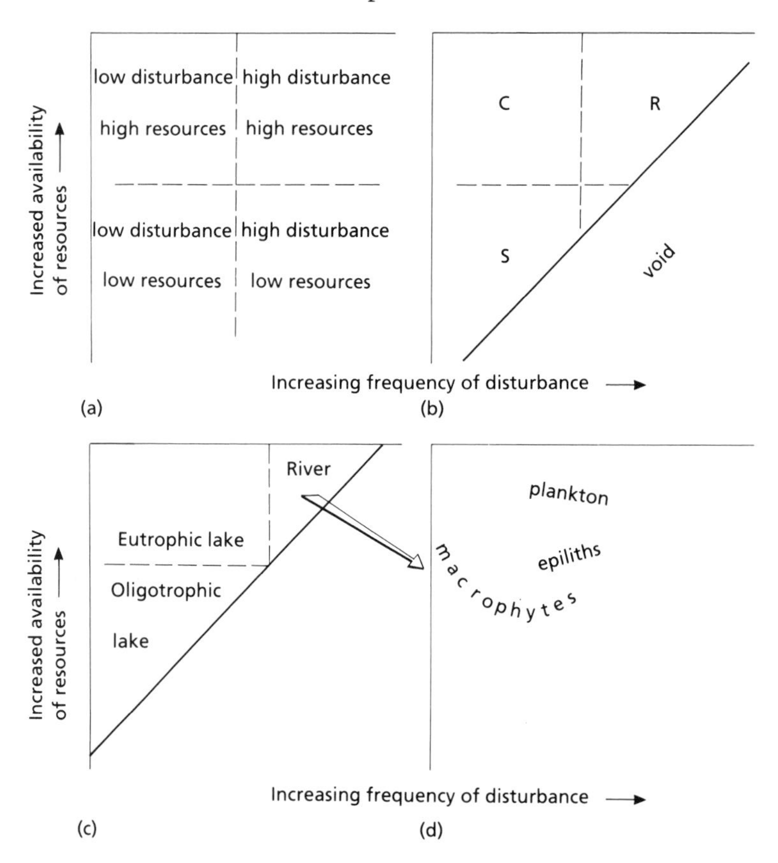

Fig. 2.10 (a) Theoretical contingencies of the incidence of the distinctive properties of the aquatic habitats of algae, as defined by axes representing available resources and disturbance frequency. (b) The primary growth and life-history strategies (colonist, stress-tolerant or ruderal—*C*, *S* or *R*) required by algae best adapted to the particular habitats (based on Grime 1979). (c) A proposed subdivision of aquatic habitats, with rivers generally occupying highly disturbed, resource-rich category; even so, within the '*RIVER*' apex, the relative success of particular algal types may be discussed (d).

resources and high disturbance leaves a triangular representation of available ecological range (Fig. 2.10(b)). The location of river algae in the right-hand apex (Figs 2.10(c) and (d)) serves to emphasize the overriding constraint placed on their ecologies by disturbance factors as opposed to, say, the chronic nutrient limitations of algae in oligotrophic lakes. This is important to the understanding of fluvial ecosystems, their management and their conservation.

REFERENCES

Arnold F, Macan TT. (1969) Studies on the fauna of a Shropshire hill stream. *Field Studies* **3**: 159–84. [2.4]

Blum JL. (1956) The ecology of river algae. *Botanical Reviews* **22**: 291–341. [2.1]

Blum JL. (1960) Algal populations in flowing waters. *Special Publications of the Pymatuning Laboratory for Field Biology* **2**: 11–21. [2.1]

Boudou A, Ribeyre F. (1989) *Aquatic Exotoxicology: Fundamental Concepts and Methodologies*, Vol. I. CRC, Boca Raton. [2.1]

Carling PA. (1994) In-stream hydraulics and sediment transport. In: Calow P, Petts GE (eds) *The Rivers Handbook*, vol. 1, pp 101–25. Blackwell Scientific Publications, Oxford. [2.2, 2.3]

Connell JH. (1978) Diversity in tropical rain forests and coral reefs. *Science* **199**: 1302–10. [2.4]

Dawson FH. (1989) Ecology and management of water plants in lowland streams. *Report, Freshwater Biological Association* **57**: 43–60. [2.2]

Décamps H, Capblancq J, Tourenq JN. (1984) Lot. In: Whitton BA (ed.) *Ecology of European Rivers*, pp 207–35. Blackwell Scientific Publications, Oxford. [2.2, 2.4]

Elber F, Schanz F. (1990) Algae, other than diatoms affecting the density, species richness and diversity of diatom communities in rivers. *Archive für Hydrobiologie* **119**: 1–14. [2.4]

Greenberg AE. (1964) Plankton of the Sacramento River. *Ecology* **45**: 40–9. [2.2]

Gregory SV. (1983) Plant–herbivore interactions in stream systems. In: Barnes JR, Minshall GW (eds) *Stream Ecology*, pp 157–89. Plenum Press, New York. [2.3, 2.4]

Grime JP. (1979) *Plant Strategies and Vegetation Processes*. Wiley-Interscience, Chichester. [2.4]

Hildrew AG, Townsend CR. (1987) Organization in freshwater benthic communities In: Gee JHR, Giller PS (eds) *The Organization of Communities, Past and Present*, pp 347–71. Blackwell Scientific Publications, Oxford. [2.4]

Humphries SE, Imberger J. (1982) *The influence of the internal structure and dynamics of Burrinjuck Reservoir on phytoplankton blooms*. Centre for Water Research. Report ED. 82–023, University of Western Australia, Nedlands. [2.3]

Hydraulics Research Laboratory (1981) *The Servern Estuary:observation of tidal currents, salinities and suspended solids concentrations*. Report EX 966. Hydraulics Research Ltd., Wallingford. [2.3]

Hydraulics Research Laboratory (1985) *The hydraulic roughness of vegetation in open channels*. Report IT 281, Hydraulics Research Ltd., Wallingford. [2.2]

Hynes HBN. (1970) *The Ecology of Running Waters*. University of Liverpool Press, Liverpool. [2.1, 2.2, 2.4]

Joint IR, Pomroy AJ. (1981) Primary production in a turbid estuary. *Estuarine, Coastal and Shelf Science* **13**: 303–16. [2.3]

Kowalczewski A, Lack TJ. (1971) Primary production and respiration of the phytoplankton of the rivers Thames and Kennet at Reading. *Freshwater Biology* **1**: 197–212. [2.3]

Krogmann DW, Butalla K, Sprinkle J. (1986) Blooms of cyanobacteria in the Potomac River. *Plant Physiology* **80**: 667–71. [2.4]

Lack TJ. (1971) Quantitative studies on the phytoplankton of the rivers Thames and Kennet at Reading. *Freshwater Biology* **1**: 213–24. [2.3]

Lack TJ, Youngman RE, Collingwood RW. (1978) Observations on a spring diatom bloom in the River Thames. *Verhandlungen des internationale Vereinigung für theoretische und angewandte Limnologie* **20**: 1435–9. [2.2]

Lamberti GA, Moore JW. (1984) Aquatic insects as primary consumers. In: Resh VH, Rosenberg DM (eds) *The Ecology of Aquatic Insects*, pp 164–95. Praeger Scientific, New York. [2.3]

Leopold LB, Wolman MG, Miller JP. (1964) *Fluvial Processes in Geomorphology*. Freeman, San Francisco. [2.3]

Lund JWG. (1965) The ecology of the freshwater phytoplankton. *Biological Reviews of the Cambridge Philosophical Society* **40**: 231–93. [2.3]

Margalef R. (1960) Ideas for a synthetic approach to the ecology of running waters. *Internationale Revue des gesamten Hydrobiologie* **45**: 133–53. [2.2, 2.3, 2.4]

Meybeck M, Chapman DV, Helmer R. (1989). *Global Freshwater Ecology*. Basil Blackwell, Oxford. [2.3]

Mohammed AA, Ahmed AM, El-Otify AM. (1989) Field and laboratory studies on Nile phytoplankton in Egypt. IV. Phytoplankton of Aswan High Dam Lake (Lake Nasser). *Internationale Revue des gesamten Hydrobiologie* **74**: 549–78. [2.4]

Moss B, Balls H. (1989) Phytoplankton distribution in a floodplain lake and river system. II. Seasonal changes in the phytoplankton communities and their control by hydrology and nutrient availability. *Journal of Plankton Research* **11**: 839–67. [2.2]

Nielsen SL, Sand-Jensen K. (1990) Allometric scaling of maximal photosynthetic growth rate to surface/volume ratio. *Limnology and Oceanography* **35**: 177–81. [2.3]

Odum EP. (1969) The strategy of ecosystem development. *Science* **164**: 262–70. [2.4]

Phinney HK, McIntyre CD. (1965) Effect of temperature on metabolism of periphyton communities developed in laboratory streams. *Limnology and Oceanography* **10**: 341–4. [2.3]

Prowse GA, Talling JF. (1958). The seasonal growth and succession of plankton algae in the White Nile. *Limnology and Oceanography* **3**: 223–8. [2.2, 2.4]

Reynolds CS. (1987) The response of phytoplankton communities to changing lake environments. *Schweizerische Zeitschrift für Hydrologie* **49**: 220–36. [2.4]

Reynolds CS. (1988) Potamoplankton: paradigms, paradoxes and prognoses. In: Round FE (ed.) *Algae and the Aquatic Environment*, pp 285–311. Biopress, Bristol. [2.2, 2.4, 2.5]

Reynolds CS. (1989) Physical determinants of phytoplankton succession. In: Sommer U (ed.) *Plankton Ecology*, pp 9–56. Science-Tech Publishers, Madison. [2.3]

Reynolds CS, Glaister MS. (1989) Remote sensing of phytoplankton concentrations in a UK river. In: White SJ (ed.) *Proceedings of the NERC Workshop on Airborne Remote Sensing, 1989*, pp 131–40. Natural Environment Research Council, Swindon. [2.2, 2.3]

Reynolds CS, Glaister MS. (1993) Spatial and temporal changes in phytoplankton abundance in the upper and middle reaches of the River Severn. *Large Rivers*. [2.2]

Reynolds CS, Carling PA, Beven K. (1991) Flow in river channels: new insights into hydraulic retention. *Archiv für Hydrobiologie* **121**: 171–9. [2.2]

Reynolds CS, White ML, Clarke RT, Marker AF. (1991) Suspension and settlement of particles in flowing water: comparison of the effects of varying water depth and velocity in circulating channels. *Freshwater Biology* **24**: 23–4. [2.3]

Round FE. (1981) *The Ecology of the Algae.* Cambridge University Press, Cambridge. [2.1]

Sabater S, Muñoz I. (1990) Successional dynamics of the phytoplankton in the lower part of the River Ebro. *Journal of Plankton Research* **12**: 573–92. [2.4]

Smith IR. (1975) *Turbulence in Lakes and Rivers.* Scientific Publication No. 29. Freshwater Biological Association, Ambleside. [2.2, 2.3]

Swale EMF. (1969) Phytoplankton in two English rivers. *Journal of Ecology* **57**: 1–23. [2.2]

Symoens J-J, Kusel-Fetzmann E, Descy J-P. (1988) Algal communities in continental waters. In: Symoens JJ (ed.) *Vegetation of Inland Waters,* pp 183–221. Kluwer Academic Publishers, Dordrecht. [2.4]

Talling JF, Rzóska J. (1967) The development of plankton in relation to hydrological régime in The Blue Nile *Journal of Ecology* **55**: 637–62. [2.2]

Valentine EM, Wood JR. (1977) Longitudinal dispersion with dead zones. *Journal of the Hydraulics Division, The American Society of Civil Engineers* **103**: 975–80. [2.2]

Ven Te Chow. (1981) *Open-channel Hydraulics,* international edition. McGraw-Hill Kogakusha, Tokyo. [2.3]

Wawrik F. (1962) Zur Frage: Fuhrt der Donaustrom autochtones Plankton? *Archiv für Hydrobiologie Supplementband* **27**: 28–35. [2.2]

Whitehead PG, Hornberger GM. (1984) Modelling algal behaviour in the River Thames. *Water Research* **18**: 945–53. [2.3]

Whitton BA. (1975) Algae. In: Whitton BA (ed.) *River Ecology,* pp 81–105. Blackwell Scientific Publications, Oxford. [2.1, 2.2, 2.3]

Wiens JA. (1989) Spatial scaling in ecology. *Functional Ecology* **3**: 385–97. [2.4]

Williams LG. (1964) Possible relationships between plankton-diatom species numbers and water-quality estimates. *Ecology* **45**: 809–23. [2.3]

Young PC, Wallis SG. (1987) The aggregated dead-zone model for dispersion in rivers. *Proceedings of the Conference on Water Quality Modelling in the Inland Natural Environment,* pp 421–33. BHRA, Cranfield. [2.3]

Zacharias O. (1898) Das Potamoplankton. *Zoologische Anzeiger* **21**: 41–8. [2.2]

3: Macrophytes

A.M.FOX

3.1 INTRODUCTION

The aquatic macrophytes comprise a diverse assemblage of plants that have become adapted from terrestrial species to life wholly, or partially, in fresh water. Their roles in aquatic environments have evoked increasing interest in recent years as multipurpose utilization of aquatic habitats has intensified. While much of the research in this field has concentrated on the nuisance species that interfere with water use, concerns over the restoration and protection of natural systems have prompted further studies relating to all aquatic plants. Many aquatic habitats can be effectively managed or protected only by applying a knowledge of the biology and ecology of macrophytes. However, more emphasis has been placed on the vegetation of lakes and canals than on that of rivers.

Both terms 'aquatic' and 'macrophytes' are open to interpretation but definitions and examples will be chosen here that describe the majority of plants found in rivers. 'Aquatic' plants will be defined according to Cook (1974) as those whose photosynthetically active parts are permanently or, at least, for several months each year submerged in, or floating on, fresh water. This definition allows differentiation between the truly aquatic species and those that tolerate only occasional high flood stages or, as with many riverbank trees, are rooted in saturated substrata.

'Macrophytes' are limited to the macroscopic flora including aquatic spermatophytes (seed-bearing plants), pteridophytes (ferns and fern allies) and bryophytes (mosses and liverworts). Algae, such as the filamentous *Cladophora* spp. and macroscopic charophytes (e.g. *Chara* and *Nitella*) are often included in this definition but they have been discussed in Chapter 2.

Macrophytes are often classified by their growth habit rather than taxonomically. Various subdivisions and terminologies have been proposed but a simple four-group system is widely accepted (Fig. 3.1: Sculthorpe 1967) *Emergent macrophytes* are rooted plants with most of their leaves and stem tissue above the water surface (e.g. *Phragmites australis*, *Typha latifolia*, *Sagittaria* spp). *Floating-leaved macrophytes* are rooted plants with most of their leaf tissue at the water surface (e.g. *Nymphaea alba*, *Nymphoides peltata*). *Free-floating macrophytes* are plants not rooted to the substratum but living unattached within or upon the water (e.g. *Ceratophyllum demersum*, *Eichhornia crassipes*, *Lemna* spp.). *Submerged macrophytes* are rooted plants with most of their vegetative tissue beneath the water surface (e.g. *Ranunculus penicillatus* var. *calcareus*, *Hydrilla verticillata*).

Although this system generally works well, depending upon environmental conditions or growth stage some species are not restricted to a single category. Amphibious plants (Hutchinson 1975) at maturity may have quite distinct (heterophyllous; defined by Sculthorpe 1967) forms of submerged and emersed leaves, the proportions of which vary with water depth (Spence *et al* 1987). Other heterophyllous species show temporal segregation of growth forms. Some (e.g. *Nuphar lutea*) have submerged leaves in the juvenile stage followed later by large floating or emergent leaves.

To appreciate the specialized adaptations that are exhibited by macrophytes in rivers, it is first necessary to recognize general adaptations to the

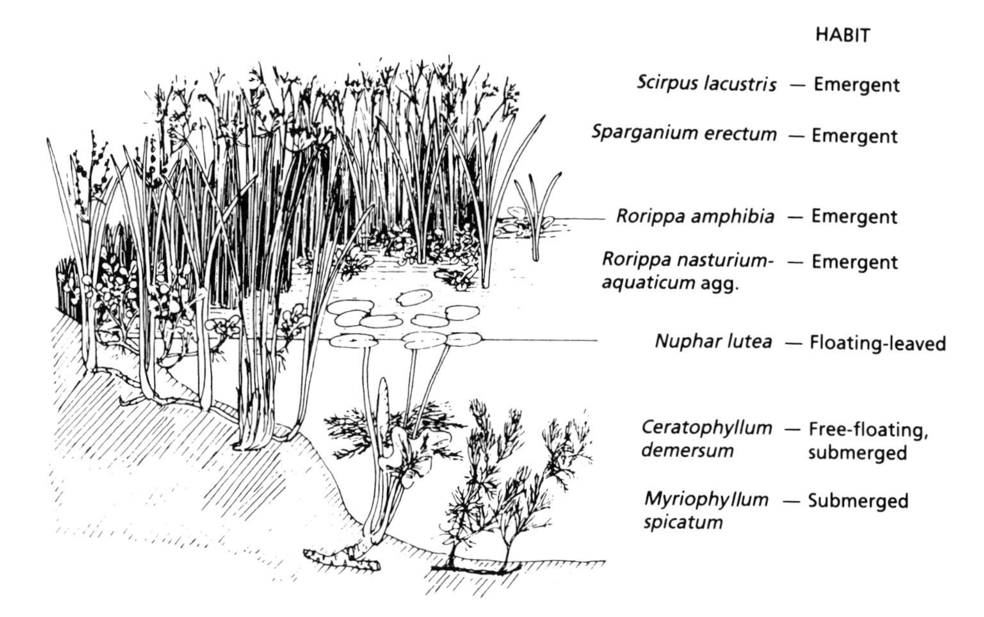

Fig. 3.1 Examples of species of the four macrophyte habits found in a lowland, clay river (modified from Haslam 1978).

aquatic environment. All freshwater macrophytes have evolved from terrestrial species, either by reduction of terrestrial characteristics or by evolving secondary adaptations. The most significant differences between the terrestrial and aquatic environments are the much slower (10 000 times) rates of gas diffusion in water compared with air (e.g. movement of carbon dioxide to leaves for photosynthesis and oxygen to roots for respiration) and the attenuation of light entering water, by reflection, absorption and scatter.

A major adaptation to these constraints is the formation of aerial leaves, as seen in the emergent, floating-leaved and surface free-floating species. More specialized adaptations required for the growth of submerged tissues include thin, dissected leaves lacking an epidermis, internal gas transfer, and specialized photosynthetic physiology (Bowes & Salvucci 1989; Boston *et al* 1989).

In this chapter, the principles involved in species distribution will be discussed first. Next, aspects of quantitative macrophyte ecology that consider growth patterns and productivity will be introduced, with a detailed treatment reserved for Chapter 9 on primary production. Finally, these principles will be incorporated into a discussion of the interactions between macrophytes and human activities, emphasizing the need to understand such principles if rational river management is to be achieved.

3.2 DISTRIBUTIONAL ECOLOGY OF MACROPHYTES

The objective of distributional ecology is to examine the factors that determine the presence or absence of a species at a particular site. Three major factors are readily identified: dispersal, tolerance of abiotic environment, and interactions with the biota. The sequence in which they should be considered if the absence of a species from a particular site is to be determined is indicated by the series of questions in Fig. 3.2.

Dispersal

It is not difficult to envisage that the rate of downstream dispersal of macrophytes within a river system is determined by the size and weight

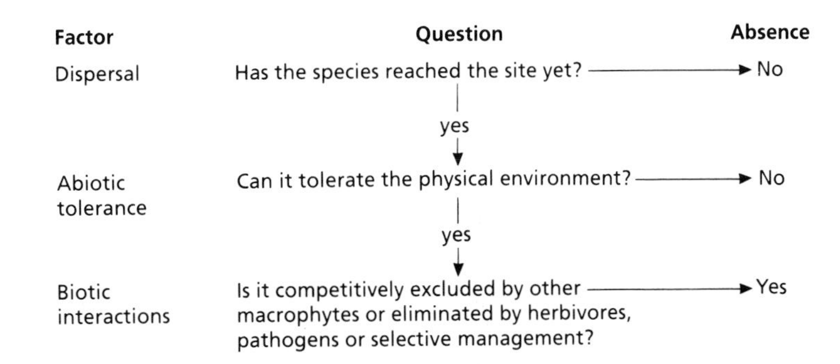

Fig. 3.2 The sequence of factors and questions to be addressed to determine why a species is absent from a particular site.

of the propagules and the force of water movement. Dispersal between river catchments or to upstream sites is less obvious and is dependent on a variety of means other than water movement.

Organs of dispersive propagation

Aquatic macrophytes in general, and submerged river plants in particular, show a tendency towards replacing sexual with vegetative reproduction (Sculthorpe 1967). For submerged macrophytes this may be related in all but a few species to the difficulty of raising the flowers above the water surface for aerial fertilization.

Organs of vegetative dispersal may include whole plants (e.g. *Lemna* spp., *Eichhornia crassipes*), shoot fragments including buds or nodes (e.g. *Ceratophyllum demersum*) or specialized organs, such as tubers (swollen stem or root sections, e.g. *Hydrilla verticillata*) and turions (dormant apices, e.g. *Hydrilla*, *Potamogeton* spp.). Spread of existing stands of macrophytes may be accomplished by other vegetative mechanisms (e.g. rhizomes, stolons) but they do not usually contribute to the colonization of new sites except under flood conditions with severe channel scouring.

Seeds are important organs of dispersal in emergent macrophytes which are less likely than other aquatic species to fragment or be wholly displaced. Flowers of emergent and surface free-floating species do not need to be modified from the terrestrial habit and are usually wind or insect pollinated (e.g. *Glyceria maxima*, *Hydrocharis morsus-ranae*). Floating-leaved species are usually fertilized in the same manner, with their chief adaptation to the aquatic environment being the production of long peduncles (flower stalks) capable of lifting the flower above the water (e.g. *Nymphaea alba*). In flowing conditions these stems need to be longer than the depth of water to accommodate changes in water level and bending as a result of water velocity.

Ironically, the evolution of the angiosperm flower as a mechanism that liberated plants from a dependence on water for gamete dispersal has been only secondarily adapted to allow pollination in water in a very few submerged species (e.g. *Vallisneria spiralis*, *Zannichellia palustris*; Sculthorpe 1967).

Mechanisms of macrophyte dispersal

In addition to the downstream movement of reproductive propagules within a river, flooding may carry propagules to adjacent lakes, canals or ditches. However, flooding usually cannot carry propagules from one watershed to another.

Wind dispersal of small seeds is common in the aquatic grasses (e.g. *Phragmites*, *Phalaris* spp.) and other emergent species (e.g. *Typha*). Transport of specialized propagules (seeds, turions or tubers) or stem fragments (e.g. *Lemna* thalli, sections of flexuous *Elodea canadensis* stem) by birds and other animals, either within the digestive tract or by external attachment, can disperse plants between river catchments. Unlike the specialized reproductive propagules that can usually tolerate some degree of digestion or desiccation (Basiouny *et al* 1978), the distance that fragments of submerged species may be transported would be limited.

Whole macrophytes or their propagules are

often transported over long distances as a result of intentional (e.g. *Eichhornia crassipes* into the USA) or accidental carriage by humans. Accidental introductions to river catchments or countries could result from transport of water, sediments, crops (e.g. rice, cotton), animal fur or wool, or with desirable macrophytes, such as aquarium plants (Schmitz *et al* 1991). The often disastrous consequences of introducing exotic weedy macrophyte species to water bodies will be discussed in the last section of this chapter.

Geographical distribution of macrophytes

The geographical distribution of a species can give an indication of how effectively it is dispersed. *Phragmites australis* and *Typha* spp., which have small windborne seeds, are widely distributed around the world (Sculthorpe 1967); however, members of the Podostemaceae family are typically endemic to small geographical areas of tropical Africa or even to single river basins. The restricted distribution in the latter group relates to the poor dispersal of their minute seeds to other suitable habitats (Willis 1914). Aquatic macrophytes are remarkable for their large proportions of ubiquitous and endemic species compared to terrestrial vegetation (Sculthorpe 1967).

Dispersal is not the only factor involved in determining geographical distribution. A narrow distribution may not be caused by poor dispersal but may result from limited tolerance of a species to certain climatic and/or geological conditions. A species on the verge of extinction or recently evolved will have a limited geographical range, regardless of its dispersive capabilities.

Tolerance of abiotic environment

The question of whether a species' incidence is determined by its dispersal (Fig. 3.2) will tend to be answered by a simple 'Yes' (the species has had the opportunity to reach the site) or 'No' (there has not been such an opportunity). The two other factors affecting species incidence (abiotic tolerance and biotic interactions), however, will also influence species abundance. A macrophyte species will be excluded from a site if it cannot tolerate the abiotic conditions to which it would be exposed throughout its normal life-cycle. For annual species (e.g. many *Najas* spp.) this would mean that even if the mature plant could tolerate a particular environment, survival would be temporary if the seedling would perish under the same conditions. Many macrophytes which reproduce vegetatively do not exhibit such susceptible stages and so may tolerate wider environmental ranges.

The absence of a species will be determined by intolerance of extremes of abiotic conditions. The abundance of a species will be related to how near the conditions are to those optimal for maximum growth rates. Limits of tolerance and growth optima may be determined for a particular species and environmental variable by correlation of species presence or abundance with measured levels of environmental variables in the field. This method requires large numbers of field samples, and direct causal relationships cannot be assumed from such data. Growth under a range of controlled environmental conditions in artificial habitats or transplants between differing natural sites may also be used to determine tolerance and growth ranges for certain variables. However, several series of experiments with multiple variables are required to determine interactions between abiotic factors (e.g. Maberly 1985).

Although it is necessary to discuss each environmental variable separately, macrophytes do not respond to each variable independently. For example, macrophyte growth may be directly influenced by substratum particle size. However, such influence cannot be easily distinguished from the effects of water velocity, which not only affects macrophytes directly but also determines substratum particle size. The effects of environmental variables may also interact if suboptimal conditions of one variable alter the tolerance of a species to another variable.

A further complication of this dynamic system is that all variables do not remain constant over time within a habitat. Short-term, seasonal, or longer-term changes (flood, drought, turbidity) can improve or reduce the suitability of a habitat for a particular species. Temporary periods of adverse conditions may be survived by the production of specialized organs of perennation (e.g. seeds, tubers, turions, rhizomes). The frequency, duration and amplitude of environmental variations will influence the tolerance of a species to an affected habitat (Westlake 1981).

Water movement

Water movement is one of the most important and specific abiotic variables influencing species composition and location of plant communities in rivers. There can be direct and indirect effects on the vegetation related to aspects of water movement such as velocity, turbulence and erosive force. Direct effects of water movement include influences on photosynthetic rates and exposure to nutrients and carbon dioxide, mechanical damage and propagule establishment. Indirect effects occur through the influence of water movement on substrata and fauna.

Metabolic rates of macrophytes are generally higher in moving water than in still water (Westlake 1967; Madsen & Søndergaard 1983). At flow rates typically found within plant beds, photosynthesis increases with water velocity in riverine *Ranunculus* species. This may result from the reduction in thickness of the boundary layers surrounding the macrophyte leaves that occurs as water velocity increases. The thinner the boundary layer, the less limiting is the rate of diffusion of dissolved carbon sources across this layer, from the main body of water to the leaf surface (Westlake 1967).

Faster flowing water will tend to be more turbulent (especially as the substrata will tend to be coarser) and this may also improve macrophyte photosynthesis by increasing the aeration of the water with atmospheric carbon dioxide. The relationship of water turbulence and concentrations of dissolved carbon dioxide explains why macrophytes unable to utilize bicarbonate carbon sources, such as many mosses, are limited in calcareous waters to areas of turbulent flow as found in highland streams, riffles, or on weirs in lowland rivers (Bain & Procter 1980).

Water movement in rivers will also expose macrophytes to a constantly replenished source of nutrients, less likely to become diminished and limiting than in static water. However, despite the metabolic stimulation resulting from the improved availability of carbon dioxide and nutrients in flowing water, there is little evidence that growth is faster in rivers than in lakes (Westlake 1975). This is probably because of increased organic metabolite losses (Wetzel 1969) resulting from the reduced boundary layer and, more significantly, the continual loss of plant material by mechanical damage.

Although the metabolic rates of some species may rise with increased water movement more than others, such factors tend to have more of an effect on total macrophyte abundance than on species selection. Some riverine species (e.g. *Ranunculus fluitans*) may only tolerate, or be competitive in, the improved metabolic conditions of flowing water. However, the tolerance of different species to the mechanical damage associated with various flow regimes probably has a greater impact on river community structure.

The most obvious mechanical influence of water movement on macrophyte communities is that free-floating species will be limited to areas, or periods, of reduced flow. This does not mean that such species cannot have major impacts on some rivers (e.g. infestations of *Eichhornia crassipes* in tropical rivers), but their role in most temperate rivers is minor compared with the rooted species.

The various types of mechanical damage that rooted river plants must tolerate—uprooting, battering, tangling, abrasion, erosion and sedimentation—have been described by Haslam (1978). The hydraulic resistance of individual plants depends on their dimensions relative to the direction of flow and their leaf, stem and branching shapes. Broad, bushy plants (e.g. *Myriophyllum spicatum*, *Rorripa* spp.) will create more resistance and be more susceptible to uprooting and battering than low, streamlined ones (e.g. *Ranunculus*, *Vallisneria* spp.). Increased anchoring strength reduces the susceptibility of a plant to uprooting and sediment erosion, and is greatest in plants with well-developed root and rhizome systems (e.g. *Sparganium erectum*). Tissue robustness determines susceptibility to abrasive damage and whether a plant will tear apart as water velocity increases. Battering damage is also influenced by the turbulence of flow and, when severe, stems of long flexuous plants may become tangled.

Laboratory experiments have been conducted on individual, and pieces of, macrophytes to try to determine hydraulic resistance and the susceptibility of different species to these various types of damage (Haslam 1978). In the field, however, the shape of plant beds may be more important than of individual stems (Pitlo & Dawson 1990). Water velocity is significantly reduced within beds of many riverine macrophyte species (Losee & Wetzel 1988; Getsinger *et al* 1990) with inner stems being protected from sudden increased water velocities and turbulence. Thus, species that have individual stems with high hydraulic resistance (e.g. *Elodea canadensis*) may form beds with an overall streamlined shape. Such species could not colonize sites with constantly high water velocity or turbulence but might survive short periods of such spate conditions once beds were established. This indicates how important the periodicity of extreme flow conditions may be and, hence, how the stability of flow regimes (as determined by the catchment geology) can influence macrophyte community structure.

Water velocity and turbulence also influence the establishment of macrophyte propagules. Unless they are sheltered from flow by being carried into backwaters, deep depressions or plant beds, buoyant or delicate propagules will be swept downstream or mechanically damaged before they can root and grow in most rivers. Very small propagules may become trapped in the boundary layers and dead water downstream of suitable obstacles, such as rocks and plant beds (*cf.* invertebrate habitats; Hynes 1970). All the dangers of mechanical damage discussed above must then be tolerated by the new plant as it grows out of the sheltered area. Although the hydraulic resistance of small shoots is less than of mature ones, their tissues are usually more delicate and their roots less secure.

Association of particular macrophyte species with certain ranges of water velocity has frequently been based on qualitative observations. The observations of Butcher (1933), for example, have often been cited (e.g. Hynes 1970; Westlake 1975). More recently, quantitative data have been collected in large-scale field surveys of Britain and Europe (Haslam 1978, 1987; Holmes 1983) with correlations made between the presence or abundance of species and flow regime.

Substratum

The geology underlying a watershed has a major influence on the physical and chemical characteristics of its river channels and lake basins, and, hence, on macrophyte colonization. Catchment geology affects patterns of rainfall runoff and the stability of flow regimes. For example, chalk rivers have little seasonal fluctuation in discharge rates and provide stable conditions for macrophyte growth. Rivers rising on erosion-resistant rocks have spatey flow regimes, dependent upon seasonal rainfall distribution and support only those macrophytes tolerant of mechanical damage or able to rapidly recolonize a flood-scoured site.

On a more local scale, variations in the resistance of underlying rocks will determine erosional channel features such as riffles and pools. Variable water depth and flow rates between such features can influence the communities of macrophytes over distances of just a few metres.

Geology also determines the general type of substratum in the river channel, whether resistant rocks, the firm gravel of limestone rivers or the easily-eroded and silty deposits of clay catchments. Secondarily, local water velocity defines the characteristics of river substrata, with fine sediments and organic matter being eroded or deposited accordingly.

Pearsall (1920) first documented the associations between certain macrophytes and substratum-type in the Cumbrian lakes, particularly with reference to silting rates at the mouths of inflowing streams. Subsequent investigations of substratum–macrophyte relations have concentrated on lacustrine species, and on nutrient availability rather than physical characteristics of the sediments (Barko & Smart 1981b).

River surveys (Butcher 1933; Haslam 1978; Holmes 1983) have indicated that many macrophyte species are associated with sediment of particular particle size. Do these species have a requirement for certain substratum characteristics, or is that relationship coincidental to their selection by water movement and the causal relationship between flow rates and sediment type? Carefully controlled experiments (e.g. Barko & Smart 1986) would be needed to answer such a question.

Substratum particle size influences macrophyte colonization with respect to root depth and stability. Deep rooting patterns (e.g. *Sparganium erectum*) associated with thick layers of fine sediments stabilize plants during spate conditions when the upper soil is disturbed but are susceptible to uprooting in sustained increases in flow. Species associated with coarser particles (e.g. *Ranunculus* spp.) tend to have networks of curling roots that can securely anchor to individual stones and consolidate the sediment for greater stability in faster flowing rivers (Haslam 1978).

The effects of macrophytes on the substratum, other than rooting consolidation, include the production of organic material and sufficient slowing of water movement within plant beds to allow the sedimentation of suspended materials. This sediment accumulation may permit the colonization of species indicative of finer sediments and lesser flow. For example, in shallow chalk streams the accumulation of silt in *Ranunculus penicillatus* var. *calcareus* beds permits the colonization of the emergent *Rorripa nasturtium-aquaticum* (Dawson *et al* 1978).

Light

The attenuation of light in clear water is a major constraint to submerged macrophyte growth. Light availability principally affects the abundance of submerged vegetation, but since some species are better adapted to low light intensities than others (e.g. *Hydrilla*; Bowes *et al* 1977) it also has some influence on community structure. Factors that increase light attenuation in water proportionately reduce the depth of plant colonization.

The presence of suspended solids (turbidity) in moving water has the most influence on light attenuation. Prolonged periods of increased water velocity and factors in the catchment that elevate sediment loading can reduce plant growth, particularly in deep waters. Water colour is also determined by characteristics of the river catchment. Heavily forested catchments and rivers with a high proportion of runoff from bogs and swamps may have brown, tannin-stained water that significantly reduces light penetration even if there is no turbidity.

In slow-moving water, shading by floating or floating-leaved macrophytes or phytoplankton, may limit the growth of submerged species. Epiphytic algae can significantly reduce photosynthetic rates of macrophytes that are susceptible to their colonization (Sand-Jensen & Borum 1984). Shading by riparian trees reduces the abundance of all types of macrophytes in narrow river channels (Canfield & Hoyer 1988).

Temperature

While some macrophyte species with a worldwide distribution do not appear to have any special temperature requirements (e.g. *Phragmites australis*, *Lemna minor*), other species are limited to certain climatic zones (e.g. cool: *Sparganium emersum*; warm: *Hydrilla verticillata*; Sculthorpe 1967; Barko & Smart 1981a). River plants do not appear to be affected by this any differently from lacustrine macrophytes but the source of a river's water may have a local influence on temperature. Spring-fed rivers will tend to be of more constant temperature throughout the year compared with rivers that are more dependent upon catchment runoff. Near the springs species may occur that require warmer (in the higher latitudes) or cooler (in the tropics) conditions than the vegetation occurring further downstream or in nearby rivers that are not spring-fed.

Water chemistry

The aspect of water chemistry that has most influence on macrophyte community structure is the complex relationship between pH, the predominant form of dissolved inorganic carbon (DIC), water hardness and calcium concentration (Butcher 1933; Spence 1967). Some species of aquatic plants are found only in acidic waters (e.g. *Potamogeton polygonifolius*) where carbon dioxide is the predominant source of DIC. Such species may be intolerant of more alkaline habitats because of an inability to use bicarbonate ions (Spence & Maberly 1985). The occurrence of certain species has been correlated with water calcium concentrations (Haslam 1978) or has been associated with catchment geologies that have a significant influence on water hardness and pH (Holmes 1983).

The correlation of macrophytes with nutrient (particularly phosphorus and nitrogen) concentrations is complicated by the ability of many species to obtain minerals either by foliar uptake from the water or from the substratum via their roots (Denny 1980). Many lowland rivers (especially temperate chalkstreams) have been shown to have an excess of available nutrients so that other factors must be responsible for limiting plant growth in these habitats (e.g. Ladle & Casey 1971; Canfield & Hoyer 1988).

Other aspects of water chemistry may have localized influences on macrophyte community composition. For example, elevated salinity may exclude intolerant species (e.g. *Hydrilla verticillata*; Haller *et al* 1974) from river/estuary boundaries and the outflow of saltwater springs or mining operations.

Water level fluctuations

The effect of catchment geology on the stability of river discharges has been mentioned in relation to macrophyte colonization. Some rivers that are not spring-fed may be susceptible to summer droughts near their sources. This limits all but the emergent vegetation, to species that produce propagules capable of tolerating periodic drying (e.g. seeds and tubers). Rivers that undergo continuous gradual changes in water levels tend to have a broader fringe of emergent vegetation than rivers that are either very stable or that are subject to sudden and brief water level fluctuations and storm erosion.

Interactions with biota

Once a species has reached a site that it can tolerate, its survival there depends upon its interactions with other biota. Severe competition between macrophyte species may result in the exclusion of one species from a site by another. Under natural conditions, herbivores and plant pathogens, while reducing macrophyte abundance, will not eliminate their plant hosts. However, a reduction in vigour of macrophytes susceptible to grazing or pathogens might lead to competitive exclusion by less susceptible plant species.

Competition

Grime (1979) identified four resources for which plants compete: light, space, nutrients and water. By definition, aquatic macrophytes do not need to compete for water resources and in lowland rivers, at least, nutrients are often available in excess of requirements (Ladle and Casey 1971). Competition for light is probably the dominating factor in the relationships between river macrophytes.

Morphological features that allow one plant to modify the environment to the detriment of others provide greatest competitive advantages. For example, the shading of low-lying submerged plants by tall, canopy-forming species or by macrophytes of other habits. The production of allelopathic substances that inhibit the growth of other species may be regarded as a competitive characteristic. Most evidence for the release of such compounds has been produced under laboratory conditions (Leather & Einhellig 1986) and conclusive examples of this phenomenon have not been identified in rivers.

Grime (1979) listed 18 competitive characteristics for terrestrial plants, and those appropriate for aquatic macrophytes have been identified (Van *et al* 1978; Murphy *et al* 1990). Characteristics relevant to river plants include: canopy formation; use of bicarbonate for DIC; use of carbon dioxide in the air; early seasonal and daily growth resulting from a low light compensation point; low root/shoot ratio; and high litter production.

Most competition studies with macrophytes have been between major weed species and the native plants that they displace (e.g. *Hydrilla* and *Vallisneria*; Haller & Sutton 1975). De Wit replacement series have been carried out with floating and lacustrine submerged species (e.g. McCreary *et al* 1983; Agami & Reddy 1989). Emergent plants have been transplanted along gradients of changing species (Grace & Wetzel

1981) but few competition experiments have been carried out in flowing water.

Some species (usually those regarded as 'weedy') do appear to be able to dominate rivers to the exclusion of others. As predicted by Grime's triangular strategy theory, competitive exclusion to one or two species usually occurs in the least disturbed and physically stressed environments, such as nutrient-enriched lowland rivers (e.g. *Potamogeton pectinatus*; Holmes 1983).

River macrophytes play an important role in relation to other biota by providing habitat diversity (Westlake *et al* 1972). The large surface area of plants and local reductions in flow provide a substratum and shelter for epiphytic algae, invertebrates, fish and their eggs. Since some of these organisms are found in varying densities on different plant species (Soszka 1975; Miller *et al* 1989), a diversity of river macrophytes is usually desirable to promote a variety of other biota. Maintenance of macrophyte diversity by the prevention of competitive exclusion by species such as *Potamogeton pectinatus* is often a major objective of river vegetation management.

Herbivory and plant pathogens

Herbivory of submerged river plants is not extensive. Some snails (e.g. Sphaeridae), insects (e.g. Elmidae) and fish (e.g. cyprinid species) will scrape algal epiphytes off macrophytes, and other invertebrates and fish feed on detritus composed largely of dead plant material (Hynes 1970). Often crayfish and temperate fish, such as *Rutilus* sp., will eat fresh vegetation in addition to their diet of invertebrates but there are few species that are exclusively herbivorous (e.g. *Tilapia* spp. and Grass Carp, *Ctenopharyngodon idella*).

As with the tropical herbivorous fish, some plant pathogens have been investigated with respect to their potential as biocontrol agents of weedy macrophytes (Charudattan 1990). However, relatively little is known about the incidence of pathogens affecting temperate riverine macrophytes.

Synthesis of the distributional ecology of macrophytes

By following the series of questions in Fig. 3.2 and by examining the environmental requirements and competitive characteristics of a macrophyte species, it should be possible both to determine why that species is absent from a particular site and to predict suitable habitats for its growth. Trying to explain or anticipate detailed spatial and temporal changes in species composition and relative abundances is more complex because of the many interrelationships between macrophytes and their environment. Detailed studies of some lowland rivers have resulted in models (Fig. 3.3) that attempt to predict some

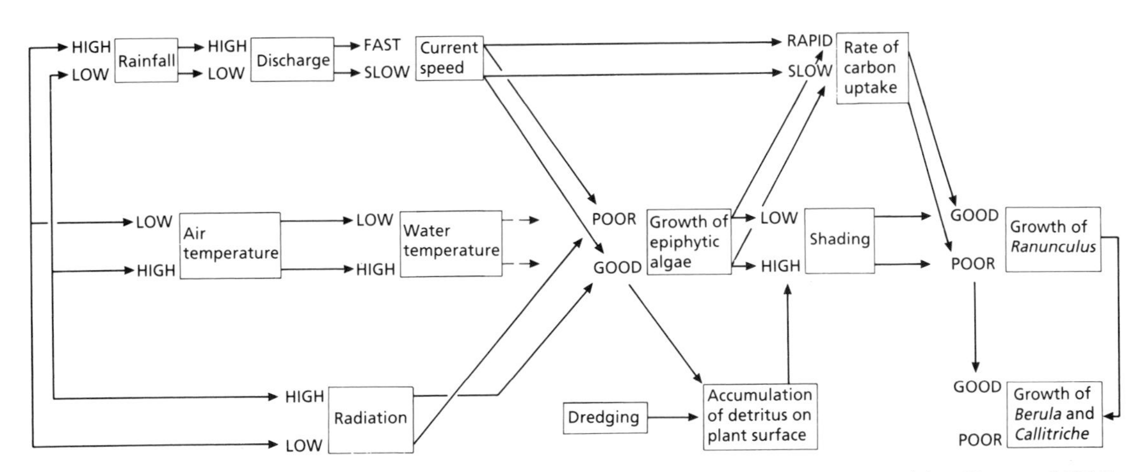

Fig. 3.3 A qualitative model of factors influencing the growth of *Ranunculus* at Bangor, UK (after Ham *et al* 1981).

simple changes in species dominance (Ham *et al* 1981). However, the production of more quantitative models for describing the dynamics of river communities is a far more daunting task requiring the interaction of researchers from a variety of physical and biological disciplines.

The dynamic interrelationships of plants, flow and sediments result in a mosaic of macrophyte species that changes by the month (Fig. 3.4), season or year (Butcher 1933; Ladle & Casey 1971). The unstable physical characteristics of river channels preclude the succession leading to climax vegetation often observed in small lakes and deltas. Rivers maintain vegetation typical of early succession due to periodic scouring. These may be seasonal events, washing out the summer's accumulation of vegetation, or the infrequent major spates that change the whole morphology of the channel. Only in the isolation of ox-bow lakes can the traditional concept of hydroseral succession proceed.

3.3 QUANTITATIVE ECOLOGY OF MACROPHYTES

In discussing the general quantitative ecology of macrophytes, the abundance and productivity of aquatic species and habitats are considered in relation to variable environmental factors. Temporal variations are considered with regard to seasonal development and growth patterns, and spatial variations are discussed in comparisons of the productivity of different types of aquatic vegetation and river sites. The difficulties of quantifying vegetation in flowing water provide a major constraint on this aspect of macrophyte ecology.

Seasonal growth patterns

In temperate and subtropical climates, seasonal changes in day length and temperature undoubtedly play an important role in regulating macrophyte growth patterns, just as they do for terrestrial plants. Rainfall patterns, and hence river discharge and erosive forces, may be just as significant. Many of the examples mentioned will assume the typical temperate pattern of low summer rainfall and higher rainfall in winter and/or spring and autumn.

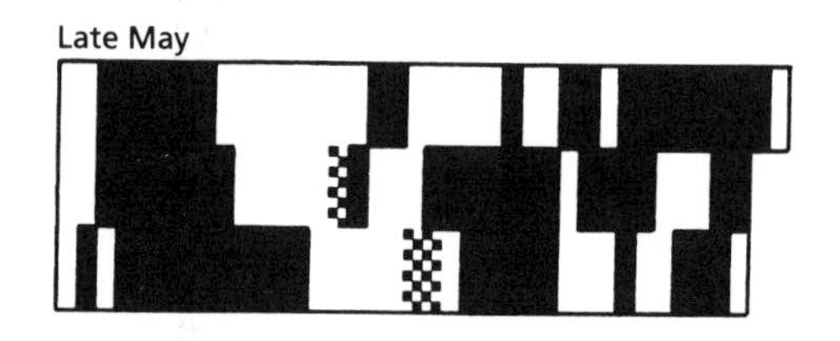

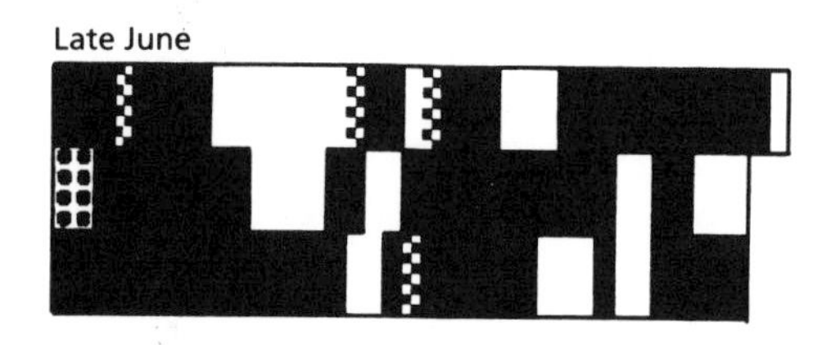

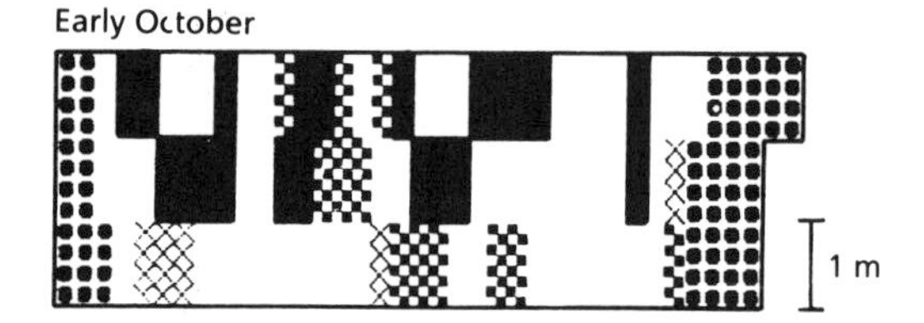

Ranunculus penicillatus var. *calcareus*.

Potamogeton pectinatus.

Other submerged species, e.g. *Myriophyllum spicatum*.

Filamentous algae, e.g. *Cladophora glomerata*.

Emergent species, e.g. *Phalaris arundinacae*.

Fig. 3.4 Maps of permanent transects in the lowland, clay/chalk stream River Windrush, UK, showing changes in species dominance throughout the summer of 1984. The dominant macrophytes in each 0.25 × 1.0-m rectangle based on surface coverage are indicated.

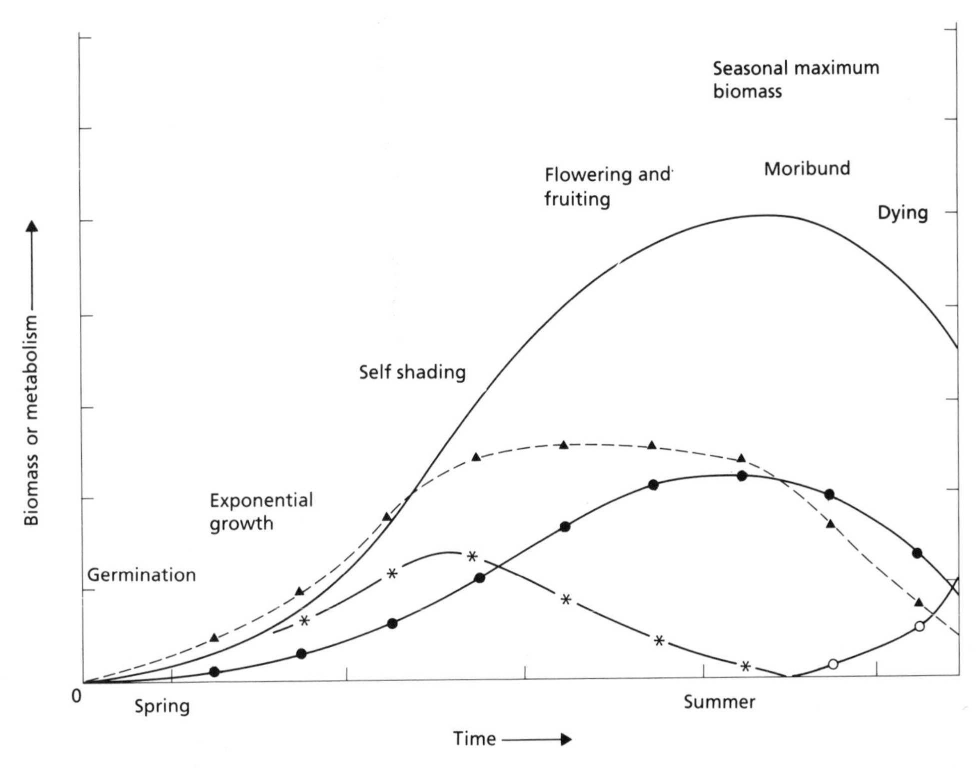

Fig. 3.5 Hypothetical growth and metabolism curves for an annual plant. – Biomass; ▲ current gross productivity; * current net productivity; ● current respiration rate; ○ death losses (after Westlake 1965).

Westlake (1965) described the typical growth pattern for an annual plant assuming negligible overwintering biomass (Fig. 3.5). The timing of the summer maximum biomass will vary with species or even with biotype within a species (e.g. *Ranunculus penicillatus* var. *calcareus*; Dawson 1980a). The timing and severity of the first autumn storms will influence the postflowering wash-out.

This pattern is appropriate for many perennial submerged species with the exception that growth in the spring starts from an overwintering biomass which may vary annually, depending upon the previous summer's production and winter discharges. Detailed studies of the production of *Ranunculus penicillatus* var. *calcareus* (Dawson 1976) revealed such a pattern, with self-shading within large beds slowing the exponential growth rates of early summer. To raise the insect-pollinated flowers above the water surface, this species forms thick buoyant flowering stems that are more brittle than the vegetative stems. Losses of these fragile stems after flowering initiates the late summer reduction in biomass, with 75% of the shoots being fragmented and lost within 3 months.

Such growth patterns are evident from biomass samples of most submerged river plants because underground biomass is small (0.5% of maximum *Ranunculus* biomass; Dawson 1976) and moribund tissues are immediately removed. This is not the case for many emergent and floating-leaved species that possess substantial rhizome systems (e.g. *Typha*, *Phragmites*, *Nuphar*) or other perennating organs (e.g. bulb-like root stock of *Sagittaria*) which may comprise much of the overwintering biomass. Emergent species are not as affected by seasonal increases in water discharge as submerged species but are more susceptible to winter frost damage.

Some of the problems of estimating the abundance and productivity of macrophytes, such as whether to sample biomass from both above and below ground, are not unique to quantitative studies of aquatic plants (Westlake 1963). Biomass sampling in rivers does require some specialized techniques to prevent the downstream loss of harvested material. Dredges or grab samplers operated from boats or by divers may be necessary in deep water (Sliger *et al* 1990). In shallow rivers, material removed from quadrats can be collected in surrounding boxes or nets (e.g. Lambourn sampler; Hiley *et al* 1981). Dry weights should be measured because it is difficult to ensure a uniform removal of surface water. Also the water content of different macrophytes varies considerably, with ranges of 85–92% of tissue weight being water for submerged species and 70–85% for emergent species (Wetzel 1983).

Estimates of vegetation cover either by remote sensing (e.g. aeroplane or balloon photography) or ground mapping tend to be less time consuming than biomass sampling. Random or systematic point sampling within the channel, or by use of grapnels from the bank, provides estimates of relative species abundances. Use of permanent transects to map plant beds can indicate changes in the placement of species over time. The division of transects into rectangles, from which the dominant species is recorded (see Fig. 3.4; Wright *et al* 1981) can speed up mapping procedures without great loss of information. The size of sampling units should reflect the level of diversity at a site. The patchy and dynamic distribution of submerged species encountered in diverse river communities (Butcher 1933) is best indicated by mapping techniques, and such patterns have to be assessed in the choice of size, number and placement of biomass samples.

Dawson (1976) showed that the percentage cover of *Ranunculus* was not a good indicator of its biomass in a shallow stream. While the *Ranunculus* biomass was halved over a 4-year period, the cover of the stream bed did not change significantly. Volume estimates of bed size rather than just those expressed in terms of area might improve the relationship with biomass but would be almost as time consuming to collect (e.g. using a recording fathometer; Maceina *et al* 1984).

Productivity of macrophyte communities

The continual losses of tissue from macrophytes in the river channel (Dawson 1980b) and accumulation of litter by emergent species complicate the estimation of productivity using repeated biomass measurements. Estimates have been achieved by tagging individual plants and leaves, and measuring changes in their size (Odum 1957; Dawson 1976), but such methods are highly intensive and small-scale. Indirect methods have included measurement of photosynthetic substrates and products (Odum 1956) but these studies are often restricted to specialized areas (e.g. near a spring mouth or weir) and are usually short term.

As a result, the productivity of river macrophytes has not received the same attention as terrestrial, wetland or lacustrine vegetation. However, a knowledge of some of the environmental factors influencing macrophyte growth can help in predicting the relative productivities of various lotic sites.

Submerged species are less productive than emergent and surface-floating plants (Westlake 1963; Wetzel 1983) because of the slow diffusion of photosynthetic gases and the reduced intensity of incident light. By virtue of warmer and longer growing seasons, macrophytes in tropical rivers are more productive than temperate ones. Within a climatic zone, differences between rivers or reaches of a river depend upon all the abiotic features outlined above, as well as growth characteristics of the composite species.

3.4 INTERACTIONS BETWEEN MACROPHYTES AND HUMAN ACTIVITIES

As with many biological disciplines, the most intensively studied species of macrophytes are the aquatic weeds; by definition, those that interfere with human activities. Murphy (1988) reviewed the major aquatic weed species of the world, and many of the problems that they cause for humans.

Conversely, human activities related to the use of natural resources, power production and flood control often affect rivers and associated macrophytes. The important roles that macrophytes play in lotic habitats can be summarized as: primary production; photosynthetic production of oxygen; substratum for algae; habitat for invertebrates and fish eggs; nutrient cycling to and from sediments; stabilization of river beds and banks. In view of these roles, activities that reduce or drastically alter the composition of macrophyte communities must be acknowledged.

Macrophytes affecting human activities

Flooding risks

The ways in which macrophytes influence water flow in rivers tend to be related to the abundance and habit of the vegetation rather than to species composition. As the biomass of macrophytes in a channel increases so does the hydraulic roughness (Dawson 1978; Carling 1994) which results in a decrease of mean water velocity. If the river discharge is constant, a reduction in velocity will cause an increase in water depth which may overflow the river banks, or raise the water table and flood or waterlog what is commonly valuable agricultural land.

The deposition of suspended sediments within macrophyte beds reduces the effective cross-section of the channel available for discharge. This can result in channelization of rapid flow between plant beds which will locally scour the substratum and may undercut river banks. Under spate conditions, these narrow channels cannot carry all the additional water, causing flooding.

Interference with fishing and navigation

Tall emergent vegetation along river banks can prevent access for shoreline fishing. Dense growths of submerged plants, particularly those that have reached the water surface, interfere both with fishing itself and the use of boats to get to areas inaccessible from the shore. Perennial submerged species can spoil the gravel spawning beds of salmonid fish (e.g. *Potamogeton pectinatus*; Caffrey 1990). High densities of photosynthesizing macrophytes can cause large diel fluctuations in oxygen which can stress fish. In non-turbulent waters, under prolonged hot and cloudy conditions, when photosynthesis does not exceed respiration, oxygen depletion can result in fish mortality (Brooker *et al* 1977).

Some of the greatest obstacles to commercial fishing and navigation have occurred on tropical rivers as a result of floating vegetation, such as *Eichhornia crassipes* and *Salvinia molesta*. Sudd formation (floating islands) is a major problem to navigation in parts of Africa and India, and results from dense mats of floating vegetation on top of which emergent and even woody species can grow (Sculthorpe 1967).

Reduction in water supply

The continuous loss of fragments from submerged vegetation near irrigation or drinking water intakes can interfere with water supply by the clogging of pumps and filters. In tropical climates the loss of water by the transpiration of emergent and some floating vegetation exceeds the evaporative loss from an equivalent area of unvegetated water surface (Rao 1988), and in arid environments such losses may significantly reduce water supply.

Habitat for disease organisms

Several human diseases are carried by intermediate hosts that favour aquatic habitats vegetated by macrophytes. Examples include: schistosomiasis (bilharzia) carried by aquatic snails; filariasis carried by *Mansonia* mosquitoes; and malaria carried by *Anopheles* mosquitoes. The intermediate hosts are either dependent upon certain macrophytes for completion of their life-cycle (e.g. *Mansonia* attaches to roots of *Pistia stratiotes*) or inhabit stagnant water resulting from the obstruction of water-courses by vegetation. The damming of many rivers and construction of irrigation canals has particularly increased the incidence of such diseases in many tropical areas of the world (Brown & Deom 1973).

Human activities affecting macrophytes

Weed management

To reduce the problems created by aquatic weeds,

various forms of plant management are employed throughout the world. Usually the objectives of responsible weed management are to reduce the biomass of problem species rather than the total elimination of vegetation. Management of some fisheries or conservation areas may not require reductions in total vegetation so much as increasing species diversity, or disturbing the environment to retain a colonizing flora.

Chemical control is less likely to be used for submerged weed management in rivers than in lentic systems because of the difficulties of maintaining a period of contact, sufficient for a phytotoxic effect, between the herbicide and plant tissues. Herbicides used in drainage and irrigation canals (e.g. acrolein, xylene) are not selective and so are not suitable for use in rivers where fauna and some vegetation are desired. Fewer herbicides are permitted to be applied to aquatic sites than terrestrial ones because of the sensitivity of the habitats and the concerns of residues reaching irrigation and potable water supplies.

Selective placement of the contact herbicide diquat has been achieved in rivers when a gel-like formulation with alginate is applied to submerged weed beds (Barrett & Murphy 1982; Fox & Murphy 1990), but this product is less satisfactory for large-scale management. Contact herbicides that do not damage underground organs tend to cause temporary reductions in vegetation. Prolonged (10–12 weeks) applications of fluridone in Florida, USA have resulted in the selective removal of the submerged weed *Hydrilla* from several kilometres of river (Haller *et al* 1990). Being a systemic herbicide such management should have a longer-lasting effect than contact herbicides. In the long term, appropriate herbicide applications to rivers, that reduce the biomass of a dominant species such as *Hydrilla*, are unlikely to have detrimental effects on the habitat. Systemic herbicides, such as glyphosate and dalapon, may be used extensively on emergent species. Repeated use here does have potential for permanent community changes although there is evidence that overall diversity is not reduced (Wade 1980).

Physical removal of weeds, while appearing to have an advantage over herbicides in removing nutrients from the system, is not very efficient in this objective because of the high proportion of water that must be moved. Whether weed removal is manual or mechanical, results tend to be short-lived with multiple harvests needed in a growing season, particularly of species that have basal meristems (e.g. grasses) or underground perennating organs. Removal of all plant material by dredging will have a more drastic short-term effect on the habitat if repeated regularly but, in the long term, communities may remain fairly stable (Wade 1978). It has been suggested that the characteristic submerged macrophyte communities of chalkstreams in the south of England have evolved as a result of their regular, physical management over several centuries (Westlake 1979).

The use of herbivorous fish (e.g. *Ctenopharyngodon idella*) for biological control of weeds in rivers has been limited because of the problems of confining the fish to reaches where management is needed. In tropical and subtropical areas, insects have been used with varying degrees of efficacy to control floating plants on rivers. Two notable successes have been *Agasicles* beetles released on alligatorweed (*Alternanthera philoxeroides*) in the south-eastern USA (Maddox *et al* 1971) and *Cyrtobagous* weevils on *Salvinia molesta* in Australia (Room *et al* 1981). Grazing of domestic animals on emergent vegetation commonly occurs, but over prolonged periods this may select for unpalatable species and trampling can accelerate bank erosion.

Macrophyte management by environmental manipulation has been proposed in the planting of trees on river banks to shade narrow river channels (Dawson & Haslam 1983). Regulation of water levels may reduce macrophyte growth but that is not usually the objective of such manipulation in rivers.

Pollution

Macrophytes are not used as commonly as invertebrate species as indicators of pollution in rivers. While reductions in the abundance of macrophytes that occasionally occur downstream of point sources of pollution can sometimes be ascribed to heavy metal or detergent toxicity, increased turbidity and smothering of plants by silt and mining spoils are more common causes (Hynes 1960). General increases in turbidity may result from catchment deforestation, poor agricultural practices and intensive boat usage. Macrophytes may also be shaded out by dense populations of sewage fungus, filamentous or epiphytic algae downstream of organic pollution.

Acidification of poorly buffered rivers either from point sources, such as coal mines, or acid rain deposition is likely to result in changes in the species composition of vegetation rather than rapid changes in productivity (Stokes 1986; Weatherley & Ormerod 1989). Heat pollution and changes in salinity also alter species composition (e.g. *Egeria densa* in British power-station outfall; Sculthorpe 1967).

Eutrophication is unlikely to increase macrophyte productivity in many lowland rivers that already have sufficient phosphorus and nitrogen for optimum growth rates. Nutrient inputs may increase macrophyte productivity in oligotrophic rivers but evidence of such changes is not always easy to find as it depends upon the availability of pre-enrichment data. Eutrophic conditions are usually associated with a loss of macrophyte diversity as highly competitive species predominate (e.g. *Potamogeton pectinatus*).

River engineering

Engineering projects, such as the building of bridges or channel straightening, will temporarily reduce macrophytes in the disturbed and silted areas. The eventual return of the original flora will depend upon any permanent changes in depth, substratum and flow, and upon species recruitment from upstream.

A change from riverine to lacustrine flora will occur immediately upstream of dam construction. As upstream water movement is reduced, floating species often dominate in hot climates, frequently disrupting the purpose of the dam (e.g. *Salvinia molesta* behind the Kariba irrigation dam in Zimbabwe; Sculthorpe 1967). The regulation of flow downstream may have an effect on the composition of submerged species and can reduce the width of the emergent fringe.

Introduction of exotic macrophytes

In the tropics the introduction of exotic macrophytes has probably been the human activity most responsible for severe weed problems in natural water bodies. Released from the constraints of natural competitors and herbivores that kept them in balance in their native countries, exotic species can rapidly dominate the flora, excluding native plants and reducing diversity. Floating (e.g. *Eichhornia crassipes*, *Salvinia molesta*) and canopy-forming submerged species (e.g. *Hydrilla verticillata*) have created problems in rivers throughout the world (Sculthorpe 1967; Haller 1978). In temperate climates exotic species have more often become established in the slower flows of lakes and canals (e.g. *Elodea canadensis* in British canals; Murphy *et al* 1982; *Crassula helmsii* potentially spreading from lakes to streams; Dawson 1983). However, some serious river infestations have occurred (e.g. *Myriophyllum spicatum* in Tennessee Valley, USA; Bates *et al* 1985).

Efforts to control existing exotic weed problems already strain the river management resources in many tropical countries. The prevention of spread between water bodies and prevention of the deliberate or accidental importation of new species must become an important objective of regulators who are concerned about river management.

ACKNOWLEDGEMENTS

The author is most grateful to Bill Haller and George Bowes for their helpful comments. Published with the approval of the Department of Agronomy and Florida Agricultural Experiment Station as Journal Series Number R-01196.

REFERENCES

Agami M, Reddy KR. (1989) Inter-relationships between *Salvinia rotundifolia* and *Spirodela polyrhiza* at various interaction stages. *Journal of Aquatic Plant Management* **27**: 96–102. [3.2]

Bain JT, Procter CF. (1980) The requirement of aquatic bryophytes for free CO_2 as an inorganic carbon source: some experimental evidence. *New Phytologist* **86**: 393–400. [3.2]

Barko JW, Smart RM. (1981a) Comparative influences of light and temperature on the growth and metabolism of selected submersed freshwater macrophytes. *Ecological Monographs* **51**: 219–35. [3.2]

Barko JW, Smart RM. (1981b) Sediment-based nutrition of submersed macrophytes. *Aquatic Botany* **10**: 339–52. [3.2]

Barko JW, Smart RM. (1986) Sediment-related mechanisms or growth limitation in submerged macrophytes. *Ecology* **67**: 1328–40. [3.2]

Barrett PRF, Murphy KJ. (1982) The use of diquat-alginate for weed control in flowing waters. *Proceedings of European Weed Research Society, 6th Symposium on Aquatic Weeds, 1982*, 200–8. European Weed Research Society Wageningen. [3.4]

Basiouny FM, Haller WT, Garrard LA. (1978) Survival of hydrilla (*Hydrilla verticillata*) plants and propagules after removal from the aquatic habitat. *Weed Science* **26**: 502–4. [3.2]

Bates AL, Burns ER, Webb DH. (1985) Eurasian water-milfoil (*Myriophyllum spicatum*) in the Tennessee Valley: an update on biology and control. *Proceedings of 1st International Symposium on Watermilfoil* (Myriophyllum spicatum) *and Related Halagoraceae spp.*, pp 104–15. Aquatic Plant Management Society, Washington, DC. [3.4]

Boston HL, Adams MS, Madsen JD. (1989) Photosynthetic strategies and productivity in aquatic systems. *Aquatic Botany* **34**: 27–57. [3.1]

Bowes G, Salvucci ME. (1989) Plasticity in the photosynthetic carbon metabolism of submersed aquatic macrophytes. *Aquatic Botany* **34**: 232–66. [3.1]

Bowes G, Van TK, Garrard LA, Haller WT. (1977) Adaptation to low light levels by hydrilla. *Journal of Aquatic Plant Management* **15**: 32–5. [3.2]

Brooker MP, Morris DL, Hemsworth RJ. (1977) Mass mortalities of adult salmon (*Salmo salar*) in the river Wye, 1976. *Journal of Applied Ecology* **14**: 409–17. [3.4]

Brown AWA, Deom JO. (1973) Summary: health aspects of man-made lakes. In: Ackerman W, White GR, Worthington GB (eds) *Man-made Lakes: Their Problems and Environmental Effects*, pp 755–64. William Byrd Press, Richmond, Virginia. [3.4]

Butcher RW. (1933) Studies on the ecology of rivers. I. On the distribution of macrophytic vegetation in the rivers of Britain. *Journal of Ecology* **21**: 58–91. [3.2, 3.3]

Caffrey JM. (1990) Problems relating to the management of *Potamogeton pectinatus* L. in Irish rivers. *Proceedings of the European Weed Research Society, 8th Symposium on Aquatic Weeds, 1990*, pp 61–8. European Weed Research Society, Wageningen. [3.4]

Canfield DE Jr, Hoyer MV. (1988) Influence of nutrient enrichment and light availability on the abundance of aquatic macrophytes in Florida streams. *Canadian Journal of Fisheries and Aquatic Sciences* **45**: 1467–72. [3.2]

Carling PA. (1994) In-stream hydraulics and sediment transport. In: Calow P, Petts GE (eds) *The Rivers Handbook*, vol. 1, pp 101–25. Blackwell Scientific Publications, Oxford. [3.4]

Charudattan R. (1990) Biological control of aquatic weeds by means of fungi. In: Pieterse AH, Murphy KJ (eds) *Aquatic Weeds*, pp 186–201. Oxford University Press, Oxford. [3.2]

Cook CDK. (1974) *Water Plants of the World*. Dr W. Junk b.v., Publishers, The Hague. [3.1]

Dawson FH. (1976) The annual production of the aquatic macrophyte *Ranunculus penicillatus* var. *calcareus* (R.W. Butcher) CDK Cook. *Aquatic Botany* **2**: 51–73. [3.3]

Dawson FH. (1978) The seasonal effects of aquatic plant growth on the flow of water in a stream. *Proceedings of the European Weed Research Society, 5th Symposium on Aquatic Weeds, 1978*, pp 71–8. [3.4]

Dawson FH. (1980a) Flowering of *Ranunculus penicillatus* (Dun) Bab. var. *calcareus* (R.W. Butcher) C.D.K. Cook in the River Piddle (Dorset, England). *Aquatic Botany* **9**: 145–57. [3.3]

Dawson FH. (1980b) The origin, composition and downstream transport of plant material in a small chalk stream. *Freshwater Biology* **10**: 419–35. [3.3]

Dawson FH. (1983) *Crassula helmsii* (T. Kirk) Cockayne: is it an aggressive alien aquatic plant in Britain? *Biological Conservation* **42**: 247–72. [3.4]

Dawson FH, Haslam SM. (1983) The management of river vegetation with particular reference to shading effects of marginal vegetation. *Landscape Planning* **10**: 147–69. [3.4]

Dawson FH, Castellano E, Ladle M. (1978) Concept of species succession in relation to river vegetation and management. *Verhandlungen der Internationalen Vereinigung für Theoretische und Angewandte Limnologie* **20**: 1429–34. [3.2]

Denny P. (1980) Solute movement in submerged angiosperms. *Biological Reviews* **55**: 65–92. [3.2]

Fox AM, Murphy KJ. (1990) The efficacy and ecological impacts of herbicides and cutting regimes on the submerged plant communities of four British rivers. I. A comparison of management efficacies. *Journal of Applied Ecology* **27**: 520–40. [3.4]

Getsinger KD, Green WR, Westerdahl HE. (1990) Characterization of water movement in submersed plant stands. *Miscellaneous Paper A-90-5*. US Army Engineer Waterways Experiment Station, Vicksburg, Mississippi. [3.2]

Grace JB, Wetzel RG. (1981) Habitat partitioning and competitive displacement in cattails (*Typha*): experimental field studies. *The American Naturalist* **118**: 463–74. [3.2]

Grime JP. (1979) *Plant Strategies and Vegetation*

Processes. John Wiley & Sons, New York. [3.2]
Haller WT. (1978) Hydrilla: a new and rapidly spreading aquatic weed problem. *Extension Circular S-245*. Agricultural Experiment Station/IFAS/Gainesville, Florida/Agronomy Department, University of Florida. [3.4]
Haller WT, Sutton DL. (1975) Community structure and competition between hydrilla and vallisneria. *Hyacinth Control Journal* **13**: 48–50. [3.2]
Haller WT, Sutton DL, Barlowe WC. (1974) Effects of salinity on growth of several aquatic macrophytes. *Ecology* **55**: 891–4. [3.2]
Haller WT, Fox AM, Shilling DG. (1990) Hydrilla control program in the upper St. Johns River, Florida, USA. *Proceedings of the European Weed Research Society, 8th Symposium on Aquatic Weeds, 1990*, pp 111–16. European Weed Research Society, Wageningen. [3.4]
Ham SF, Wright JF, Berrie AD. (1981) Growth and recession of aquatic macrophytes on an unshaded section of the River Lambourn, England, from 1971 to 1976. *Freshwater Biology* **11**: 381–90. [3.2]
Haslam SM. (1978) *River Plants*. Cambridge University Press, Cambridge. [3.1, 3.2]
Haslam SM. (1987) *River Plants of Western Europe*. Cambridge University Press, Cambridge. [3.2]
Hiley PD, Wright JF, Berrie AD. (1981) A new sampler for stream benthos, epiphytic macrofauna and aquatic macrophytes. *Freshwater Biology* **11**: 79–85. [3.3]
Holmes NTH. (1983) Typing British rivers according to their flora. *Focus on Nature Conservation*, 4. Nature Conservancy Council, Peterborough. [3.2]
Hutchinson GE. (1975) *A Treatise on Limnology, Volume III, Limnological Botany*. John Wiley & Sons, New York. [3.1]
Hynes HBN. (1960) *The Biology of Polluted Waters*. Liverpool University Press, Liverpool. [3.4]
Hynes HBN. (1970) *The Ecology of Running Waters*. University of Liverpool Press, Liverpool. [3.2]
Ladle M, Casey H. (1971) Growth and nutrient relationships of *Ranunculus penicillatus* var. *calcareus* in a small chalk stream. *Proceedings of the European Weed Research Council, 3rd Symposium on Aquatic Weeds, 1971*, pp 53–64. European Weed Research Society, Wageningen. [3.2]
Leather GR, Einhellig FA. (1986) Bioassays in the study of allelopathy. In: Putnam AR, Tang C (eds) *The Science of Allelopathy*, pp 133–45. John Wiley & Sons, New York. [3.2]
Losee RF, Wetzel RG. (1988) Water movement within submersed littoral vegetation. *Verhandlungen der Internationalen Vereinigung für Theoretische und Angewandte Limnologie* **23**: 62–6. [3.2]
Maberly SC. (1985) Photosynthesis by *Fontinalis antipyretica*. II. Assessment of environmental factors limiting photosynthesis and production. *New Phytologist* **100**: 141–55. [3.2]
McCreary NJ, Carpenter SR, Chaney JE. (1983) Coexistence and interference in two submersed freshwater perennial plants. *Oecologia* **59**: 393–6. [3.2]
Maceina MJ, Shireman JV, Langeland KA, Canfield DE Jr. (1984) Prediction of submersed plant biomass by use of a recording fathometer. *Journal of Aquatic Plant Management* **22**: 35–8. [3.3]
Maddox DM, Andres LA, Hennessey RD, Blackburn RD, Spencer NR. (1971) Insects to control alligatorweed. An invader of aquatic ecosystems in the United States. *Bioscience* **21**: 985–91. [3.4]
Madsen TV, Søndergaard M. (1983) The effects of current velocity on the photosynthesis of *Callitriche stagnalis* Scop. *Aquatic Botany* **15**: 187–93. [3.2]
Miller AC, Beckett DC, Bacon EJ. (1989) The habitat value of aquatic macrophytes for macroinvertebrates: benthic studies in Eau Galle Reservoir, Wisconsin. *Proceedings of the 23rd Annual Meeting, Aquatic Plant Control Research Program Miscellaneous Paper A-89-1*, pp 190–201. US Army Engineer Waterways Experiment Station, Vicksburg, Mississippi. [3.2]
Murphy KJ. (1988) Aquatic weed problems and their management: a review. I. The worldwide scale of the aquatic weed problem. *Crop Protection* **7**: 232–48. [3.4]
Murphy KJ, Eaton JW, Hyde TM. (1982) The management of aquatic plants in navigable canal systems used for amenity and recreation. *Proceedings of the European Weed Research Society, 6th Symposium on Aquatic Weeds, 1982*, pp 141–51. European Weed Research Society, Wageningen. [3.4]
Murphy KJ, Rørslett B, Springuel I. (1990) Strategy analysis of submerged lake macrophyte communities: an international example. *Aquatic Botany* **36**: 303–23. [3.2]
Odum HT. (1956) Primary production in flowing waters. *Limnology and Oceanography* **1**: 102–17. [3.3]
Odum HT. (1957) Trophic structure and productivity of Silver Springs, Florida. *Ecological Monographs* **27**: 55–112. [3.3]
Pearsall WH. (1920) The aquatic vegetation of the English lakes. *Journal of Ecology* **8**: 164–201. [3.2]
Pitlo RH, Dawson FH. (1990) Flow resistance by aquatic weeds. In: Pieterse AH, Murphy KJ (eds) *Aquatic Weeds*, pp 74–84. Oxford University Press, Oxford. [3.2]
Rao AS. (1988) Evapotranspiration rates of *Eichhornia crassipes* (Mart.) Solms, *Salvinia molesta* DS Mitchell and *Nymphaea lotus* (L.) Willd. Linn. in a humid tropical climate. *Aquatic Botany* **30**: 215–22. [3.4]
Room PM, Harley KLS, Forno IW, Sands DPA. (1981) Successful biological control of the floating weed salvinia. *Nature* **294**: 78–80. [3.4]
Sand-Jensen K, Borum J. (1984) Epiphyte shading and its effects on photosynthesis and diel metabolism of

Lobelia dortmanna L. during the spring bloom in a Danish lake. *Aquatic Botany* **20**: 109–19. [3.2]

Schmitz DC, Nelson BV, Nall L, Schardt JD. (1991) Exotic aquatic plants in Florida: a historical perspective and review of the present aquatic plant regulation program. In: Center TD, Doren RF, Hofstetter RL, Myers RL, Whiteaker LD (eds) *Proceedings of the Symposium on Exotic Pest Plants, Miami, Florida, 1988*, pp 303–26. US Dept. of Interior, National Park Service Natural Resources Publication Office, Denver, Colorado. [3.2]

Sculthorpe CD. (1967) *The Biology of Aquatic Vascular Plants*. Edward Arnold, London. [3.1, 3.2, 3.4]

Sliger WA, Henson JW, Shadden RC. (1990) A quantitative sampler for biomass estimates of aquatic macrophytes. *Journal of Aquatic Plant Management* **28**: 100–2. [3.3]

Soszka GJ. (1975) The invertebrates on submersed macrophytes in three Masurian lakes. *Ekologia Polska* **23**: 371–91. [3.2]

Spence DHN. (1967) Factors controlling the distribution of freshwater macrophytes with particular reference to the Lochs of Scotland. *Journal of Ecology* **55**: 147–70. [3.2]

Spence DHN, Maberly SC. (1985) Occurrence and ecological importance of HCO_3^- use among aquatic higher plants. In: Lucas WJ, Berry JA (eds) *Inorganic Carbon Uptake by Aquatic Photosynthetic Organisms*, pp 125–43. American Society of Plant Physiologists, Rockville, Maryland. [3.2]

Spence DHN, Bartley MR, Child R. (1987) Photomorphogenic processes in freshwater angiosperms. In: Crawford RMM (ed.) *Plant Life in Aquatic and Amphibious Habitats*, pp 153–66. Blackwell Scientific Publications, Oxford. [3.1]

Stokes PM. (1986) Ecological effects of acidification on primary producers in aquatic systems. *Water, Air and Soil Pollution* **30**: 421–38. [3.4]

Van TK, Haller WT, Bowes G. (1978) Some aspects of the competitive biology of *Hydrilla*. *Proceedings of the European Weed Research Society, 5th Symposium on Aquatic Weeds, 1978*, pp 117–26. European Weed Research Society, Wageningen. [3.2]

Wade PM. (1978) The effect of mechanical excavators on the drainage channel habitat. *Proceedings of the European Weed Research Society, 5th Symposium on Aquatic Weeds, 1978*, pp 333–42. European Weed Research Society, Wageningen. [3.4]

Wade PM. (1980) The effects of channel maintenance on the aquatic macrophytes of the drainage channels of the Monmouthshire Levels, South Wales, 1840–1976. *Aquatic Botany* **8**: 307–22. [3.4]

Weatherley NS, Ormerod SJ. (1989) Modelling ecological impacts of the acidification of Welsh streams: temporal changes in the occurrence of macroflora and macroinvertebrates. *Hydrobiologia* **185**: 163–74. [3.4]

Westlake DF. (1963) Comparisons of plant productivity. *Biological Reviews* **38**: 385–425. [3.3]

Westlake DF. (1965) Some basic data for investigations of the productivity of aquatic macrophytes. *Memorie dell'Istituto Italiano di Idrobiologia Dott Marco de Marchi* **18** (Suppl.): 229–48. [3.3]

Westlake DF. (1967) Some effects of low-velocity currents on the metabolism of aquatic macrophytes. *Journal of Experimental Botany* **18**: 187–205. [3.2]

Westlake DF. (1975) Macrophytes. In: Whitton BA (ed.) *River Ecology*, pp 106–28. Blackwell Scientific Publications, Oxford. [3.2]

Westlake DF. (1979) The ecology of chalk streams. *Watsonia* **12**: 387. [3.4]

Westlake DF. (1981) Temporal changes in aquatic macrophytes and their environment. In: Hoestlandt H (ed.) *Symposium Dynamique des Populations et Qualité des Eaux*. Acts Symp. Institute d'Ecologie du Bassin de la Somme, Chantilly, 5–8 November 1979, pp 111–38. [3.2]

Westlake DF, Casey H, Dawson FH, Ladle M, Mann RHK, Marker AFH. (1972) The chalk-stream ecosystem. In: Kajak Z, Hillbricht-Ilkowska A (eds) *Proceedings of the IBP-UNESCO Symposium on Productivity Problems of Freshwaters, Kazimierz, Dolny, 1970*, pp 616–35. Polish Scientific Publishers, Warszawa-Krakow, Poland. [3.2]

Wetzel RG. (1969) Factors influencing photosynthesis and excretion of dissolved organic matter by aquatic macrophytes in hard-water lakes. *Verhandlungen der Internationalen Vereinigung für Theoretische und Angewandte Limnologie* **17**: 72–85. [3.2]

Wetzel RG. (1983) *Limnology*, 2nd edn. Saunders College Publishing, Philadelphia. [3.3]

Willis JC. (1914) On the lack of adaptation in the Tristichaceae and Podostemaceae. *Proceedings of the Royal Society, London, Series B* **87**: 532–50. [3.2]

Wright JF, Hiley PD, Ham SF, Berrie AD. (1981) Comparison of three mapping procedures developed for river macrophytes. *Freshwater Biology* **11**: 369–79. [3.3]

4: Heterotrophic Microbes

L. MALTBY

4.1 INTRODUCTION

This chapter is concerned with heterotrophic micro-organisms (i.e. bacteria and fungi) involved in the breakdown of detritus in rivers and streams. It will consider the types of techniques used to study these organisms (section 4.2), their ecology and distribution (sections 4.3 and 4.4) and their interactions with detritus (section 4.5). Finally, the effect of water quality on heterotrophic bacteria and fungi will be considered (section 4.6). The reason for considering techniques first is that, to investigate microbial ecology, it is necessary to obtain reliable estimates of the populations present. Although this is a modest objective, it is fraught with problems (Herbert 1990) which greatly limit our understanding of the role and importance of these organisms in freshwater systems.

4.2 TECHNIQUES

Many methods have been used to study the fungi and bacteria associated with detritus in fresh waters (Ingold 1975; Jones 1977, 1979; Poindexter & Leadbetter 1986; Austin 1988; Grigorova & Norris 1990). The aim of this section is not to provide a detailed description of each of these techniques, but rather to provide an introduction to the most commonly used ones (Table 4.1), their advantages and limitations. Methods used to enumerate microbial populations can be classified as either direct or indirect; direct methods being further divided into those that record total or viable organisms. As there is no universal growth medium for all micro-organisms, the main disadvantage of viable counts is that they underestimate microbial populations, often by large amounts (Staley & Konopka 1985). Indirect methods involve measurement of chemicals that act as indicators of biomass. Hence, a major potential limitation of these is that they require that the indicator chemical has a known, constant, ratio to cell biomass.

Fungi

Direct microscopic observations

The advantage of using direct observation to assess fungal assemblages is that it relies on the production of reproductive structures over a short incubation time and therefore only records active biomass (Bärlocher & Kendrick 1974). Generally, detrital material is collected from the field and incubated in sterile water for a few days before being microscopically examined and the spores present identified. The disadvantage of this technique is that not all active biomass will sporulate during the incubation period, or will produce spores which are easy to identify. Direct microscopic observations will favour conidial fungi that produce large characteristic spores (i.e. hyphomycetes).

Direct observations can provide information on the diversity, distribution and abundance of the mycoflora associated with detritus. Shearer and Lane (1983) described a technique for assessing the relative importance value (RIV) of fungal species based on recording the occurrence of species on discs cut from several leaves. The RIV of a species ranges from zero (absent) to two (only species present) (Shearer & Webster 1985a) and provides a measure of the abundance and distri-

Table 4.1 Methods used for studying heterotrophic microbes associated with detritus

Unit of study	Method
Fungi	Direct observations Baiting Dilution plating Particle plating Foam/water samples Hyphal biomass Chitin assay Ergosterol analysis
Bacteria	Direct counts Viable counts ^{3}H-thymidine, ^{3}H-adenine, ^{3}H-leucine incorporation
Microbial community	ATP concentration Oxygen consumption ^{14}C utilization ETS activity Microbial fatty acids

bution of a species both between and within leaves; this is important given the patchy distribution of aquatic fungi on leaves. The disadvantage of the technique is that it is time consuming and tedious. Shearer and Lane (1983) concluded that the optimum number of discs that should be examined per leaf was ten. However, the optimum number of leaves per site will be dependent on the site itself; more diverse communities require more samples.

Baiting

Two types of baits have been used: 'natural' baits (e.g. leaves, wood) for use in the field, and seeds for baiting material in the laboratory (Park 1972; Duddridge & Wainwright 1980). The most common application of baiting techniques is to collect organic material from the field and place it into containers containing sterile water and a number of baits. After 2–3 weeks' incubation, the baits are removed and either examined microscopically or plated on agar and the resulting colonies identified. This technique favours species which have motile spores including members of the genera *Pythium*, *Achlya* and *Saprolegnia* (i.e. Mastigomycotina). Park (1972) assessed the effectiveness of a range of different seed baits. He concluded that small baits, such as *Brassica* half-cotyledons, yielded a higher variety of fungi than larger baits such as hemp seeds.

Although wood baits have been used to sample fungi in the field (Shearer 1972; Willoughby & Archer 1973; Lamore & Goos 1978; Sanders & Anderson 1979; Shearer & Von Bodman 1983), most studies have been concerned with the mycoflora of decaying leaves and have therefore used leaf baits (e.g. Bärlocher & Kendrick 1974; Suberkropp & Klug 1976; Bärlocher 1980; Chamier & Dixon 1982a; Suberkropp 1984; Shearer & Webster 1985a, 1985b, 1985c). Leaves are incubated in the river, either as leaf packs or in mesh bags, for a set period of time (>2 weeks) after which the fungal species colonizing them are identified, either by direct observation or after particle plating on agar. Factors to consider when using natural baits are bait size and deployment method (i.e. bags or packs). Leaf packs attached to an anchor (e.g. Cummins *et al* 1980) are more representative of natural leaf accumulations, although they are vulnerable to consumption by macroinvertebrates and physical abrasion. To reduce these losses, leaf baits are often placed in mesh bags; but here the size of mesh needs to be considered. Whereas fine mesh (<1 mm) may be desired to exclude macroinvertebrates, which may consume the material and associated microbes, small mesh size will limit the flow of water though the bag, reducing the exchange of nutrients and oxygen and thereby influencing colonization patterns and decomposition rates (Bird & Kaushik 1981; Webster & Benfield 1986).

The effect of bait size on the number of species isolated from it has been assessed for both wood (Sanders & Anderson 1979) and leaves (Bärlocher & Schweizer 1983). In both cases there was a positive correlation between bait size and number of aquatic hyphomycetes isolated. Leaf baits used by Bärlocher and Schweizer (1983) were small (0.16–4.00 cm^2). Usually, however, whole leaves are used as baits and therefore size differences are probably less important in influencing diversity estimates.

The possible limitations of using leaf baits for ecological studies include: (i) they tend to consist of a single leaf type and are often deployed at the same time; hence they may underestimate the diversity of the fungal community; (ii) grouping leaves together, either in packs or within bags, tends to facilitate within-group colonization, again influencing diversity and distribution patterns. Despite these reservations, Shearer and Webster (1985c) found a high degree of similarity (67–87%) between the assemblages of fungi colonizing leaf baits (packs) and naturally occurring leaves. Such results suggest that leaf baits may provide a useful tool for determining the composition of leaf mycoflora.

Plating methods

There are several types of plating methods used. For dilution plating, detrital material is homogenized in sterile water and serial dilutions of the homogenate are spread over agar plates which are then incubated (Park 1972). In particle plating, small pieces of detrital material are placed directly on to agar plates. A modified particle plating technique has recently been proposed by Kirby *et al* (1990). As with dilution plating, the detritus is initially homogenized. The resulting particles are then sieved through 600 μm and 500 μm mesh sieves. Particles retained by the smaller sieve are resuspended and spread over agar plates. The advantage of this method is that the particles are small and standardized. Particle plating techniques provide information on the fungal species present on a particle but not their relative abundances. As particle size increases, both the number of species sampled per particle and the similarity between particles will increase, until eventually all species will be present on all particles. Therefore, with large particles it may not be possible to calculate the relative abundances of species in a sample nor the diversity of the asssemblage. If, on the other hand, particles were so small that they were colonized by a single species, sampling a large number of particles would provide an accurate and sensitive measurement of abundance and therefore diversity. Kirby *et al* (1990) found that particles in the size range 500–600 μm were colonized by an average of 1.12 species and the maximum number of species colonizing a given particle was two.

A limitation of plating techniques is that not all the fungal species that develop on the agar plates will have been active on the detritus. For example, Kaushik and Hynes (1968) plated leaf discs, which had previously been incubated in river water, directly on to agar plates that were then incubated at room temperature. All the fungi that grew on the plates were hyphomycetes typical of terrestrial, rather than aquatic, habitats. Inactive propagules may be removed from detritus by surface sterilization with mercuric chloride or serial washing (Kirby 1987).

Foam/water samples

The branched and sigmoid conidia of aquatic hyphomycetes (Fig. 4.1) are easily trapped in foam commonly observed in fast-flowing rivers and streams. The fungal spores present in foam samples have been used to provide information on the distribution and diversity of river mycoflora (e.g. Iqbal & Webster 1973, 1977; Sanders & Anderson 1979; Bärlocher & Rosset 1981; Wood-Eggenschwiler & Bärlocher 1983; Aimer & Segedin 1985a; Shearer & Webster 1985a, 1985b, 1985c). However, the presence of spores in river foam is not proof that the species is aquatic (i.e. can grow and sporulate under water), nor are all species equally represented in foam samples. Foam samples are biased in favour of highly

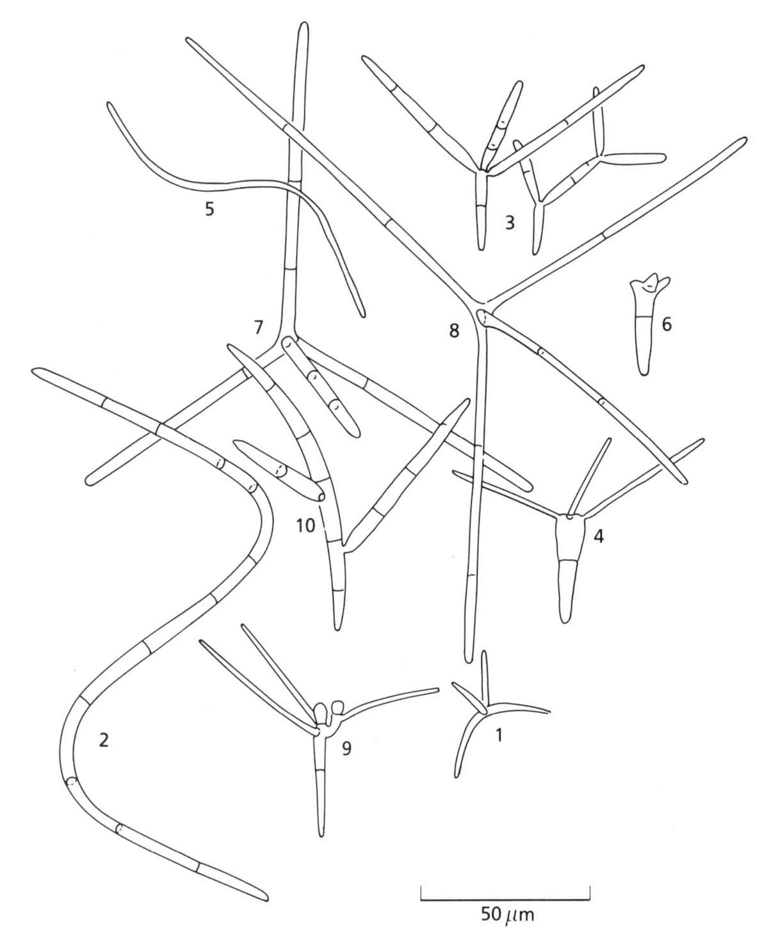

Fig. 4.1 Conidia of some common aquatic hyphomycetes. **1**: *Alatospora acuminata*; **2**: *Anguillospora longissima*; **3**: *Articulospora tetracladia*; **4**: *Clavariopsis aquatica*; **5**: *Flagellospora curvula*; **6**: *Heliscus lugdunensis*; **7**: *Lemonniera aquatica*; **8**: *Tetracladium elegans*; **9**: *Tetrachaetum marchalianum*; **10**: *Tricladium splendans* (After Ingold 1975).

branched spores, which are more easily removed from suspension by air bubbles than less branched spores (Iqbal & Webster 1973).

An alternative method of sampling water-borne spores is filtration. Water samples are filtered through millipore filters, the volume filtered being dependent on spore density (Iqbal & Webster 1973; Shearer & Lane 1983; Aimer & Segedin 1985b). In rivers of high spore density a 250-ml water sample may be satisfactory; however, in those of low spore density it may be necessary to filter up to 5 litres of water (Webster & Descals 1981). To prevent germination, spores should be fixed and stained in the field (e.g. by Lugol's iodine, lactophenol–fuchsin or cotton blue or trypan blue in either lactic acid or lactophenol (Iqbal & Webster 1973; Suberkropp & Klug 1976; Akridge & Koehn 1987; Chamier 1987; Findlay & Arsuffi 1989)). The water samples are filtered and the millipore filters rendered transparent by heating at 60°C for 1 h.

There are a number of problems with the filtration technique (Shearer & Lane 1983); they include:

1 Blocking of filters when sampling waters high in suspended solids.

2 Variability between samples (Shearer & Webster 1985c, but see Iqbal & Webster 1973).

3 Difficulty in identifying spores due to either obstruction by debris or deformation during filtering. Many species have similar conidial morphologies.

4 Relationship between the number of conidia produced by a species and its actual occurrence and activity is generally unknown (Suberkropp 1991).

5 Selectivity, i.e. spores of different sizes and branching patterns may be removed differentially.
6 Number of conidia produced by many species is dependent on flow rate; therefore, in slow-flowing systems, species that produce few fungi may go undetected.
7 Origin of spores. The propagule pool receives contributions from many sources, e.g. decaying leaves, wood or macrophytes and spores may have been transported considerable distances. The presence of spores in a site is not evidence of active growth; in fact, several studies have noted a discrepancy between the species list obtained from filtering of water samples and that obtained by direct examination of organic matter (Iqbal & Webster 1977; Chamier & Dixon 1982a; Bärlocher & Schweizer 1983; Shearer & Lane 1983; Shearer & Webster 1985c).

Hyphal biomass

The problem of measuring fungal biomass in detritus is that it is embedded within an opaque solid structure. Several approaches have been developed to overcome this; homogenizing the detritus and either: (i) embedding it within a thin film of agar (Jones & Mollison 1948; Frankland *et al.* 1978; Bengtsson 1982); or (ii) staining it with a fluorescent dye and filtering on to black filters (Paul & Johnson 1977; van Frankenhuyzen & Geen 1986; Findlay & Arsuffi 1989); or (iii) chemically clearing the detritus and measuring the hyphae within intact substrate (Hering & Nicholson 1964; Bärlocher & Kendrick 1974).

Newell and Hicks (1982) compared the effectiveness of these three approaches for assessing fungal biovolume in dead leaves. They concluded that agar-films provided the best method for direct estimation of quantities of mycelium. However, this method does have limitations, including difficulty in identifying fungal fragments in litter homogenate; inability of homogenization to release all fungal hyphae; and problems in recognizing fungal reproductive structures making fungal identification difficult. The problem of separating fungal mycelium from plant tissue may be overcome by staining fungal material. A commonly used stain is aniline blue which causes mycelia to fluoresce when excited by ultraviolet light. However, the degree to which hyphae fluoresce varies both between and within species (Newell & Hicks 1982; Findlay & Arsuffi 1989), thereby introducing variability in estimates of hyphal abundance.

Hyphal dimensions, and hence biovolume, have usually been measured either directly (Newell & Hicks 1982) or from photographs (Findlay & Arsuffi 1989), although it is possible to use image-analysis techniques (Morgan *et al* 1991). Biovolume estimates may be converted to biomass using appropriate conversion factors (Newell & Statzell-Tallman 1982).

Indirect methods

There are several indirect methods for estimating fungal biomass in plant material (Whipps *et al* 1982) including measuring chitin (Ride & Drysdale 1972; Cochran & Vercellotti 1978) or ergosterol (Seitz *et al* 1979; Lee *et al* 1980).

Chitin assay Hydrolysis of chitin, a major constituent of most fungal cell walls (Hudson 1986), produces hexosamines such as glucosamine that can be assayed colorimetrically (Ride & Drysdale 1972; Jones 1979) or by gas chromatography (Hicks & Newell 1983). As the amount of glucosamine per unit weight of mycelium varies between species and between different ages of mycelium, this can reliably be used only to assess the fungal biomass of material colonized by fungi of known age and composition. Another disadvantage is that glucosamine polymers are also present in some bacterial cell walls (Gottschalk 1986; West *et al* 1987).

Ergosterol analysis Ergosterol (ergosta-5,7,22-trien-3β-ol) is the prominent membrane-bound sterol in most fungi but is absent from vascular plant material and bacteria. It has been used to estimate fungal biomass in terrestrial (Matcham *et al* 1985; West *et al* 1987), marine (Lee *et al* 1980; Newell & Fallon 1991) and, more recently, freshwater systems (Gessner & Schwoerbel 1991; Sinsabaugh *et al* 1991). It can be extracted from detrital material by saponification with methanol and quantified by high-performance liquid chromatography with an ultraviolet detector

(Grant & West 1986; Newell *et al* 1988). Because ergosterol has a high rate of mineralization upon the death of fungi, it is indicative of living biomass. This is in contrast to chitin, which is more resistant to breakdown.

Selectivity of techniques

All the techniques described above are selective and, therefore, several must be used to produce a comprehensive list of fungi present in a sample. For example, direct microscopic examination of detritus is selective for those groups with large distinctive spores (i.e. aquatic hyphomycetes), whereas particle plating may be selective towards different genera depending on the incubation conditions used. Incubating material on nutrient-rich agar at room temperature will favour terrestrial hyphomycetes, whereas incubating it on nutrient-poor agar at lower temperatures will favour aquatic hyphomycetes (Bärlocher & Kendrick 1974; Suberkropp & Klug 1976; Chergui & Pattee 1988). A list of the media used to culture aquatic fungi is presented in Table 4.2. To reduce bacterial contamination, lactic acid or antibiotics (e.g. penicillin, streptomycin) are frequently added to culture media (Kaushik & Hynes 1968; Webster & Descals 1981; Chamier *et al* 1984).

Conidial frequencies in water samples have been extensively used to study aquatic hyphomycete communities (e.g. Iqbal & Webster 1973, 1977; Bärlocher & Rosset 1981; Shearer &

Table 4.2 Media used for culturing aquatic fungi

Medium	Reference
Cellulose	Park (1972, 1974b)
Cooke agar	Kirby *et al* (1990)
Cornmeal agar	Duddridge & Wainwright (1980)
Czapek agar	Kaushik & Hynes (1968)
Goos' medium	Webster *et al* (1976)
Glucose-peptone/Rose Bengal/aureomycin agar	Park (1974b)
Glucose–yeast extract–starch medium	Kirby (1987)
Inorganic salts agar	Suberkropp & Klug (1976) Suberkropp (1984)
Malt extract agar	Abel & Bärlocher (1984) Bärlocher & Kendrick (1974) Butler & Suberkropp (1986) Chamier *et al* (1984) Kirby (1987) Kirby *et al* (1990) Rosset & Bärlocher (1985) Thornton (1963)
Malt extract broth	Bärlocher & Kendrick (1973)
Potato dextrose agar	Godfrey (1983)
Starch–casein–nitrate medium	Suberkropp & Klug (1976)
Water agar	Bärlocher & Kendrick (1974)
Water agar + leaf material	Bärlocher & Kendrick (1974) Chamier *et al* (1984)
Yeast extract–glucose broth	Arsuffi & Suberkropp (1984)

Webster 1985a, 1985b). However, except for the most common species, there is little agreement between the abundance of species in the propagule pool and their abundance on leaf litter (Shearer & Lane 1983; Shearer & Webster 1985c).

Although there is good agreement between the types of aquatic hyphomycetes colonizing leaf baits and naturally occurring leaves (Shearer & Lane 1983), baiting in the laboratory with seeds tends to be selective towards aquatic Oomycetes (e.g. *Pythium*, *Achlya* and *Saprolegnia*) which have motile zoospores (Park 1972).

Bacteria

Direct microscopic counts

Bacteria present on detrital material may be observed, *in situ*, using scanning electron microscopy (e.g. Suberkropp & Klug 1974; Armstrong & Bärlocher 1989; Groom & Hildrew 1989). Alternatively, they can be dislodged from detritus by sonification (Chamier 1987; Bott & Kaplan 1990) or homogenization (Suberkropp & Klug 1976; van Frankenhuyzsen & Geen 1986; Henebry & Gorden 1989). However, the efficiency of removal of attached bacteria from particles by sonification ranges from 20% to 50% (Benner *et al* 1988), and homogenization results in samples of high detritus content which can cause problems when using epifluorescence microscopy. The dislodged bacterial cells can be stained with a fluorescent dye, usually acridine orange, and filtered on to black 0.2-μm pore size filters before being counted using epifluorescent microscopy. If acridine orange preparations are excited with blue light at 470 nm, debris fluoresces red, orange or yellow whereas bacteria fluoresce green.

Epifluorescence direct counting can also be used for estimating the size and biomass of bacteria (Fry 1990). In order to estimate biomass, bacteria are measured and biovolume:carbon conversion factors applied. The measurement of bacterial cells is not without problems (Fry 1988) and conversion factors can vary by a factor of two (Benner *et al* 1988). Image analysis techniques can be used both to count and measure bacteria in epifluorescence images (Bjørnsen 1986).

Difficulties may be encountered when using acridine orange for direct counts especially if the sample contains a large amount of detritus since this may pick up the stain and autofluoresce. Porter and Feig (1980) described the use of a DNA-specific fluorescing stain, 4′,6-diamidino-2-phenylindole (DAPI), which can be used for epifluorescent counts. When exposed to ultraviolet light (365 nm) the DNA–DAPI complex fluoresces bright blue whereas unbound DAPI and DAPI bound to non-DNA material may fluoresce a weak yellow. This technique has been used successfully for studying stream bacteria (Kaplan & Bott 1983, 1985; Bott & Kaplan 1990). However, Edwards (1987) found the technique to be unsuitable for studying bacteria in a blackwater river. DAPI is specific for A–T-rich DNA and may not therefore be suitable for all types of bacterial assemblages.

Viable counts

Not all the bacterial cells recorded by direct microscopic counts will be viable. Viable counting procedures involve providing bacteria with a nutrient source and recording the number of active cells (i.e. those that have grown and/or divided) after incubation. Viable counts can be obtained either by plating samples on to agar plates and recording the number of colonies produced (i.e. colony counts) or by 'most probable number' (MPN) procedures (Jones 1979; Herbert 1990). To obtain MPN counts, samples are serially diluted and aliquots of each dilution are inoculated into, or on to, a growth medium. Most MPN methods use liquid media which become turbid due to growth of bacterial cells.

The results obtained from viable count methods are dependent upon a number of factors, including dilution procedures, incubation conditions and duration, and the type of medium used. Jones (1979) suggests the use of casein–peptone–starch (CPS) medium for freshwater bacteria, although other media including peptone–yeast extract–glucose (PYG) agar (Suberkropp & Klug 1976) and nutrient agar (Baker & Farr 1977; Chamier *et al* 1984) have been used to study bacteria associated with detritus. The type of medium used will influence both the diversity and abundance of bacteria isolated from environmental samples.

For example, Baker and Farr (1977) consistently isolated one to three times more bacteria from stream water using CPS medium than nutrient agar. Other problems with viable plate counts include: (i) bacteria remain aggregated and therefore individual colonies are derived from many rather than single cells; (ii) antagonistic interaction occurs between proximate bacterial colonies; (iii) they encourage the growth of cells that are metabolically inactive *in situ*.

Staley and Konopka (1985) concluded that with viable plate counts, the maximum recovery of heterotrophic bacteria from a variety of oligotrophic and mesotrophic habitats was 1% of the total direct count. This is in sharp contrast with results from studies using tritiated amino acids and thymidine which have concluded that at least 85% of the total direct count are metabolically active (Tabor & Neihof 1984).

Because of these limitations, Fry (1982) stated that 'viable counts should only be used with care in natural habitats in limited circumstances'. Boulton and Boon (1991) go further and state that the continued use of viable counts for the determination of bacterial abundance is 'indefensible'. Despite this, viable counts are still used (e.g. Paul *et al* 1983; Hornor & Hilt 1985; Colwell *et al* 1989).

Other methods of distinguishing between metabolically active and inactive bacteria include autoradiography (O'Carroll 1988), use of inhibitors such as 2-(*p*-iodophenyl)-3-(*p*-nitrophenyl)-5-phenyl tetrazolium chloride (INT) (Tabor & Neihof 1982), and the use of vital fluorogenic stains such as fluorescein diacetate (FDA) (Schnürer & Rosswall 1982). However, none of these methods is ideal for studying bacteria associated with detritus (Boulton & Boon 1991). For example, not all bacteria fluoresce with FDA (Fry 1990). It is especially poor for Gram-negative bacteria, which tend to be the predominant group in unpolluted rivers and streams (Chrzanowski *et al* 1984).

^{3}H-thymidine incorporation

Several radioisotopic methods have been developed to assess bacterial biomass, including ^{3}H-adenine incorporation into DNA and RNA, 5-^{3}H-uridine incorporation into RNA, ^{3}H-thymidine incorporation into DNA and ^{3}H-leucine incorporation into protein (Karl 1980; Moriarty 1986; Riemann & Bell 1990). The most commonly used isotope is ^{3}H-thymidine.

The incorporation of ^{3}H-thymidine into DNA has been used to assess bacteria in the seston of both marine (Fuhrman & Azam 1980, 1982) and freshwater (Riemann *et al* 1982; Bell *et al* 1983) systems as well as marine sediments (Moriarty & Pollard 1981). Recently its usefulness as a technique for quantifying bacteria associated with benthic detritus in river systems has been evaluated (Findlay *et al* 1984a). The technique has been criticized on the grounds that compounds other than DNA are labelled. For example, ^{3}H-thymidine may be catabolized and the label incorporated into protein (Staley & Konopka 1985); it may also be incorporated into RNA (O'Carroll 1988). However, these problems can be overcome by using pulse-label techniques (Moriarty 1986; 1990). Using short incubation times (e.g. 10 min at 25°C, 20 min at 15°C) reduces the chance that thymidine will be degraded and the label incorporated into other macromolecules. Another approach is to isolate and purify the DNA before assaying radioactivity. Several methods have been used to extract DNA including: acid–base hydrolysis, phenol–chloroform extraction and enzymatic fractionation. However, the effectiveness of these methods depends on the type of material being analysed and this must therefore be considered when selecting an extraction procedure (Torréton & Bouvy 1991).

^{3}H-thymidine incorporation has been used to evaluate bacterial communities on riverine detritus in both the laboratory (e.g. Findlay *et al* 1984b, 1986; Findlay & Arsuffi 1989) and the field (e.g. Edwards & Meyer 1986; Palumbo *et al* 1987; Peters *et al* 1987). Palumbo and colleagues (1987) used this technique to assess microbial communities on leaf litter in streams and found it to be a more sensitive method than either oxygen consumption or ATP concentration.

Microbial community

ATP content

ATP is present in all living material. Its use as a means of estimating total microbial biomass was initially proposed by Holm-Hansen and Booth in 1966, but since then it has been widely used in both laboratory (Federle *et al* 1982; Suberkropp *et al* 1983; Peters *et al* 1987; Findlay & Arsuffi 1989) and field (Forbes & Magnuson 1980; Rounick & Winterbourn 1983; Taylor & Roff 1984; Palumbo *et al* 1987) based studies. Although several methods can be used to assay ATP (Karl 1980), the most commonly used techniques are those based on the luciferin–luciferase reaction. This involves a number of assumptions including: (i) ATP is not adsorbed on to detrital material, and (ii) there is a constant ratio of ATP to total cell carbon for all microbial taxa. The first assumption is possibly false and the second assumption is certainly false. ATP is not just associated with living biomass but occurs as free dissolved ATP (D-ATP) in both marine and freshwater environments (Karl 1980). The extent to which D-ATP is adsorbed on to detrital material, and the implications of this for the use of ATP as an indicator of microbial biomass, have not been fully investigated.

Frequently a carbon:ATP ratio of 250:1 is used (e.g. Kaplan & Bott 1983; Findlay & Arsuffi 1989). This was an average value derived by Holm-Hansen and Booth (1966) from a wide range of microbial cells. However, carbon:ATP ratios vary both within and between species and may range from less than 100:1 to more than 2000:1 (Jones 1979; Karl 1980). Therefore the application of the 250:1 ratio to unidentified field samples at best can provide only a crude indication of microbial biomass. ATP-biomass estimates will be most accurate when the natural microbial assemblage approaches a monoculture of a known species for which a specific carbon:ATP ratio can be determined.

Oxygen consumption

Microbial respiration has frequently been used as an index of community biomass and activity and is the basis of the 'biochemical oxygen demand' (BOD) test. It is usually measured by quantifying the oxygen uptake of a sample either by using chemical techniques (e.g. Groom & Hildrew 1989), an oxygen electrode (e.g. Findlay & Arsuffi 1989; Sinsabaugh & Linkins 1990) or a Gilson Differential Respirometer (e.g. Palumbo *et al* 1987; Elwood *et al* 1988; Garden & Davies 1988; Petersen *et al* 1989).

Microbial communities contain both heterotrophic and autotrophic organisms. Although oxygen is utilized by all these organisms during aerobic respiration, it is also generated by autotrophs during photosynthesis. Therefore, to obtain a measure of respiration alone, samples must be incubated in the dark.

^{14}C utilization

When organic compounds are mineralized by heterotrophic micro-organisms, the carbon they contain is either assimilated into microbial biomass or respired and converted to carbon dioxide. The activity of the heterotrophic microbial community can therefore be estimated by measuring the rate at which organic carbon, supplied as ^{14}C-glucose or ^{14}C-acetate, is either converted to carbon dioxide (Sorokin & Kadota 1972) or assimilated (Wright & Hobbie 1966). Hall *et al* (1990) reviewed studies using ^{14}C to estimate microbial activity in fresh waters. Recent studies using this technique to assess heterotrophic microbial activity associated with detritus include Carpenter *et al* (1983), Fairchild *et al* (1983) and Peters *et al* (1987).

Electron transport system activity

Electron transport system (ETS) activity is a characteristic of respiring cells and can be measured by quantifying the reduction of 2-(*p*-iodophenyl)-3-(*p*-nitrophenyl)-5-phenyl tetrazolium chloride (INT) to its formazan derivative, a water-insoluble precipitate which can be observed microscopically or quantified spectrophotometrically (Tabor & Neihof 1982; Trevors 1984). For example, Elwood *et al* (1988) used ETS activity to assess the microbial community on leaf litter. Leaf discs were incubated in a 0.25% solution of INT for 30 min, after which the formazan derivative was extracted in 90% acetone and its absorbance read at 520 nm on a spectrophotometer. Whereas Trevors (1984) found a strong correlation between ETS activity and oxygen consumption for freshwater sediments ($r > 0.74$), no such correlation was detected by Elwood *et al* (1988). This led them to conclude that the technique was an unreliable measure of microbial activity on decomposing leaves. It is possible that not all species of bacteria can reduce the tetrazolium salt or that it inhibits or suppresses respiratory activity in some species. The technique has also been used to study micro-organisms associated with biofilms (Freeman *et al* 1990; Ladd & Costerston 1990) and can be combined with epifluorescence microscopy as a means of assessing the abundance of metabolically active micro-organisms (Tabor & Neihof 1982).

Microbial fatty acids

The abundance and composition of fatty acids extracted from detritus has been suggested as a means of quantifying and characterizing benthic microbial communities (Bobbie & White 1980). Some fatty acids, such as polyenoic acids, are restricted to fungi, whereas others, for example *cis*-vaccenic acid, are restricted to bacteria. A list of fatty acids that can be used as biomarkers is given in Fry (1988). Van Frankenhuyzen and Geen (1986) used fatty acids to investigate the effect of pH on the microbial community of leaf litter in streams. They found the technique to be less satisfactory than direct microscopic counts.

4.3 FUNGI

The fungi found in fresh waters include members of the Mastigomycotina (e.g. *Pythium*, *Saprolegnia*), Deuteromycotina (e.g. *Lemonniera*, *Helicoon*), Ascomycotina (e.g. *Penicillium*, *Fusarium*) and Zygomycotina (e.g. *Mucor*). Although no fungi have been found to produce basidiocarps in freshwater, conidial basidiomycetes do occur (Hudson 1986). In a study of fungi associated with leaves in streams, Bärlocher and Kendrick (1974) identified 55 different species, of which 48 were deuteromycetes (29 species were terrestrial hyphomycetes, 14 species were aquatic hyphomycetes and 5 species were coelomycetes), the remaining being members of the Zygomycotina and Mastigomycotina. A similar number of species was recorded by Lamore and Goos (1978) from woody substrates. Of the 59 species they identified, 40 were deuteromycetes, 15 were ascomycetes and four were zygomycetes.

Aquatic hyphomycetes

The term 'aquatic hyphomycetes' has traditionally been used to describe conidial aquatic fungi abundant on organic material in streams and rivers. As some members of this group are neither exclusively aquatic (Park 1974a; Singh & Musa 1977; Bandoni 1981) nor hyphomycetes, some being anamorphs of ascomycetes and basidiomycetes, Webster and Descals (1981) have proposed that they be referred to as Ingoldian fungi, after C.T. Ingold who pioneered their study in the 1940s. However, the use of the term 'aquatic hyphomycetes' is widespread and is retained here.

Since the initial studies by Ingold (Ingold 1942), more than 150 species of aquatic hyphomycetes have been described. They tend to be concentrated on coarse, as opposed to fine, particulate organic matter and are particularly abundant on submerged decaying leaves (Bärlocher & Kendrick 1981). As a group they are widely distributed, and many species (e.g. *Alatospora acuminata*, *Anguillospora longissima*) have a global distribution (Nilsson 1964). Recent studies into the ecology and distribution of aquatic hyphomycetes have been conducted in Europe (Bärlocher & Rosset 1981; Chamier & Dixon 1982a; Wood-Eggenschwiler & Bärlocher 1983; Czeczuga & Próba 1987; Chergui & Pattee 1988), New Zealand (Aimer & Segedin 1985a, 1985b), Australia (Thomas *et al* 1989), India (Sridhar & Kaveriappa 1982, 1989; Gupta & Mahrotra 1989; Mer & Sati 1989), North America (Suberkropp 1984; Akridge & Koehn 1987), Hawaii (Ranzoni 1979), Puerto Rico (Padgett 1976), North Africa (Chergui 1990) and the subarctic (Müller-Haeckel & Marvanová 1979).

The conidia of aquatic hyphomycetes are large and are predominantly tetraradiate (e.g. *Alatospora*, *Tetrachaetum*, *Lemonniera*) or sigmoid (e.g. *Anguillospora*, *Flagellospora*, *Lunulospora*) in shape (Ingold 1975; Fig. 4.1). The tetraradiate shape is thought to be an adaptation to a flowing water environment, increasing the effectiveness with which the conidia become trapped on underwater surfaces (Webster 1981). It may also enhance their dispersal in surface films of water such as might occur between leaves in terrestrial habitats (Bandoni 1974, 1975).

Seasonal fluctuations in conidial numbers have been recorded in many parts of the world; peak conidial numbers occurring shortly after peak leaf fall (Iqbal & Webster 1973, 1977; Müller-Haekel & Marvanová 1979; Bärlocher & Rosset 1981; Chamier & Dixon 1982a; Aimer & Segedin 1985a, 1985b; Shearer & Webster 1985b; Thomas *et al* 1989). The conidial numbers of most species mirror the availability of substrate, but a few show distinctive patterns of seasonal abundance not explained in terms of substrate availability. In such cases, temperature is invoked as the causal factor.

Although negative relationships between temperature and species richness have been observed in the field (Mer & Sati 1989), evidence from laboratory studies on the effect of temperature on growth and sporulation of individual species is far from convincing.

The effect of temperature on the seasonal occurrence of aquatic hyphomycetes at several sites in a North American stream was investigated by Suberkropp (1984). Two groups of fungi were identified: those present only in summer (e.g. *Lunulospora curvula*, *Flagellospora penicillioides*, *Triscelophorus monosporus* and *Clavatospora tentacula*) and a winter group (e.g. *Alatospora acuminata*, *Flagellospora curvula* and *Lemonniera aquatica*) which declined in importance during the summer. During the summer, water temperatures in the study sites ranged between 9° and 24°C, whereas in winter the temperature range was between 0° and 15°C. Experiments investigating the effect of temperature on growth concluded that members of the summer group had an optimum for growth of 25–30°C and minimum for growth between 5° and 10°C, whereas those of the winter group had an optimum for growth of 20°C and continued to grow at 1°C. Although these data may provide an explanation for the absence of the summer assemblage of aquatic hyphomycetes during the colder winter months, they do not explain the decline of the winter assemblage. Also, they do not explain the decline of the summer assemblage at sites where the temperature did not fall below 9–10°C.

Typical 'summer' and 'winter' species also occur in British streams; e.g. conidia of *Lunulospora curvula* were detected in the River Creedy only from August to November in contrast to *Tricladium chaetocladium* (incorrectly identified as *T. gracile*, Webster *et al* 1976) which was detected only from December to April (Iqbal & Webster 1973, 1977). The possibility that these patterns were due to the two species having different temperature optima was investigated by Webster *et al* (1976). When grown in single-species culture in the laboratory, the optima for growth and sporulation of *L. curvula* was found to be higher (25°C) than that for *T. chaetocladium* (20°C). Interestingly, when these two species were grown together in mixed culture, the optimum temperature for sporulation decreased to 5°C for *T. chaetocladium* and 10°C for *L. curvula*, even though in single-species culture *T. chaetocladium* failed to sporulate at 5°C. It would therefore appear that the temperature responses of these species can be modified by interspecific interactions, although the mechanism by which this occurs is unclear. The possible importance of interspecific interactions in determining temporal occurrence is suggested by the observation that if leaves are already colonized by summer species, the ability of winter species to colonize them is reduced (Suberkropp 1984).

In general, the majority of species found in temperate regions have an optimum temperature for growth in the laboratory of between 15° and 25°C, and continue to grow when the temperature drops below 5°C, (Thornton 1963; Suberkropp & Klug 1981). Those species that are classified as 'tropical species' (Nilsson 1964), or are common in the summer in temperate regions, have higher optimum temperatures and do not grow below 5°C (e.g. *Clavatospora tentacula*, *Flagellospora penicilliodies*, *Lunulospora curvula*).

Spatial variation in aquatic hyphomycete abundances has been observed both between (e.g. Iqbal & Webster 1973, 1977; Wood-Eggenschwiler & Bärlocher 1983) and within (e.g. Shearer & Webster 1985a) rivers. Whereas major differences in riparian vegetation (e.g. change from moorland to woodland) are correlated with changes in aquatic hyphomycete diversity (Webster 1981; Shearer & Webster 1985a), despite the apparent substrate specificity of some species (e.g. Bengtsson 1982, 1983), changes in riparian vegetation within a habitat type (e.g. deciduous woodland) are not (Wood-Eggenschwiler & Bärlocher 1983; Aimer & Segedin 1985; Thomas *et al* 1989).

Aero-aquatic hyphomycetes

Aero-aquatic hyphomycetes are distinguished by the fact that they generally sporulate only when exposed to moist air and produce conidia which trap air (Fisher 1977; Fig. 4.2). They are commonly found in slow-flowing or static water and many appear adapted to exist in hypoxic conditions (Webster 1981). However, as aero-aquatics grow well (Fisher & Webster 1979) and reproduce under well-aerated conditions, their distribution may be limited more by their poor competitive ability than by physiological constraints (Fisher & Webster 1981).

Aero-aquatics are found on both leaves (deciduous and coniferous) and wood (Lamore & Goos 1978; Fisher 1979; Jones 1981), and common species include *Helicodendron fractum* and *Helicoon sessile*.

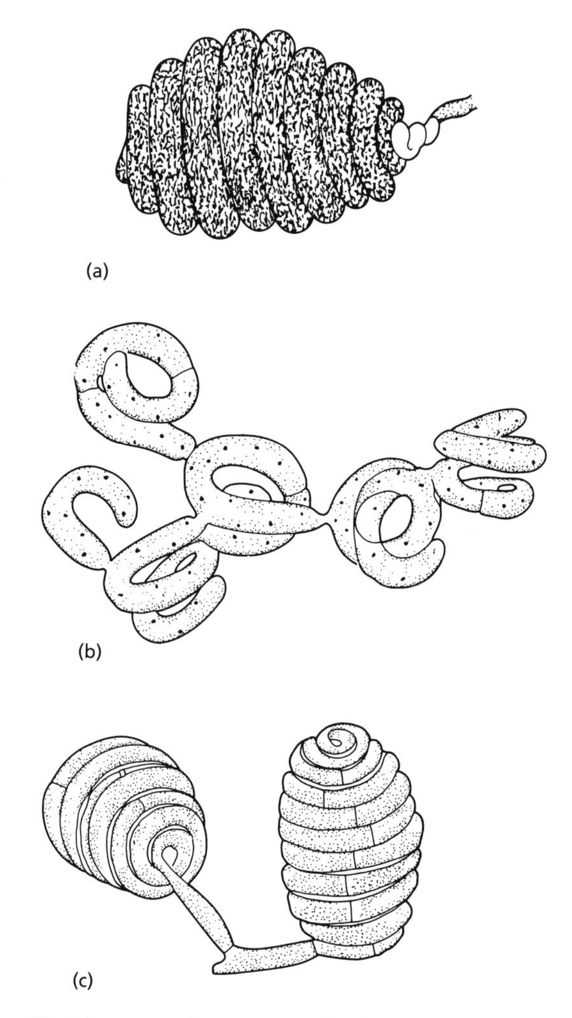

Fig. 4.2 Spores of aero-aquatic hyphomycetes. (a) *Helicoon richonis*; (b) *Helicodendron triglitziense*; (c) *Helicodendron conglomeratum* (after Webster 1980).

Enzymatic capabilities

As aquatic hyphomycetes can utilize dissolved organic matter (e.g. Thornton 1963; Bengtsson 1982), to what extent are they capable of degrading the detrital material to which they are attached? Although they colonize both leaf and wood substrates (Willoughby & Archer 1973; Sanders & Anderson 1979), most studies have concentrated on the mycoflora of decaying leaves (e.g. Suberkropp & Klug 1980, 1981; Jones 1981; Chamier & Dixon 1982b; Fisher *et al* 1983; Suberkropp *et al* 1983; Chamier *et al* 1984; Butler & Suberkropp 1986; Abdullah & Taj-Aldeen 1989).

In general, plant cell walls consist of approximately 35% (w/w) peptic polysaccharides, 24% (w/w) hemicellulose and 23% (w/w) cellulose (Darvill *et al* 1980). Cell-wall-degrading enzymes of aquatic hyphomycetes were reviewed recently by Chamier (1985) and a summary of the enzymatic capabilities of some aquatic hyphomycetes is presented in Table 4.3. As a group, these fungi can utilize a wide range of substrates and the majority of species can degrade at least one of the major components of plant cell walls. In addition to utilizing plant cells as a carbon source, fungi may also be saprophytic on other fungi. All of the 14 species of aquatic hyphomycetes tested by Chamier (1985) could utilize lamarin and a number could break down chitin. Both lamarin and chitin are constituents of fungal cell walls. Interestingly, those species with weak cellulase

Table 4.3 Enzymatic capabilities of aquatic hyphomycetes

Species	Substrate							Reference
	Cellulose	Xylan	Starch	Protein	Tannic acid	Pectin	Lignin	
Alatospora acuminata	√	√	–	x	–	√	–	beh
Anguillospora crassa	√	–	√	√	√	–	–	ag
Anguillospora furtiva	–	–	–	–	–	–	√	f
Anguillospora longissima	–	–	√	x	√	–	x	ag
Anguillospora pseudolongissima	√	√	–	√	–	√	–	e
Articulospora tetracladia	√	√	–	–	–	√	x	cdg
Articulospora inflata	√	√	–	x	–	√	–	e
Clavariopsis aquatica	√	√	–	√	–	√	x	beh
Clavatospora tentacula	√	–	–	–	–	–	–	h
Dactyleila aquatica	x	–	x	√	x	–	√	a
Dendrospora tenella	–	–	–	–	–	–	x	f
Dimorphospora foliicola	–	–	–	–	–	–	√	f
Filosporella annelidica	√	√	–	–	–	√	–	b
Flagellospora curvula	√	√	–	√	–	√	–	beh
Flagellospora penicilloides	√	–	–	–	–	–	–	h
Gyoerffyella rotula	–	–	–	–	–	–	√	f
Heliscus lugdunensis	√	√	x	√	√	√	x	abefgh
Lemonniera aquatica	√	√	√	√	√	√	x	acdefgh
Lemonniera cornuta	–	–	–	–	–	–	x	f
Lemonniera terrestris	√	–	–	√	–	√	–	e
Lunulospora curvula	√	–	√	√	√	–	x	afh
Margaritispora aquatica	–	–	–	–	–	–	x	f
Mycocentrospora angulata	√	√	–	–	–	√	–	cd
Scorpiosporium gracile	–	–	–	–	–	–	x	f
Tetrachaetum elegans	√	√	–	x	–	√	–	cde
Tetracladium marchalianum	√	√	–	√	–	√	–	deh
Tetracladium setigerum	x	–	√	√	√	√	√	acf
Tricladium giganteum	–	–	–	–	–	–	√	f
Tricladium splendans	√	√	–	–	–	√	–	cdg
Varicosporium elodeae	√	√	√	√	x	√	–	acdg

(√, detected; x, not detected; –, not assessed). a, Abdullah & Taj-Aldeen (1989); b, Butler & Suberkropp (1986); c, Chamier & Dixon (1982b); d, Chamier *et al* (1984); e, Suberkropp *et al* 1983); f, Fisher *et al* (1983); g, Jones (1981); h, Suberkropp & Klug (1981).

activity had greater laminase and chitinase activity than those species with high cellulase activity, suggesting that they may utilize both fungal and plant material as an energy source.

There is some experimental evidence in support of the utilization of wood by hyphomycetes. Fisher *et al* (1983) concluded that 15 of the 23 aero-aquatics, and six of the 16 aquatic hyphomycetes tested could degrade lignin, and Jones (1981) demonstrated that four hyphomycetes, *Anguillospora crassa*, *Anguillospora longissima*, *Lunulospora curvula* and *Tricladium splendans*, caused significant weight loss of wood under laboratory conditions.

Fungi associated with leaf litter

When leaves from deciduous trees enter rivers and streams in the autumn they are already colonized by a number of fungi, including *Alternaria*, *Aureobasidium*, *Cladosporium*, *Epicoccum* and *Humicola* (Bärlocher & Kendrick 1975). Once submerged, the mycoflora of the leaf changes to one dominated by aquatic hyphomycetes (Bärlocher & Kendrick 1975; Suberkropp & Klug 1976; Chamier & Dixon 1982a). Some members of the terrestrial mycoflora of leaves have the ability to degrade leaf material when it is submerged. For example, *Cladosporium* and *Epicoccum* have strong enzymatic activity towards cellulose, xylan and pectin (Chamier *et al* 1984) and can degrade both alder leaves and filter paper in water (Godfrey 1983). The decrease in abundance of the resident mycoflora once the leaf material becomes submerged is thought to be due to their inability to grow at low water temperatures (Bärlocher & Kendrick 1974).

Colonization of a leaf by aquatic hyphomycetes is initially random, with hyphae growing mainly on the leaf surface. After a few weeks, the leaves become covered with a mosaic of fungal colonies (Fig. 4.3; Shearer & Lane 1983; Chamier *et al* 1984). Although habitat selection has been reported for some species of aquatic hyphomycetes (Bengtsson 1983), resource specificity is generally low and the same species tend to be present on different leaf types (Bärlocher & Kendrick 1974; Suberkropp & Klug 1980; Shearer & Lane 1983; Butler & Suberkropp 1986). There is also little evidence of a succession of aquatic hyphomycetes colonizing leaf litter. Instead, the community appears to be dominated by four to eight species that arrive early and persist (Bärlocher 1982). Genera commonly isolated from leaf litter include *Alatospora*, *Anguillospora*, *Articulospora*, *Cladosporium*, *Clavatospora*, *Epicoccum*, *Flagellospora*, *Fusarium*, *Lemonniera*, *Tetracladium*, *Tricladium* and *Varicosporium*.

Most studies on the decomposition of vascular plant material in flowing waters have concentrated on aquatic hyphomycetes colonizing leaf litter from deciduous trees; however, other species of fungi (e.g. *Pythium*; Park & McKee 1978; Park 1980) and substrate types (e.g. conifer needles; Bärlocher & Michaelides 1978; Bärlocher & Oertli 1978; Bärlocher & Kendrick 1981) may be important in the economy of stream systems (see Chapter 8).

Fungi associated with wood

A significant proportion of allochthonous material enters streams and rivers as twigs, branches and logs. In a study of wood-inhabiting fungi in a North American stream, Lamore and Goos (1978) identified 59 species of fungi, of which 15 were ascomycetes, 36 were hyphomycetes and four were zygomycetes. Ascomycetes such as *Ceratastomella*, *Massarina* and *Nectria lugdunensis* were also common colonizers of wood in a British stream, although again the dominant group was the aquatic hyphomycetes which accounted for 23 of the 37 species identified (Willoughby & Archer 1973). Submerged wood decomposes more slowly than leaves (Shearer & von Bodman 1983; Chergui & Patee 1991) and, in contrast to leaf material, there is some evidence of species succession (Shearer & Webster 1991).

Common colonizers of wood in flowing waters include *Anguillospora longissima*, *Anguillospora crassa*, *Aureobasidium pullulans*, *Camposporium pellucidium*, *Clavariopsis aquatica*, *Dactylella aquatica*, *Fusarium*, *Geotrichum candidum*, *Helicoon sessile*, *Heliscus lugdunensis*, *Massarina* and *Tricladium splendens* (Willoughby & Archer 1973; Lamore & Goos 1978; Jones 1981; Shearer & Webster 1991).

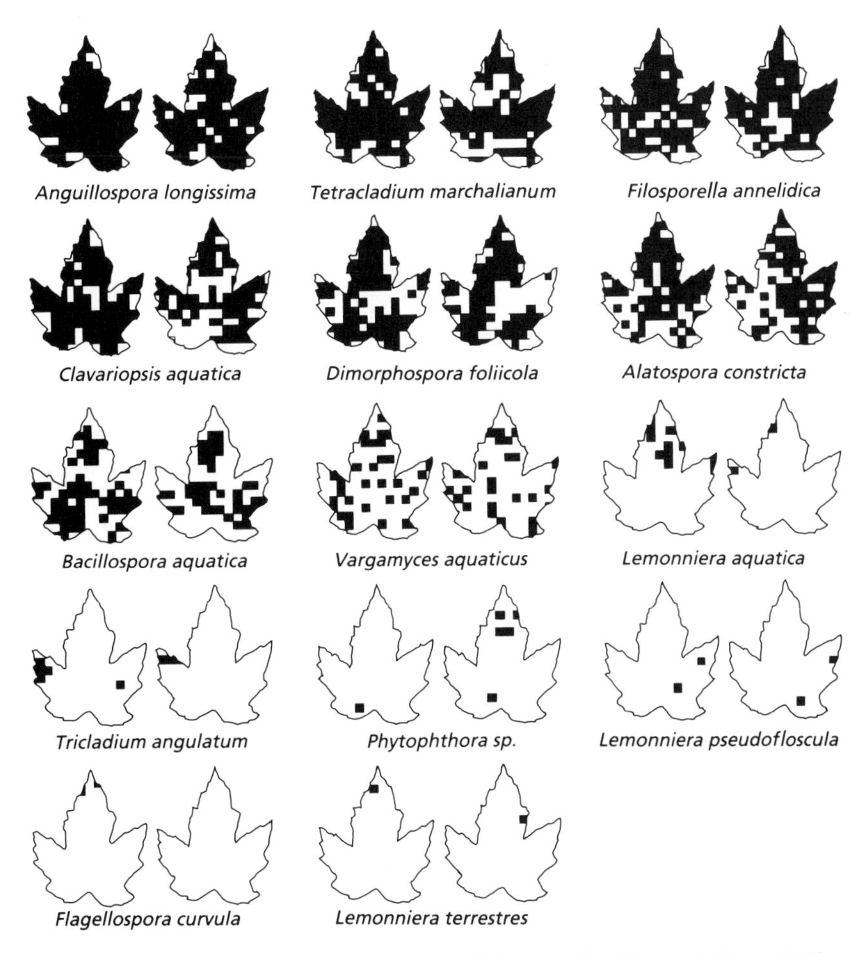

Fig. 4.3 Distribution of aquatic hyphomycetes on silver maple leaves (after Shearer & Lane 1983).

4.4 BACTERIA

Little is known about the species composition and variability of natural bacterial communities, principally because conventional identification requires pure cultures and less than 1% of natural bacteria are cultivatable (Lee & Fuhrman 1990). Jones (1987) stated that 'since we can only isolate, readily, approximately 0.25% of the bacteria seen in freshwater and sediments, there is little hope of assessing the diversity of a microbial community at present'.

Freshwater bacteria may occur either as free-living or attached cells suspended in the water column or associated with sediments. The bacterial flora of detritus consists of rods, cocci, filamentous and mycelial forms as well as myxobacteria (Fenchel & Jørgensen 1977; Armstrong & Bärlocher 1989). In unpolluted streams, Gram-negative rods predominate, although stalked bacteria such as *Hyphomicrobium*, *Caulobacter* and *Gallionella* are often present. For example, Baker and Farr (1977), studying the bacterial flora of two chalk streams in southern England, found that 87% of the bacteria isolated from water samples were Gram-negative rods dominated by *Pseudomonas*, *Flavobacterium*, *Acinetobacter*, *Moraxella*, *Xanthominas* and *Aeromonas*. In nutrient-rich waters, the numbers of *Flavobacterium* and *Achromobacter* species decrease and the bacterial flora becomes dominated by members of the Pseudomonadaceae, Bacillaceae and Enterobacteriaceae (Rheinmeimer 1985).

Suberkropp and Klug (1976) conducted a detailed study of the bacteria associated with leaves during their processing in a North American woodland stream. Bacterial colonization of leaves increased rapidly during the first 2 weeks of exposure to river water and then continued to increase exponentially during the 6-month study period (Fig. 4.4). Although a wide variety of bacterial colonies was observed during the study, no general trend or correlation of colony type with stage of decomposition was detected. The most common genera isolated from leaf material were *Flavobacterium*, *Flexibacter*, *Pseudomonas*, *Acinetobacter*, *Achromobacter*, *Chromobacterium*, *Serriata*, *Bacillus*, *Alcaligenes*, *Cytophaga*, *Sporocytophaga* and *Arthrobacter*.

Whereas the number of bacterial cells in the seston of rivers may reach 10^9-10^{11} cells l^{-1} (e.g. Edwards 1987), not all of these cells will be active. Studies on the Ogeechee River in North America have distinguished two groups of bacteria: a small group of actively growing cells attached to particles and a much larger group of slowly growing or inactive free-living cells (Edwards & Meyer 1986). A similar observation was made by Kirchman (1983) for a pond system where less than 10% of the total bacterial population was attached to particles at any one time. There is a strong correlation between discharge and bacterial density in the Ogeechee River, suggesting that many of the free-living bacteria are inactive soil bacteria washed in from adjacent land (Edwards 1987; Carlough & Meyer 1989), although no positive correlation between bacterial numbers and discharge was detected by Goulder (1980) for the River Hull (UK). Such inwashing of bacteria may explain the differences in seasonal patterns observed for different systems. In the Ogeechee River the number of suspended bacteria increased in autumn and remained high during winter before decreasing in spring and summer (Carlough & Meyer 1989). In contrast, in the River Hull, numbers of suspended bacteria were highest in summer and lowest in winter (Goulder 1980). Both of these patterns differ from that recorded for bacteria associated with leaf litter by Suberkropp and Klug (1976). Here, bacterial numbers increased exponentially from the time when litter entered the system (November) until the point when the litter had decomposed, 5–7 months later (i.e. April–June). In streams where the major

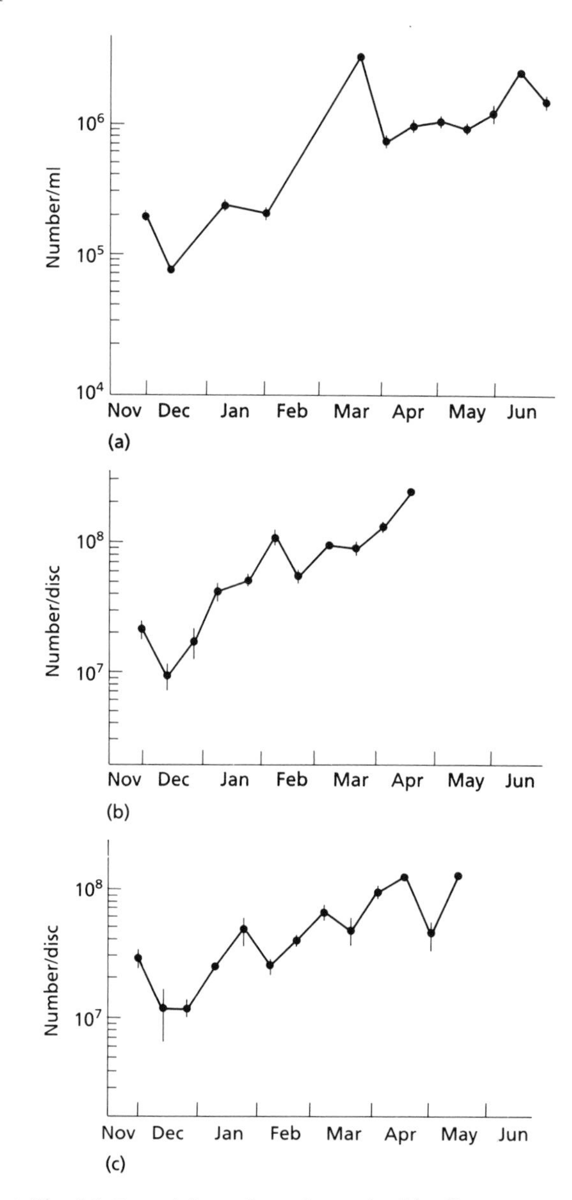

Fig. 4.4 Bacterial numbers, determined by direct microscopic counts, in the water (a) and associated with oak (b) and hickory leaves (c) (after Suberkropp & Klug 1976).

source of suspended bacteria is surfaces and sediments within the stream, bacterial abundance is positively correlated with distance from the source (Goulder 1984; Rimes & Goulder 1986).

Bacteria are an important component of biofilms coating stone surfaces. The number of bacteria in biofilms is often greater than suspended bacterial populations (Geesey *et al* 1978; Lock *et al* 1984) and may reach densities of 10^7–10^8 cells cm^{-2} (Croker & Meyer 1987; Bärlocher & Murdoch 1989). These bacterial communities are diverse, consisting of rods, cocci and spiral-shaped cells (Armstrong & Bärlocher 1989).

The question that must be addressed is whether bacteria associated with particulate detritus are actively involved in its decomposition or are simply using it as a substrate for attachment. In order to decompose vascular plant material, microbes must have the ability to break down cell wall materials such as cellulose, cellulobiose and lignin. Although freshwater bacteria can utilize a range of substrates, including proteins (e.g. *Pseudomonas*), urea (e.g. *Proteus vulgaris*), uric acid (e.g. *Vibrio*) and starch (e.g. *Bacillus*, *Clostridium*), cellulose is utilized mainly by myxobacteria (e.g. *Cytophaga*, *Sporocytophaga*) (Suberkropp & Klug 1976; Rheinmeimer 1985).

Several studies have shown that bacteria in freshwaters can utilize dissolved organic matter (DOM) (Geesey *et al* 1978; Haack & McFeters 1982; Rounick & Winterbourn 1983; Winterbourn *et al* 1986). DOM may originate from many sources (Fig. 4.5; Miller 1987), and is an important component of the energy budget of many stream systems (see Chapter 8). DOM may be leached from detritus or be released by the enzymatic activity of micro-organisms associated with it. Non-detrital DOM may be released by algae and macrophytes.

Heterotrophic bacteria are the major utilizers of DOM in freshwater systems (see section 4.5) and Chapter 8). Their biomass is positively correlated with dissolved organic carbon concentration and responds rapidly to changes in its availability (Kaplan & Bott 1983; Henebry & Gorden 1989).

4.5 BREAKDOWN OF ORGANIC MATTER

The decomposition of particulate organic material in fresh waters has been reviewed by several

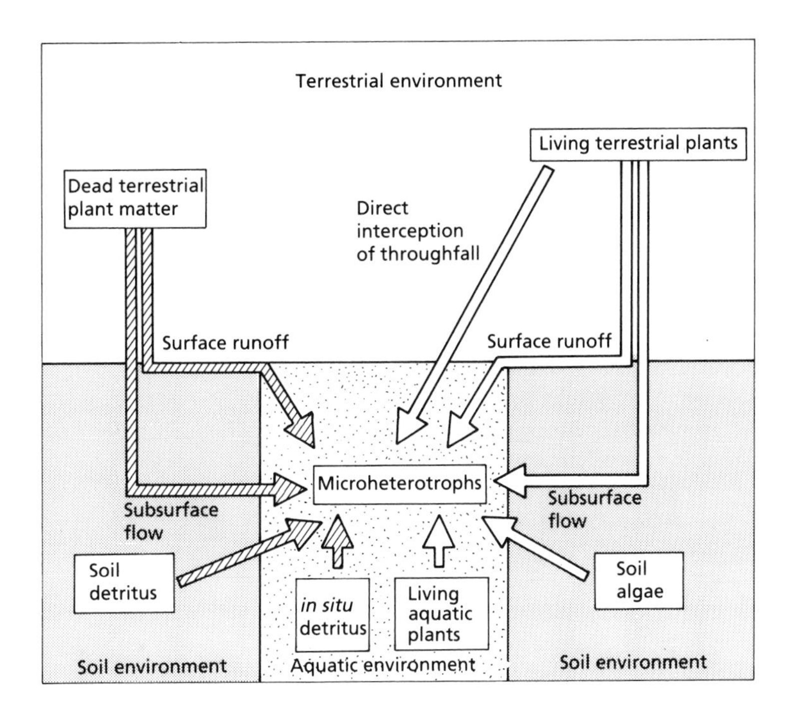

Fig. 4.5 Sources of detrital (hatched arrows) and non-detrital (open arrows) dissolved organic matter (DOM) (after Miller 1987).

authors, including Willoughby (1974), Zdanowski (1977), Anderson and Sedell (1979), Webster and Benfield (1986), and is the subject of Chapter 8.

In lotic environments, cellulose is broken down mainly by myxobacteria and higher fungi. Because of the ability of their hyphae to penetrate the interior of leaves, Bärlocher and Kendrick (1976) concluded that aquatic hyphomycetes were the most important group involved in the degradation of leaf litter. More recent studies of the enzymatic capabilities of these fungi support this view (section 4.3). Amongst bacteria, members of the genera *Cytophaga* and *Sporocytophaga* are the predominant cellulose decomposers (Rheinheimer 1985). Members of the genus *Cytophora* also have the ability to 'tunnel' through plant cell walls (Porter *et al* 1989). Observational studies and laboratory experiments using bacteriocides suggest that fungi are more important than bacteria in the decomposition of leaf litter in flowing water systems (e.g. Triska 1970; Kaushik & Hynes 1971; Suberkropp & Klug 1976). However, bacteria increase in abundance during the latter stages of leaf decomposition and are the major group of micro-organisms on fine particulate organic matter (Mason 1976; Suberkropp & Klug 1976; Chamier *et al* 1984; Sinsabaugh & Linkins 1990). They may also play an important role in the decomposition of conifer needles and more recalcitrant leaves (Iversen 1973; Davis & Winterbourn 1977).

Bacteria are the major group utilizing DOM. Benthic bacteria associated with biofilms appear to be more important than suspended bacteria in the processing of DOM (Lock & Hynes 1976; Dahm 1981; McDowell 1985). Of the suspended bacteria, attached bacteria account for the majority of the heterotrophic activity (Edwards & Meyer 1986). This is in contrast to open water systems (i.e. lakes, oceans) where free-living bacteria are mainly responsible for DOM uptake (Azam *et al* 1990).

4.6 EFFECT OF POLLUTANTS ON BACTERIA AND FUNGI

The distribution and activity of heterotrophic micro-organisms is influenced by a range of pollutants, including cadmium (Giesy 1978), copper (Leland & Carter 1985), zinc (Hornor & Hilt 1985), acid mine drainage (Carpenter *et al* 1983; Maltby & Booth 1991), pesticides (Milner & Goulder 1986; Chandrashekar & Kaveriappa 1989; Cuffney *et al* 1990), organic enrichment (Suberkropp *et al* 1988) and acidity (Traaen 1980; Chamier 1987; Palumbo *et al* 1987a; Thompson & Bärlocher 1989).

Hardness, pH

Bärlocher and Rosset (1981), in a study of the mycoflora of rivers in West Germany and Switzerland, recorded a greater diversity of aquatic hyphomycetes in softwater than in hardwater streams. These differences, which were related to pH (and associated variables) rather than to riparian vegetation, were also apparent in a subsequent, more extensive, study (Wood-Eggenschwiler & Bärlocher 1983). In an attempt to explain the observed distribution patterns, Rosset and Bärlocher (1985) studied the effects of pH, calcium and bicarbonate ions on the growth of aquatic hyphomycetes isolated from hardwater and softwater streams. When tested over the range pH 3–9, all species of fungi investigated grew best at pH values of between 4 and 5. However, the ecological relevance of these results is put into question by the observation that several of the fungi tested exhibited little or no growth at pH values corresponding to those of their native stream. Although species-specific effects on growth were observed for bicarbonate ions, these were not related to the distribution of the species in the field. Calcium stimulated growth of all species tested.

Several studies have noted an accumulation of organic material in acidified waters (Friberg *et al* 1980; Traaen 1980), although there are exceptions (Schindler 1980). Such accumulations are assumed to be due to reductions in microbial and/or faunal processing rates. Although reductions in the microbial activity of leaf litter in acid streams has been recorded (Chamier 1987; Palumbo *et al* 1987b), the results from field and laboratory studies on the effect of pH on aquatic fungi are variable. Whereas some studies have found a negative effect of low pH on fungal abundance (e.g. Hall *et al* 1980; McKinley & Vestal 1982) others have recorded a positive effect (e.g. van Frankenhuyzen *et al* 1985). Thompson and Bärlocher (1989) found that, whereas weight loss of leaf discs was positively correlated with pH and temperature in the field, in a laboratory stream there was a negative correlation between weight loss and pH over the same pH range. One possible reason for the apparent contradiction between laboratory and field results is that under natural conditions, low pH is frequently accompanied by high concentrations of aluminium. Aluminium is concentrated by leaf litter and may be toxic to the associated microbes (Chamier *et al* 1989).

Changes in pH also alter the structure and functioning of bacterial communities. The metabolic activity of both suspended bacteria and that associated with leaf litter is lower in acidic (pH 4.5–5.7) than in circumneutral (pH 6.4–7.8) streams (Rimes & Goulder 1986; Mulholland *et al* 1987). Low pH stress can also result in morphological changes of bacterial cells (Rao *et al* 1984).

The effect of changes in pH on the structure of bacterial communities is illustrated by the study performed by Guthrie *et al* (1978). The structure of a bacterial community was determined before and after the addition of fly ash. The addition of fly ash reduced the pH from 6.5 to 4.6 and resulted in a change in both the number of bacterial species present and their relative abundances (Table 4.4).

Table 4.4 Dominance of bacterial groups before and after acidification (from Guthrie *et al* 1978)

pH 6.5	pH 4.6
Bacillus	*Pseudomonas*
Sarcina	*Flavobacterium*
Achromobacter	*Chromobacterium*
Flavobacterium	*Bacillus*
Pseudomonas	*Brevibacterium*

Metals

Alterations in fungal assemblages and reduced decomposition rates have been observed below mine effluent discharges (Carpenter *et al* 1983; Maltby & Booth 1991). For example, Carpenter *et al* (1983) found a significant reduction in the decomposition rate of leaves in a site receiving acid-mine drainage. Although bacterial biomass was not reduced at the polluted site, heterotrophic activity was reduced, owing to a reduction in either bacterial or fungal activity. The effect of coal-mine discharges on aquatic fungi has recently been investigated in British rivers (Maltby & Booth 1991; Bermingham S, unpublished data). Results of these studies suggest that the group of fungi most susceptible to mine discharges are the aquatic hyphomycetes, whose diversity reduces by almost 50% (Fig. 4.6).

Reductions in diversity may occur as a result of either the lethal (death) or sublethal (reductions in growth, reproduction) effects of pollutants. Heavy metals have been shown to inhibit both the growth and sporulation of aquatic fungi (Duddridge & Wainwright 1980; Abel & Bärlocher 1984; Bermingham S, unpublished data). Abel and Bärlocher (1984) compared the sensitivity to cadmium of aquatic hyphomycetes from different sites. Sporulation was ten times more sensitive than growth and, although toxicity was greater for softwater hyphomycete communities, this was due to differences in water chemistry rather than genetic differences between the fungi. Of the five species studied in detail, *Heliscus lugdunensis* and *Tetracladium marchalianum* were the least sensitive, *Alatospora acuminata* was the most sensitive, and *Clavariopsis aquatica* and *Flagellospora curvula* were intermediate.

Metal-induced changes in microbial communities in favour of tolerant species and strains has been observed for heterotropic bacteria (Hornor & Hilt 1985). Bacteria from zinc-polluted sites were more tolerant of zinc stress than those from clean sites (Fig. 4.7) and selection for tolerant bacterial communities was achieved after 20 days' exposure to zinc in artificial stream systems (Colwell *et al* 1989).

Organic enrichment

Suberkropp *et al* (1988) studied the aquatic hyphomycete community above and below the

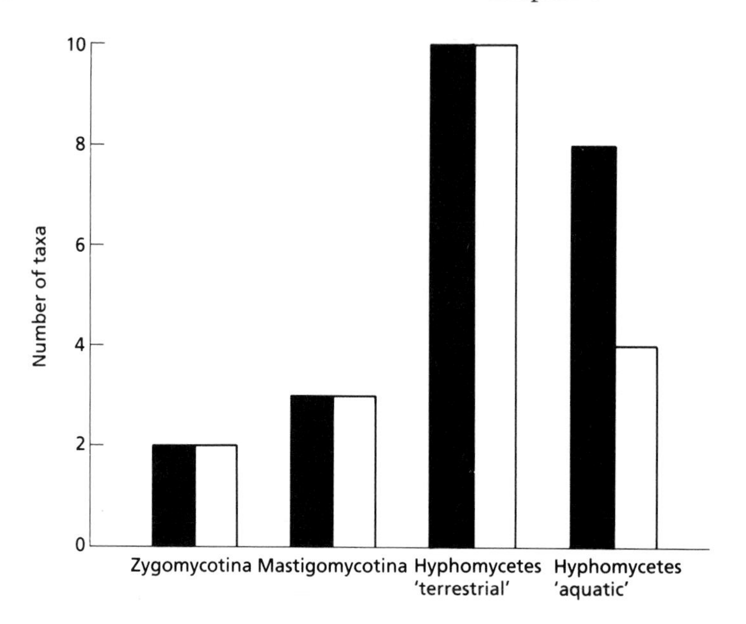

Fig. 4.6 Frequency of fungal taxa on leaf litter collected above and below an effluent discharge from a disused coal mine (■ upstream; □ downstream) (data from Maltby & Booth 1991).

effluent outfall of a sewage treatment works. The effluent resulted in an elevation in biochemical oxygen demand, nitrate, nitrite, ammonia and phosphate; however, it only caused a slight depression in oxygen concentration. The total number and species composition of aquatic hyphomycetes was similar above and below the effluent input. In contrast, Czeczuga and Próba (1987), studying the River Narew and its tributaries (Poland), found a strong negative correlation between fungal species richness and nitrate concentration. The nitrate values measured during their study ranged from 0.05 to 2.5 mg l^{-1}.

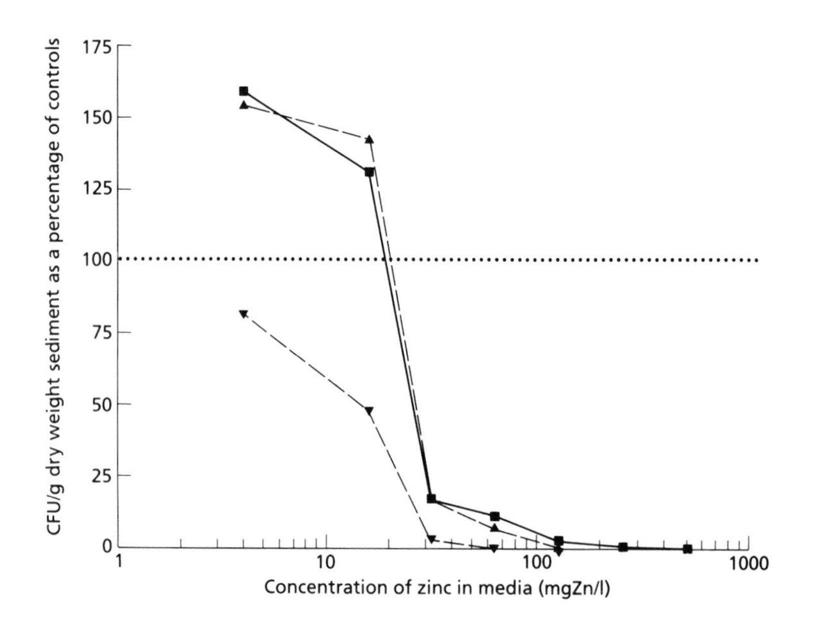

Fig. 4.7 Effect of zinc on the heterotrophic bacterial communities from three sites on the Cedar Run, Virginia, which were exposed to varying degrees of zinc contamination site (CFU, colony forming units) (data from Hornor & Hilt 1985). ■——■ High contamination site (3124.6 μg Zn/g); ▲– – – –▲ medium contamination site (291.4 μg Zn/g); ▼----▼ low contamination site (109.0 μg Zn/g).

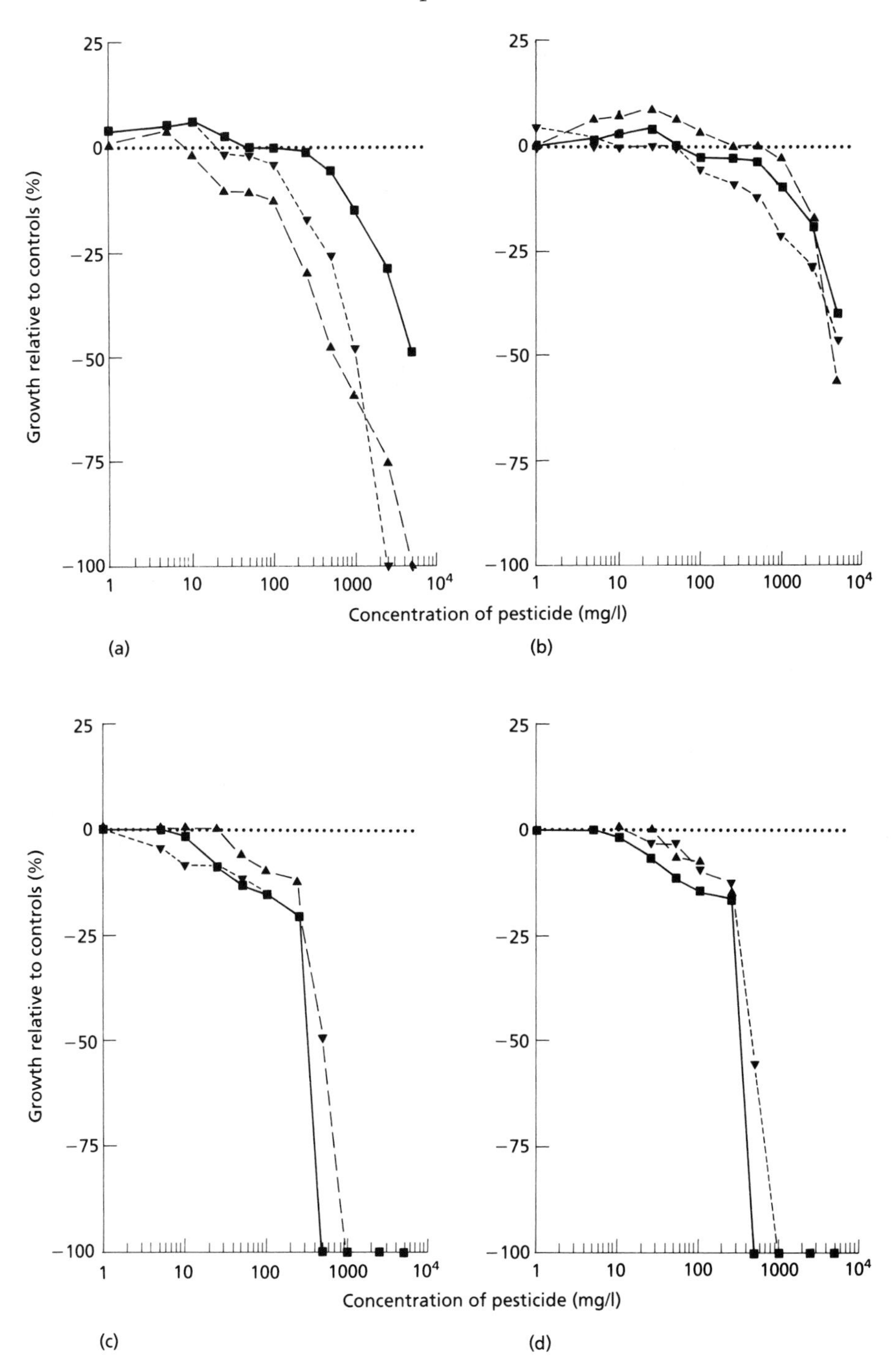

Fig. 4.8 Effect of four pesticides on the growth of three species of aquatic hyphomycetes. (a) Paraquat; (b) 2,4-DB; (c) Mancozeb; (d) Captafol. ■——■ *Flagellospora penicillioides*; ▲- - -▲ *Lunulospora curvula* and ▼----▼ *Phalangispora constricta* (data from Chandrashekar & Kaveriappa 1989).

Pesticides

Chandrashekar and Kaveriappa (1989) investigated the effect of four pesticides on the growth of three species of aquatic hyphomycetes: *Flagellospora penicillioides*, *Lunulospora curvula* and *Phalangispora constricta*. The two herbicides tested, paraquat and 2,4-dichlorophenoxybutyric acid (2,4-DB), stimulated growth whereas the two fungicides tested, mancozeb and captafol, either had no effect or inhibited growth (Fig. 4.8). Although species differed in their response to these chemicals, their relative sensitivities were not consistent across all toxicants. For example, whereas *P. constricta* was the most tolerant to the two fungicides, it was the most susceptible to the two herbicides. Other pesticides which have been studied include DDT (Dalton *et al* 1970), diquat (Fronda & Kendrick 1985) and methoxychlor (Cuffney *et al* 1990).

4.7 SUMMARY

Studies on heterotrophic micro-organisms associated with detritus have concentrated upon fungi, particularly aquatic hyphomycetes. Part of the reason for this is the generally held belief that aquatic hyphomycetes play a major role in the decomposition of coarse particulate organic matter, in particular leaves from deciduous trees, and are therefore important in the functioning of allochthonous-based systems. Despite this concentration of effort, relatively little information is available on the ecology and ecophysiology of specific species. Even less is known about other members of the leaf microflora or of the micro-organisms associated with other forms of detritus. There is an obvious need for research into these areas, especially in view of the fact that recent research has highlighted the possible importance of dissolved organic matter in river systems and its utilization by bacteria.

To understand fully the ecology of heterotrophic microbes, it is necessary to be able to sample and identify them. Although techniques are available for studying microbial communities, most of them have limitations. For example, it is possible to quantify total fungal biomass and to identify the species present in a particular fungal assemblage. However, techniques are not available to quantify the biomass of specific individual species within that assemblage. Similarly, because of taxonomic problems, most ecological studies of bacteria use the bacterial, or microbial, community as the unit of study. Techniques under development which may address some of these problems include gene probes (Steffan *et al* 1989), immunofluorescence (Currin *et al* 1990) and DNA hybridization (Lee & Fuhrman 1990).

Until techniques are developed to overcome these problems, the question of the role and importance of specific heterotrophic micro-organisms will remain unanswered. Such information is necessary if we are to predict the impact that changes in water quality and land use will have on the structure and functioning of river systems.

REFERENCES

Abdullah SK, Taj-Aldeen SJ. (1989) Extracellular enzymatic activity of aquatic and aero-aquatic conidial fungi. *Hydrobiologia* **174**: 217–23. [4.3]

Abel TH, Bärlocher F. (1984) Effects of cadmium on aquatic hyphomycetes. *Applied and Environmental Microbiology* **48**: 245–51. [4.2, 4.6]

Aimer RD, Segedin BP. (1985a) Some aquatic hyphomycetes from New Zealand streams. *New Zealand Journal of Botany* **23**: 273–99. [4.2, 4.3]

Aimer RD, Segedin BP. (1985b) Fluctuation in spore numbers of aquatic hyphomycetes in a New Zealand stream. *Botanical Journal of the Linnean Society* **91**: 61–6. [4.2, 4.3]

Akridge RE, Koehn RD. (1987) Amphibious hyphomycetes from San Marcos River in Texas. *Mycologia* **79**: 228–33. [4.2, 4.3]

Anderson NH, Sedell JR. (1979) Detritus processing by macroinvertebrates in stream ecosystems. *Annual Review of Entomology* **24**: 351–77. [4.5]

Armstrong SM, Bärlocher F. (1989) Adsorption and release of amino acids from epilithic biofilms in streams. *Freshwater Biology* **22**: 153–9. [4.2, 4.4]

Arsuffi TL, Suberkropp K. (1984) Leaf processing capabilities of aquatic hyphomycetes: interspecific differences and influence on shredder feeding preferences. *Oikos* **42**: 144–54. [4.2]

Austin B. (1988) *Methods in Aquatic Bacteriology*. John Wiley & Sons, Chichester. [4.2]

Azam F, Cho BC, Smith DC, Simon M. (1990) Bacterial cycling of matter in the pelagic zone of aquatic ecosystems. In: Tilzer MM, Serruya C (eds) *Large Lakes, Ecological Structure and Function*, pp 477–88.

Springer-Verlag, Berlin. [4.5]
Baker JH, Farr IS. (1977) Origins, characterization & dynamics of suspended bacteria in two chalk streams. *Archiv für Hydrobiologie* **80**: 308–26. [4.2, 4.4]
Bandoni RJ. (1974) Mycological observations on the aqueous films covering decaying leaves and other litter. *Transactions of the Mycological Society of Japan* **15**: 309–15. [4.3]
Bandoni RJ. (1975) Significance of the tetraradiate form in dispersal of terrestrial fungi. *Reports of Tottori Mycology Institute* **12**: 105–13. [4.3]
Bandoni RJ. (1981) Aquatic hyphomycetes from terrestrial litter. In: Wicklow DT, Carroll GC (eds) *The Fungal Community. Its Organization and Role in the Ecosystem*, pp 693–708. Marcel Dekker, New York. [4.3]
Bärlocher F. (1980) Leaf-eating invertebrates as competitors of aquatic hyphomycetes. *Oecologia* **47**: 303–6. [4.2]
Bärlocher F. (1982) On the ecology of Ingoldian fungi. *BioScience* **32**: 581–6. [4.3]
Bärlocher F, Kendrick B. (1973) Fungi and food preferences of *Gammarus pseudolimnaeus*. *Archiv für Hydrobiologie* **72**: 501–16. [4.2]
Bärlocher F, Kendrick B. (1974) Dynamics of the fungal populations on leaves in a stream. *Journal of Ecology* **62**: 761–91. [4.2, 4.3]
Bärlocher F, Kendrick B. (1975) Assimilation efficiency of *Gammarus pseudolimnaeus* (Amphipoda) feeding on fungal mycelium or autumn-shed leaves. *Oikos* **26**: 55–9. [4.3]
Bärlocher F, Kendrick B. (1976) Hyphomycetes as intermediaries of energy flow in streams. In: Jones EBG (ed.) *Recent Advances in Aquatic Mycology*, pp 435–46. Elek Science, London. [4.5]
Bärlocher F, Kendrick B. (1981) Role of aquatic hyphomycetes in the trophic structure of streams. In: Wicklow DT, Carroll GC (eds) *The Fungal Community. Its Organization and Role in the Ecosystem*. pp 743–60. Marcel Dekker, New York. [4.3]
Bärlocher F, Michaelides J. (1978) Colonization and conditioning of *Pinus resinosa* needles by aquatic hyphomycetes. *Archiv für Hydrobiologie* **81**: 462–74. [4.3]
Bärlocher F, Murdoch JH. (1989) Hyporheic biofilms—a potential food source for interstitial animals. *Hydrobiologia* **184**: 61–7. [4.4]
Bärlocher F, Oertli JJ. (1978) Colonization of conifer needles by aquatic hyphomycetes. *Canadian Journal of Botany* **56**: 57–62. [4.3]
Bärlocher F, Rosset J. (1981) Aquatic hyphomycete spora of two Black Forest and two Swiss Jura streams. *Transactions of the British Mycological Society* **76**: 479–83. [4.2, 4.3, 4.6]
Bärlocher F, Schweizer M. (1983) Effects of leaf size and decay rate on colonization by aquatic hyphomycetes. *Oikos* **41**: 205–10. [4.2]
Bell RT, Ahlgren GM, Ahlgren I. (1983) Estimating bacterioplankton production by measuring [^{3}H] thymidine incorporation in a eutrophic Swedish lake. *Applied and Environmental Microbiology* **45**: 1709–21. [4.2]
Bengtsson G. (1982) Patterns of amino acid utilization by aquatic hyphomycetes. *Oecologia* **55**: 355–63. [4.2, 4.3]
Bengtsson G. (1983) Habitat selection in two species of aquatic hyphomycetes. *Microbial Ecology* **9**: 15–26. [4.3]
Benner R, Lay J, K'nees E, Hodson RE. (1988) Carbon conversion efficiency for bacterial growth on lignocellulose: implications for detritus-based food webs. *Limnology and Oceanography* **33**: 1514–26. [4.2]
Bird GA, Kaushik NK (1981) Coarse particulate organic matter in streams. In: Lock MA, Williams DD (eds) *Perspectives in Running Water Ecology*, pp 41–68. Plenum Press, New York. [4.2]
Bjørnsen PK. (1986) Automatic determination of bacterioplankton biomass by image analysis. *Applied and Environmental Microbiology* **51**: 1199–204. [4.2]
Bobbie RJ, White DC. (1980) Characterization of benthic microbial community structure by high resolution gas chromatography of fatty acid methyl esters. *Applied and Environmental Microbiology* **39**: 1212–22. [4.2]
Bott TL, Kaplan LA. (1990) Potential for protozoan grazing of bacteria in streambed sediments. *Journal of the North American Benthological Society* **9**: 336–45. [4.2]
Boulton AJ, Boon PI. (1991) A review of methodology used to measure leaf litter decomposition in lotic environments: time to turn over an old leaf? *Australian Journal of Marine and Freshwater Research* **42**: 1–43. [4.2]
Butler SK, Suberkropp K. (1986) Aquatic hyphomycetes on oak leaves: comparison of growth, degradation and palatability. *Mycologia* **78**: 922–8. [4.2, 4.3]
Carlough LA, Meyer JL. (1989) Protozoans in two southeastern blackwater rivers and their importance to trophic transfer. *Limnology and Oceanography* **34**: 163–77. [4.4]
Carpenter J, Odum WE, Mills A. (1983) Leaf litter decomposition in a reservoir affected by acid mine drainage. *Oikos* **41**: 165–72. [4.2, 4.6]
Chamier A-C. (1985) Cell-wall-degrading enzymes of aquatic hyphomycetes: a review. *Botanical Journal of the Linnean Society* **91**: 67–81. [4.3]
Chamier A-C. (1987) Effect of pH on microbial degradation of leaf litter in seven streams of the English Lake District. *Oecologia* **71**: 491–500. [4.2, 4.6]
Chamier A-C, Dixon PA. (1982a) Pectinases in leaf degradation by aquatic hyphomycetes I: the field study. The colonization pattern of aquatic hypho-

mycetes on leaf packs in a Surrey stream. *Oecologia* **52**: 109–15. [4.2, 4.3]

Chamier A-C, Dixon PA. (1982b) Pectinases in leaf degradation by aquatic hyphomycetes: the enzymes and leaf maceration. *Journal of General Microbiology* **128**: 2469–83. [4.3]

Chamier A-C, Dixon PA, Archer SA. (1984) The spatial distribution of fungi on decomposing alder leaves in a freshwater stream. *Oecologia* **64**: 92–103. [4.2, 4.3, 4.5]

Chamier A-C, Sutcliffe DW, Lishman JP. (1989) Changes in Na, K, Ca, Mg and Al content of submersed leaf litter, related to ingestion by the amphipod *Gammarus pulex* (L.). *Freshwater Biology* **21**: 181–9. [4.6]

Chandrashekar KR, Kaveriappa KM. (1989) Effects of pesticides on the growth of aquatic hyphomycetes. *Toxicology Letters* **48**: 311–15. [4.6]

Chergui H. (1990) The dynamics of aquatic hyphomycetes in an eastern Moroccan stream. *Archiv für Hydrobiologie* **118**: 341–52. [4.3]

Chergui H, Pattee E. (1988) The dynamics of Hyphomycetes on decaying leaves in the network of the River Rhone (France). *Archiv für Hydrobiologie* **114**: 3–20. [4.2, 4.3]

Chergui H, Pattee E. (1991) The breakdown of wood in the side arm of a large river: preliminary observations. *Verhandlungen Internationale Vereinigung für Theoretische und Angewandt Limnologie* **24**: 1785–1788. [4.3]

Chrzanowski TH, Crotty RD, Hubbard JG, Welch P. (1984) Applicability of the fluorescein diacetate method of detecting active bacteria in freshwater. *Microbial Ecology* **10**: 179–85. [4.2]

Cochran TW, Vercellotti JR. (1978) Hexosamine biosynthesis and accumulation by fungi in liquid and solid media. *Carbohydrate Research* **61**: 529–43. [4.2]

Colwell FS, Hornor SG, Cherry DS (1989) Evidence of structural and functional adaptations in epilithon exposed to zinc. *Hydrobiologia* **171**: 79–90. [4.2, 4.6]

Croker MT, Meyer JL. (1987) Interstitial dissolved organic carbon in sediments of a southern Appalachian headwater stream. *Journal of the North American Benthological Society* **6**: 159–67. [4.4]

Cuffney TF, Wallace B, Lugthart GJ. (1990) Experimental evidence quantifying the role of benthic invertebrates in organic matter dynamics of headwater streams. *Freshwater Biology* **23**: 281–99. [4.6]

Cummins KW, Spengler GL, Ward GM, Speaker RM, Ovink RM, Mahan DC *et al* (1980) Processing of confined and naturally entrained leaf litter in a woodland stream ecosystem. *Limnology and Oceanography* **25**: 952–7. [4.2]

Currin CA, Paerl HW, Suba GK, Alberte RS. (1990) Immunofluorescence detection and characterization of N_2-fixing microorganisms from aquatic environments. *Limnology and Oceanography* **35**: 59–71. [4.7]

Czeczuga B, Próba D. (1987) Studies of aquatic fungi. VII Mycoflora of the upper part of the River Narew and its tributaries in a differentiated environment. *Nova Hedwigia* **44**: 151–61. [4.3, 4.6]

Dahm PR. (1981) Pathways and mechanisms for removal of dissolved organic carbon from leaf leachate in streams. *Canadian Journal of Fisheries and Aquatic Sciences* **38**: 68–76. [4.5]

Dalton SA, Hodkinson M, Smith KA. (1970) Interactions between DDT and river fungi. *Applied Microbiology* **20**: 662–6. [4.6]

Darvill A, McNeil M, Albersheim P, Delmer DP (1980) The primary cell walls of flowering plants. In: Tolbert NE (ed.) *The Plant Cell*, pp 92–157. Academic Press, New York. [4.3]

Davis SF, Winterbourn MJ. (1977) Breakdown and colonization of *Nothofagus* leaves in a New Zealand stream. *Oikos* **28**: 250–5. [4.5]

Duddridge JE, Wainwright M. (1980) Heavy metal accumulation by aquatic fungi and reduction in viability of *Gammarus pulex* fed Cd^{2+} contaminated mycelium. *Water Research* **14**: 1605–11. [4.2, 4.6]

Edwards RT. (1987) Sestonic bacteria as a food source for filtering invertebrates in two southeastern blackwater rivers. *Limnology and Oceanography* **32**: 221–34. [4.2, 4.4]

Edwards RT, Meyer JL. (1986) Production and turnover of planktonic bacteria in two southeastern blackwater rivers. *Applied and Environmental Microbiology* **52**: 1317–23. [4.2, 4.4, 4.5]

Elwood JW, Mulholland PJ, Newbold JD. (1988) Microbial activity and phosphorus uptake on decomposing leaf detritus in a heterotrophic stream. *Verhandlungen Internationale Vereinigung für Theoretische und Angewandt Limnologie* **23**: 1198–208. [4.2]

Fairchild JF, Boyle TP, Robinson-Wilson E, Jones JR. (1983) Microbial action in detritus leaf processing and the effects of chemical perturbation. In: Fontaine TD, Bartell SM (eds) *Dynamics of Lotic Ecosystems*, pp 437–56. Ann Arbor Science, Michigan. [4.2]

Federle TW, McKinley VL, Vestal JR. (1982) Physical determinants of microbial colonization and decomposition of plant litter in an arctic lake. *Microbial Ecology* **8**: 127–38. [4.2]

Fenchel TM, Jørgensen BB. (1977) Detrital food chains of aquatic ecosystems: the role of bacteria. *Advances in Microbial Ecology* **1**: 1–58. [4.4]

Findlay SEG, Arsuffi TL. (1989) Microbial growth and detritus transformations during decomposition of leaf litter in a stream. *Freshwater Biology* **21**: 261–9. [4.2]

Findlay SEG, Meyer JL, Edwards RT. (1984a) Measuring bacterial production via rate of incorporation of [^{3}H] thymidine into DNA. *Journal of Microbiological*

Methods **2**: 57–72. [4.2]
Findlay S, Meyer JL, Smith PJ. (1984b) Significance of bacterial biomass in the nutrition of a freshwater isopod (*Lirceus* sp.). *Oecologia* **63**: 38–42. [4.2]
Findlay S, Meyer JL, Smith PJ. (1986) Incorporation of microbial biomass by *Peltoperla* sp. (Plecoptera) and *Tipula* sp. (Diptera). *Journal of the North American Benthological Society* **5**: 306–10. [4.2]
Fisher PJ. (1977) New methods for detecting and studying saprophytic behaviour of aero-aquatic hyphomycetes from stagnant water. *Transactions of the British Mycological Society* **68**: 407–11. [4.3]
Fisher PJ. (1979) Colonization of freshly abscissed and decaying leaves by aero-aquatic hyphomycetes. *Transactions of the British Mycological Society* **73**: 99–102. [4.3]
Fisher PJ, Webster J. (1979) Effect of oxygen and carbon dioxide on growth of four aero-aquatic hyphomycetes. *Transactions of the British Mycological Society* **72**: 57–61. [4.3]
Fisher PJ, Webster J. (1981) Ecological studies on aero-aquatic hyphomycetes. In: Wicklow DT, Carroll GC (eds) *The Fungal Community. Its Organization and Role in the Ecosystem*, pp 709–30. Marcel Dekker, New York. [4.3]
Fisher PJ, Davey RA, Webster J. (1983) Degradation of lignin by aquatic and aero-aquatic hyphomycetes. *Transactions of the British Mycological Society* **80**: 166–8. [4.3]
Forbes AM, Magnuson JJ. (1980) Decomposition and microbial colonization of leaves in a stream modified by coal ash effluent. *Hydrobiologia* **76**: 263–7. [4.2]
Frankland JC, Lindley DK, Swift MJ. (1978) A comparison of two methods for the estimation of biomass in leaf litter. *Soil Biology and Biochemistry* **10**: 323–33. [4.2]
Freeman C, Lock MA, Marxsen J, Jones SE. (1990) Inhibitory effects of high molecular weight dissolved organic matter upon metabolic processes in biofilms from contrasting rivers and streams. *Freshwater Biology* **24**: 159–66. [4.2]
Friberg F, Otto C, Svensson BS. (1980) Effects of acidification on the dynamics of allochthonous leaf material and benthic invertebrate communities in running waters. In: Drablos D, Tollan A. (eds) *Proceedings of the International Conference on the Ecological Impact of Acid Precipitation, Norway 1980*, pp 304–5. SNSF project. [4.6]
Fronda A, Kendrick B. (1985) Uptake of diquat by four aero-aquatic hyphomycetes. *Environmental Pollution* **37**: 229–44. [4.6]
Fry JC. (1982) The analysis of microbial interactions and communities *in situ*. In: Bull AT, Slater JH (eds) *Microbial Interactions and Communities*, Vol. 1, pp 103–52. Academic Press, London. [4.2]
Fry JC (1988) Determination of Biomass. In: Austin B (ed.) *Methods in Aquatic Bacteriology*, pp 27–72. John Wiley & Sons, New York. [4.2]
Fry JC. (1990) Direct methods and biomass estimation. *Methods in Microbiology* **22**: 41–85. [4.2]
Fuhrman JA, Azam F. (1980) Bacterioplankton secondary production estimates for coastal waters of British Columbia, Antarctica and California. *Applied and Environmental Microbiology* **39**: 1085–95. [4.2]
Fuhrman JA, Azam F. (1982) Thymidine incorporation as a measure of heterotrophic bacterioplankton production in marine surface waters: evaluation and field results. *Marine Biology* **66**: 109–20. [4.2]
Garden A, Davies RW. (1988) Decay rates of autumn and spring leaf litter in a stream and effects on growth of a detritivore. *Freshwater Biology* **19**: 297–303. [4.2]
Geesey GG, Mutch R, Costerton JW, Green RB. (1978) Sessile bacteria: an important component of the microbial population in small mountain streams. *Limnology and Oceanography* **23**: 1214–23. [4.4]
Gessner MO, Schwoerbel J. (1991) Fungal biomass associated with decaying leaf litter in a stream. *Oecologia* **87**: 602–603. [4.2]
Giesy JP. (1978) Cadmium inhibition of leaf decomposition in an aquatic microcosm. *Chemosphere* **6**: 467–75. [4.6]
Godfrey BES. (1983) Growth of two terrestrial microfungi on submerged alder leaves. *Transactions of the British Mycological Society* **81**: 418–21. [4.2, 4.3]
Gottschalk G. (1986) *Bacterial Metabolism*, 2nd edn. Springer-Verlag, New York. [4.2]
Goulder R. (1980) Seasonal variation in heterotrophic activity and population density of planktonic bacteria in a clean river. *Journal of Ecology* **68**: 349–63. [4.4]
Goulder R. (1984) Downstream increase in the abundance and heterotrophic activity of suspended bacteria in an intermittent calcareous headstream. *Freshwater Biology* **14**: 611–19. [4.4]
Grant WD, West AW. (1986) Measurement of ergosterol, diaminopimelic acid and glucosamine in soil: evaluation as indicators of microbial biomass. *Journal of Microbial Methods* **6**: 47–53. [4.2]
Grigorova R, Norris JR. (1990) *Methods in Microbiology. Vol 22, Techniques in Microbial Ecology*. Academic Press, London. [4.2]
Groom AP, Hildrew AG. (1989) Food quality for detritivores in streams of contrasting pH. *Journal of Animal Ecology* **58**: 863–81. [4.2]
Gupta AK, Mahrotra RS. (1989) Seasonal periodicity of aquatic fungi in tanks at Kurukshetra, India. *Hydrobiologia* **173**: 219–29. [4.3]
Guthrie RK, Cherry DS, Singleton FL. (1978) Responses of heterotrophic bacterial populations to pH changes in coal ash effluent. *Water Resources Bulletin* **14**: 803–8. [4.6]
Haack TK, McFeters GA. (1982) Nutritional relation-

ships among microorganisms in an epilithic biofilm community. *Microbial Ecology* **8**: 115–26. [4.4]

Hall GH, Jones JG, Pickup RW, Simon BM. (1990) Methods to study the bacterial ecology of freshwater environments. *Methods in Microbiology* **22**: 182–209. [4.2]

Hall RJ, Likens GE, Fiance SB, Hendrey GR. (1980) Experimental acidification of a stream in the Hubbard Brook Experimental Forest, New Hampshire. *Ecology* **61**: 976–89. [4.6]

Henebry MS, Gorden RW. (1989) Summer bacterial populations in Mississippi River Pool 19: implications for secondary production. *Hydrobiologia* **182**: 15–23. [4.2, 4.4]

Herbert RA. (1990) Methods for enumerating microorganisms and determining biomass in natural environments. *Methods in Microbiology* **22**: 1–39. [4.1, 4.2]

Hering TF, Nicholson PB. (1964) A clearing technique for the examination of fungi in plant tissue. *Nature* **201**: 942–3. [4.2]

Hicks RE, Newell SY. (1983) An improved gas chromatographic method for measuring glucosamine and muramic acid concentrations. *Analytical Biochemistry* **128**: 438–45. [4.2]

Holm-Hansen O, Booth CR. (1966) The measurement of adenosine triphosphate in the ocean and its ecological significance. *Limnology and Oceanography* **11**: 510–19. [4.2]

Hornor SG, Hilt BA. (1985) Distribution of zinc-tolerant bacteria in stream sediments. *Hydrobiologia* **128**: 155–60. [4.2, 4.6]

Hudson HJ. (1986) *Fungal Biology*. Edward Arnold, London. [4.2, 4.3]

Ingold CT. (1942) Aquatic Hyphomycetes of decaying alder leaves. *Transactions of the British Mycological Society* **25**: 339–417. [4.3]

Ingold CT. (1975) *Guide to Aquatic Hyphomycetes*. Freshwater Biological Association (Publication No. 30), Cumbria. [4.2, 4.3]

Iqbal SH, Webster J. (1973) Aquatic hyphomycete spora of the River Exe and its tributaries. *Transactions of the British Mycological Society* **61**: 331–46. [4.2, 4.3]

Iqbal SH, Webster J. (1977) Aquatic hyphomycete spora of some Dartmoor streams. *Transactions of the British Mycological Society* **69**: 233–41. [4.2, 4.3]

Iversen TM. (1973) Decomposition of autumn-shed beech leaves in a springbrook and its significance for the fauna. *Archiv für Hydrobiologie* **72**: 305–12. [4.5]

Jones EBG. (1981) Observations on the ecology of lignicolous aquatic hyphomycetes. In: Wicklow DT, Carroll GC (eds) *The Fungal Community. Its Organization and Role in the Ecosystem*, pp 731–42. Marcel Dekker, New York. [4.3]

Jones JG. (1977) The study of aquatic microbial communities. In: Skinner FA, Shewan JM (eds) *Aquatic Microbiology*, pp 1–30. Academic Press, London. [4.2]

Jones JG. (1979) *A Guide to Methods for Estimating Microbial Numbers and Biomass in Fresh Water*. Freshwater Biological Association (Publication No. 39), Cumbria. [4.2]

Jones JG. (1987) Diversity of freshwater microbiology. In: Fletcher M, Gray TRG, Jones JG (eds) *Ecology of Microbial Communities*, pp 236–59. Cambridge University Press, Cambridge. [4.4]

Jones PCT, Mollison JE. (1948) A technique for the quantitative estimation of soil microorganisms. *Journal of General Microbiology* **2**: 54–69. [4.2]

Kaplan LA, Bott TL. (1983) Microbial heterotrophic utilization of dissolved organic matter in a Piedmont stream. *Freshwater Biology* **13**: 363–77. [4.2, 4.4]

Kaplan LA, Bott TL. (1985) Acclimation of stream-bed heterotrophic microflora: metabolic responses to dissolved organic matter. *Freshwater Biology* **15**: 479–92. [4.2]

Karl DM. (1980) Cellular nucleotide measurements and applications in microbial ecology. *Microbiological Reviews* **44**: 739–96. [4.2]

Kaushik NK, Hynes HBN. (1968) Experimental study on the role of autumn shed leaves in aquatic environments. *Journal of Ecology* **56**: 229–43. [4.2]

Kaushik NK, Hynes HBN. (1971) The fate of dead leaves that fall into streams. *Archiv für Hydrobiologie* **68**: 465–515. [4.5]

Kirby JJH. (1987) A comparison of serial washing and surface sterilization. *Transactions of the British Mycological Society* **88**: 559–62. [4.2]

Kirby JJH, Webster J, Baker JH. (1990) A particle plating method for analysis of fungal community composition and structure. *Mycological Research* **94**: 621–6. [4.2]

Kirchman D. (1983) The production of bacteria attached to particles suspended in a freshwater pond. *Limnology and Oceanography* **28**: 858–72. [4.4]

Ladd TI, Costerton JW. (1990) Methods for studying biofilm bacteria. *Methods in Microbiology* **22**: 285–307. [4.2]

Lamore BJ, Goos RD. (1978) Wood-inhabiting fungi of a freshwater stream in Rhode Island. *Mycologia* **70**: 1025–34. [4.2, 4.3]

Lee C, Howarth RW, Howes BL. (1980) Sterols in decomposing *Spartina alterniflora* and the use of ergosterol in estimating the contribution of fungi to detrital nitrogen. *Limnology and Oceanography* **25**: 290–303. [4.2]

Lee S, Fuhrman JA. (1990) DNA hybridization to compare species compositions of natural bacterioplankton assemblages. *Applied and Environmental Microbiology* **56**: 739–46. [4.4, 4.7]

Leland HV, Carter JL. (1985) Effects of copper on production of periphyton, nitrogen fixation and processing of leaf litter in a Sierra Nevada, California stream. *Freshwater Biology* **15**: 155–73. [4.6]

Lock MA, Hynes HBN. (1976) The fate of dissolved organic carbon derived from autumn-shed maple leaves (*Acer saccharum*) in a temperate hard water stream. *Limnology and Oceanography* **21**: 436–43. [4.5]

Lock MA, Wallace RR, Costerton JW, Ventullo RM, Charlton SE. (1984) River epilithon: towards a structural-functional model. *Oikos* **42**: 10–22. [4.4]

McDowell WH. (1985) Kinetics and mechanisms of dissolved organic carbon retention in a headwater stream. *Biogeochemistry* **2**: 329–53. [4.5]

McKinley VL, Vestal JR. (1982) Effects of acid on plant litter decomposition in an arctic lake. *Applied and Environmental Microbiology* **43**: 1188–95. [4.6]

Maltby L, Booth R. (1991) The effect of coal-mine effluent on fungal assemblages and leaf breakdown. *Water Research* **25**: 247–50. [4.6]

Mason CF. (1976) Relative importance of fungi and bacteria in the decomposition of phragmites leaves. *Hydrobiologia* **51**: 65–9. [4.5]

Matcham SE, Jordan BR, Wood DA. (1985) Estimation of fungal biomass in a solid substrate by three independent methods. *Applied Microbiology and Biotechnology* **21**: 108–12. [4.2]

Mer GS, Sati SC. (1989) Seasonal fluctuations in species composition of aquatic hyphomycetous flora in a temperate freshwater stream of central Himalaya, India. *Internationale Revue de Gesamten Hydrobiologie* **74**: 433–7. [4.3]

Miller JC. (1987) Evidence for the use of non-detrital dissolved organic matter by microheterotrophs on plant detritus in a woodland stream. *Freshwater Biology* **18**: 483–94. [4.4]

Milner CR, Goulder R. (1986) The abundance, heterotrophic activity and taxonomy of bacteria in a stream subject to pollution by chlorophenols, nitrophenols and phenoxyalkanoic acids. *Water Research* **20**: 85–90. [4.6]

Morgan P, Cooper CJ, Battersby NS, Lee SA, Lewis ST, Machin TM, Graham SC, Watkinson RJ. (1991) Automated image analysis method to determine fungal biomass in soils and on solid matrices. *Soil Biology and Biochemistry* **23**: 609–16. [4.2]

Moriarty DJW. (1986) Measurement of bacterial growth rates in aquatic systems from rates of nucleic acid synthesis. *Advances in Microbial Ecology* **9**: 245–92. [4.2]

Moriarty DJW. (1990) Techniques for estimating bacterial growth rates and production of biomass in aquatic ecosystems. *Methods in Microbiology* **22**: 212–34. [4.2]

Moriarty DJW, Pollard PC. (1981) DNA synthesis as a measure of bacterial productivity in seagrass sediments. *Marine Ecology Progress Series* **5**: 151–6. [4.2]

Mulholland PJ, Palumbo AV, Elwood JW, Rosemond AD. (1987) Effects of acidification on leaf decomposition in streams. *Journal of the North American Benthological Society* **6**: 147–58. [4.6]

Müller-Haeckel A, Marvanová L. (1979) Periodicity of aquatic hyphomycetes in the subarctic. *Transactions of the British Mycological Society* **73**: 109–16. [4.3]

Newell SY, Arsuffi TL, Fallon RD. (1988) Fundamental procedures for determining ergesterol content of decaying material by liquid chromatography. *Applied and Environmental Microbiology* **54**: 1876–9. [4.2]

Newell SY, Fallon RD. (1991) Toward a method for measuring instantaneous fungal growth rates in field samples. *Ecology* **72**: 1547–59. [4.2]

Newell SY, Hicks RE. (1982) Direct-count estimates of fungal and bacterial biovolume in dead leaves of smooth cordgrass (*Spartina alterniflora* Loisel). *Estuaries* **5**: 246–60. [4.2]

Newell SY, Statzell-Tallman A. (1982) Factors for conversion of fungal biovolume values to biomass, carbon and nitrogen: variation with mycelial ages, growth conditions and stains of fungi from a salt marsh. *Oikos* **39**: 261–8. [4.2]

Nilsson S. (1964) Freshwater hyphomycetes. Taxonomy, morphology and ecology. *Symbolae Botanicae Upsalienses* **XVIII**(2): 1–130. [4.3]

O'Carroll K. (1988) Assessment of bacterial activity. In: Austin B (ed.) *Methods in Aquatic Bacteriology*, pp 347–66. John Wiley & Sons, Chichester. [4.2]

Padgett DE. (1976) Leaf decomposition by fungi in a tropical rainforest stream. *Biotropica* **8**: 166–78. [4.3]

Palumbo AV, Bogle MA, Turner RR, Elwood JW, Mulholland PJ. (1987a) Bacterial communities in acidic and circumneutral streams. *Applied and Environmental Microbiology* **53**: 337–44. [4.6]

Palumbo AV, Mulholland PJ, Elwood JW. (1987b) Microbial communities on leaf material protected from macroinvertebrate grazing in acidic and circumneutral streams. *Canadian Journal of Fisheries and Aquatic Sciences* **44**: 1064–70. [4.2, 4.6]

Park D. (1972) Methods of detecting fungi in organic detritus in water. *Transactions of the British Mycological Society* **58**: 281–90. [4.2]

Park D. (1974a) Aquatic hyphomycetes in non-aquatic habitats. *Transactions of the British Mycological Society* **63**: 183–7. [4.3]

Park D. (1974b) Accumulation of fungi by cellulose exposed in a river. *Transactions of the British Mycological Society* **63**: 437–47. [4.2]

Park D. (1980) A two-year study of numbers of cellulolytic *Pythium* in river water. *Transactions of the British Mycological Society* **74**: 253–8. [4.3]

Park D, McKee W. (1978) Cellulolytic *Pythium* as a component of the river mycoflora. *Transactions of the British Mycological Society* **71**: 251–9. [4.3]

Paul EA, Johnson RL. (1977) Microscopic counting and

adenosine 5′-triphosphate measurement in determining microbial growth in soils. *Applied and Environmental Microbiology* **34**: 263–9. [4.2]

Paul RW, Benfield EF, Cairns J. (1983) Dynamics of leaf processing in a medium-sized river In: Fontaine TD, Bartell SM (eds) *Dynamics of Lotic Ecosystems*, pp 403–23. Ann Arbor Science, Michigan. [4.2]

Peters GT, Webster JR, Benfield EF. (1987) Microbial activity associated with seston in headwater streams: effects of nitrogen, phosphorus and temperature. *Freshwater Biology* **18**: 405–13. [4.2]

Petersen RC, Cummins KW, Ward GM. (1989) Microbial and animal processing of detritus in a woodland stream. *Ecological Monographs* **59**: 21–39. [4.2]

Poindexter JS, Leadbetter ER. (1986) *Bacteria in Nature. Vol. 2. Methods and Special Applications in Bacterial Ecology*. Plenum Press, New York. [4.2]

Porter KG, Feig YS. (1980) The use of DAPI for identifying and counting aquatic microflora. *Limnology and Oceanography* **25**: 943–8. [4.2]

Porter D, Newell SY, Lingle WL. (1989) Tunneling bacteria in decaying leaves of a seagrass. *Aquatic Botany* **35**: 395–401. [4.5]

Ranzoni FW. (1979) Aquatic hyphomycetes from Hawaii. *Mycologia* **71**: 786–95. [4.3]

Rao SS, Paolini D, Leppard GG. (1984) Effects of low-pH stress on the morphology and activity of bacteria from lakes receiving acid precipitation. *Hydrobiologia* **114**: 115–21. [4.6]

Rheinheimer G. (1985) *Aquatic Microbiology*. John Wiley & Sons, Chichester. [4.4, 4.5]

Ride JP, Drysdale RB. (1972) A rapid method for the chemical estimation of filamentous fungi in plant tissue. *Physiological Plant Pathology* **2**: 7–15. [4.2]

Riemann B, Bell RT. (1990) Advances in estimating bacterial biomass and growth in aquatic systems. *Archiv für Hydrobiologie* **118**: 385–402. [4.2]

Riemann B, Fuhrman J, Azam F. (1982) Bacterial secondary production in freshwater measured by (^{3}H)-thymidine incorporation method. *Microbial Ecology* **8**: 101–14. [4.2]

Rimes CA, Goulder R. (1986) Suspended bacteria in calcareous and acid headstreams: abundance, heterotrophic activity and downstream change. *Freshwater Biology* **16**: 663–51. [4.4, 4.6]

Rosset J, Bärlocher F. (1985) Aquatic hyphomycetes: influences of pH, Ca^{2+} and HCO_3^- on growth *in vitro*. *Transactions of the British Mycological Society* **84**: 137–45. [4.2, 4.6]

Rounick JS, Winterbourn MJ. (1983) The formation, structure and utilization of stone surface organic layers in two New Zealand streams. *Freshwater Biology* **13**: 57–72. [4.2, 4.4]

Sanders PF, Anderson JM. (1979) Colonization of wood blocks by aquatic hyphomycetes. *Transactions of the British Mycological Society* **73**: 103–7. [4.2, 4.3]

Schindler DW. (1980) Experimental acidification of a whole lake: a test of the oligotrophication hypothesis. In: Drabløs D, Tolan A (eds) *Proceedings International Conference on the Ecological Impact of Acid Precipitation, Norway 1980*, pp 340–4. SNSF Project. [4.6]

Schnürer J, Rosswall T. (1982) Fluorescein diacetate hydrolysis as a measure of total microbial activity in soil and litter. *Applied and Environmental Microbiology* **43**: 1256–61. [4.2]

Seitz LM, Saver DB, Burroughs R, Mohr HE, Hubbard JD. (1979) Ergosterol as a measure of fungal growth. *Phytopathology* **69**: 1202–3. [4.2]

Shearer CA. (1972) Fungi of the Chesapeake Bay and its tributaries. III The distribution of wood-inhabiting Ascomycetes and fungi imperfecti of the Patuxent River. *American Journal of Botany* **59**: 961–9. [4.2]

Shearer CA, Lane LC. (1983) Comparison of three techniques for the study of aquatic hyphomycete communities. *Mycologia* **75**: 498–508. [4.2, 4.3]

Shearer CA, Webster J. (1985a) Aquatic hyphomycete communities in the River Teign. I. Longitudinal distribution patterns. *Transactions of the British Mycological Society* **84**: 489–501. [4.2, 4.3]

Shearer CA, Webster J. (1985b) Aquatic hyphomycete communities in the River Teign. II. Temporal distribution patterns. *Transactions of the British Mycological Society* **84**: 503–7. [4.2, 4.3]

Shearer CA, Webster J. (1985c) Aquatic hyphomycete communities in the River Teign. III. Comparison of sampling techniques. *Transactions of the British Mycological Society* **84**: 509–18. [4.2]

Shearer CA, von Bodman SB. (1983) Patterns of occurrence of Ascomycetes associated with decomposing twigs in a mid western stream. *Mycologia* **75**: 518–30. [4.2]

Shearer CA, Webster J. (1991) Aquatic hyphomycete communities in the river Teign. IV Twig colonization. *Mycological Research* **95**: 413–20. [4.2]

Singh N, Musa TM. (1977) Terrestrial occurrence and the effect of temperature on growth, sporulation and spore production of some tropical aquatic hyphomycetes. *Transactions of the British Mycological Society* **68**: 103–6. [4.3]

Sinsabaugh RL, Golladay SW, Linkins AE. (1991) Comparison of epilithic and epixylic biofilm development in a boreal river. *Freshwater Biology* **25**: 179–87. [4.2]

Sinsabaugh RL, Linkins AE. (1990) Enzymic and chemical analysis of particulate organic matter from a boreal river. *Freshwater Biology* **23**: 301–9. [4.2, 4.5]

Sorokin YI, Kadota H. (1972) *Techniques for the Assessment of Microbial Production and Decomposition of Fresh Waters*. IBP Handbook No. 23, Blackwell, Oxford. [4.2]

Sridhar KR, Kaveriappa KM. (1982) Aquatic fungi of the

western Ghat forests in Karnataka. *Indian Phytopathology* **35**: 293–6. [4.3]

Sridhar KR, Kaveriappa KM. (1989) Colonization of leaves by water-borne hyphomycetes in a tropical stream. *Mycological Research* **92**: 392–6. [4.3]

Staley JT, Konopka A. (1985) Measurement of *in situ* activities of nonphotosynthetic microorganisms in aquatic and terrestrial habitats. *Annual Review of Microbiology* **39**: 321–46. [4.2]

Steffan RJ, Breen A, Atlas RM, Sayler GS. (1989) Application of gene probe methods for monitoring specific microbial populations in freshwater ecosystems. *Canadian Journal of Microbiology* **35**: 681–5. [4.7]

Suberkropp K. (1984) Effect of temperature on seasonal occurrence of aquatic hyphomycetes. *Transactions of the British Mycological Society* **82**: 53–62. [4.2, 4.3]

Suberkropp K. (1991) Relationship between growth and sporulation of aquatic hyphomycetes on decomposing leaf litter. *Mycological Research* **95**: 843–850. [4.2]

Suberkropp KF, Klug MJ. (1974) Decomposition of deciduous leaf litter in a woodland stream. I. A scanning electron microscopic study. *Microbial Ecology* **1**: 96–103. [4.2]

Suberkropp K, Klug MJ. (1976) Fungi and bacteria associated with leaves during processing in a woodland stream. *Ecology* **57**: 707–19. [4.2, 4.3, 4.4, 4.5]

Suberkropp K, Klug MJ. (1980) The maceration of deciduous leaf litter by aquatic hyphomycetes. *Canadian Journal of Botany* **58**: 1025–31. [4.3]

Suberkropp K, Klug MJ. (1981) Degradation of leaf litter by aquatic hyphomycetes. In: Wicklow DT, Carroll GC (eds) *The Fungal Community. Its Organization and Role in the Ecosystem*, pp 761–76. Marcel Dekker, New York. [4.3]

Suberkropp K, Arsuffi TL, Anderson JP. (1983) Comparison of degradative ability, enzymatic activity and palatability of aquatic hyphomycetes grown on leaf litter. *Applied and Environmental Microbiology* **46**: 237–44. [4.2, 4.3]

Suberkropp K, Michelis A, Lorch HJ, Ottow JCG. (1988) Effect of sewage treatment plant effluents on the distribution of aquatic hyphomycetes in the River Erms, Schwäbische Alb, FGR. *Aquatic Biology* **32**: 141–53. [4.6]

Tabor PS, Neihof RA. (1982) Improved method for determination of respiring individual microorganisms in natural waters. *Applied and Environmental Microbiology* **43**: 1249–55. [4.2]

Tabor PS, Neihof RA. (1984) Direct determination of activities for microorganisms of Chesapeake Bay populations. *Applied and Environmental Microbiology* **48**: 1012–19. [4.2]

Taylor BR, Roff JC. (1984) Use of ATP and carbon: nitrogen ratio as indicators of food quality of stream detritus. *Freshwater Biology* **15**: 195–201. [4.2]

Thomas K, Chilvers GA, Norris RH. (1989) Seasonal occurrence of conidia of aquatic hyphomycetes (fungi) in Lees Creek, Australian Capital Territory. *Australian Journal of Marine and Freshwater Research* **40**, 11–23. [4.3]

Thompson PL, Bärlocher F. (1989) Effect of pH on leaf breakdown in streams and in the laboratory. *Journal of the North American Benthological Society* **8**: 203–10. [4.6]

Thornton DR. (1963) The physiology and nutrition of some aquatic hyphomycetes. *Journal of General Microbiology* **33**: 23–31. [4.2, 4.3]

Torréton JP, Bouvy M. (1991) Estimating DNA synthesis from [^{3}H] thymidine incorporation: discrepancies among macromolecular extraction procedures. *Limnology and Oceanography* **36**: 299–306. [4.2]

Traaen TS. (1980) Effects of acidity on decomposition of organic matter in aquatic environments. In: Drabløs D, Tollan A (eds) *Proceedings International Conference on the Ecological Impact of Acid Precipitation, Norway 1980*, pp 340–1. SNSF Project. [4.6]

Trevors JT. (1984) The measurement of electron transport system (ETS) activity in freshwater sediments. *Water Research* **18**: 581–4. [4.2]

Triska FJ. (1970) *Seasonal distribution of aquatic hyphomycetes in relation to the disappearance of leaf litter from a woodland stream.* Unpublished PhD thesis, University of Pittsburg. [4.5]

van Frankenhuyzen K, Geen GH. (1986) Microbe-mediated effects of low pH on availability of detrital energy to a shredder, *Clistoronia magnifica* (Trichoptera, Limnephilidae). *Canadian Journal of Zoology* **64**: 421–6. [4.2]

van Frankenhuyzen K, Geen GH, Koivisto C. (1985) Direct and indirect effects of low pH on the transformation of detrital energy by the shredding caddisfly, *Clistoronia magnifica* (Banks) (Limnephilidae). *Canadian Journal of Zoology* **63**: 2298–304. [4.6]

Webster J. (1980) *Introduction to Fungi*, 2nd edn. Cambridge University Press, Cambridge. [4.3]

Webster J. (1981) Biology and ecology of aquatic hyphomycetes. In: Wicklow DT, Carroll GC (eds) *The Fungal Community. Its Organization and Role in the Ecosystem*, pp 681–91. Marcel Dekker, New York [4.3]

Webster JR, Benfield EF. (1986) Vascular plant breakdown in freshwater ecosystems. *Annual Review of Ecology and Systematics* **17**: 567–94. [4.2, 4.5]

Webster J, Descals E. (1981) Morphology, distribution and ecology of conidial fungi in freshwater habitats. In: Cole GT, Kendrick B (eds) *Biology of Conidial Fungi*, Vol 1, pp 295–355. Academic Press, New York. [4.2, 4.3]

Webster J, Moran ST, Davey RA. (1976) Growth and sporulation of *Tricladium chaetocladium* and *Lunulospora curvula* in relation to temperature. *Transactions of the British Mycological Society* **67**:

491–5. [4.2, 4.3]
West AW, Grant WD, Sparling GP. (1987) Use of ergosterol, diaminopimelic acid and glucosamine contents of soils to monitor changes in microbial populations. *Soil Biology and Biochemistry* **19**: 607–12. [4.2]
Whipps JM, Haselwandter K, McGee EEM, Lewis DH. (1982) Use of biochemical markers to determine growth, development and biomass of fungi in infected tissues, with particular reference to antagonistic and mutualistic biotrophs. *Transactions of the British Mycological Society* **79**: 385–400. [4.2]
Willoughby LG. (1974) Decomposition of litter in fresh water. In: Dickinson CH, Pugh GJF (eds) *Biology of Plant Litter Decomposition*, Vol 2, pp 659–81. Academic Press, London. [4.5]
Willoughby LG, Archer JF. (1973) The fungal spora of a freshwater stream and its colonization pattern on wood. *Freshwater Biology* **3**: 219–39. [4.2, 4.3]
Winterbourn MJ, Rounick JS, Hildrew AG. (1986) Patterns of carbon resource utilization by benthic invertebrates in two British river systems: a stable isotope study. *Archiv für Hydrobiologie* **3**: 349–61. [4.4]
Wood-Eggenschwiler S, Bärlocher F. (1983) Aquatic hyphomycetes in sixteen streams in France, Germany and Switzerland. *Transactions of the British Mycological Society* **81**: 371–9. [4.2, 4.3, 4.6]
Wright RT, Hobbie JE. (1966) Use of glucose and acetate by bacteria and algae in aquatic ecosystems. *Ecology* **43**: 447–64. [4.2]
Zdanowski MK. (1977) Microbial degradation of cellulose under natural conditions. A review. *Polskie Archiwum Hydrobiologii* **24**: 215–25. [4.5]

5: Invertebrates

K.W.CUMMINS

5.1 INTRODUCTION

Historically the invertebrates as a group have received major attention in the study of running water ecosystems. The reasons for this are clear. The macroinvertebrates (approximately >0.5 mm; Cummins 1975a) stand as the link between algae and micro-organisms, which serve as their primary food resources, and the fish (and other vertebrates), for which they are prey. Macroinvertebrates are also intermediate in turnover rates. They have replacement times greater than the very small, rapid turnover micro-organisms, and faster replacement rates than the generally larger fish with their characteristic slower turnover times. Thus, the macroinvertebrates are large enough to be observed with the unaided eye, abundant enough to be readily collected, and have lifecycles of suitable length (several weeks to 1 or 2 years) for short-term seasonal or annual field investigations. With these credentials, it is not surprising that the macroinvertebrates have been widely used to assess the prey base available to support fish populations (e.g. Waters 1988) and to evaluate water quality (e.g. Hilsenhoff 1987; Yount & Niemi 1990; Karr 1991).

Aside from the immense number of taxonomic studies on running-water invertebrates, especially on the aquatic insects (e.g. see review in Merritt & Cummins 1984), research on lotic invertebrates over the last century has focused on the patterns of distribution and abundance, and the environmental factors controlling these (e.g. Hynes 1970; Macan & Worthington 1951, Macan 1962, 1974). The study of the physical, chemical and biological (especially riparian vegetation, e.g. Corkum 1989) factors controlling the distribution and abundance of lotic invertebrates has evolved to the use of clustering techniques or principal component analyses to identify the key parameters involved (e.g. Moss *et al* 1987).

Since the mid-1970s, emphasis has shifted from the primary focus on structure, towards studies on process and function (e.g. Cummins 1974, 1988; Fig. 5.1). Examples of the areas that have received particular attention are: rates of biomass production, life-history strategies, resource partitioning, and population parameters such as competition and predator–prey interactions. (e.g. Cummins 1973, 1974, 1975a, 1988; Hynes 1975; Macan 1977; Townsend 1980; Cummins *et al* 1984; Resh & Rosenberg 1984; Minshall *et al* 1985). In these and many other areas, there has been extensive debate concerning appropriate methodology for estimating the dimensions of the various parameters involved; an example being the debate about the most suitable technique for quantifying gross and net invertebrate production (e.g. Waters 1979a, 1979b; Benke 1984; Iverson & Dall 1989).

The microinvertebrates, or meiofauna (generally <0.5 mm in size), of running waters have received much less attention than the macroinvertebrates, for most of the same reasons cited above relative to micro-organisms. In recent years, some of the long-standing interest in the benthic meiofauna of marine systems has shifted to running waters (e.g. Ward & Voelz 1990). In particular, studies are now being conducted on benthic microcrustacea in lotic systems. Of course, the two groups overlap in size, and the distinction as to what is studied is largely a matter of the mesh size selected for sorting (see below).

The review presented in this chapter is not

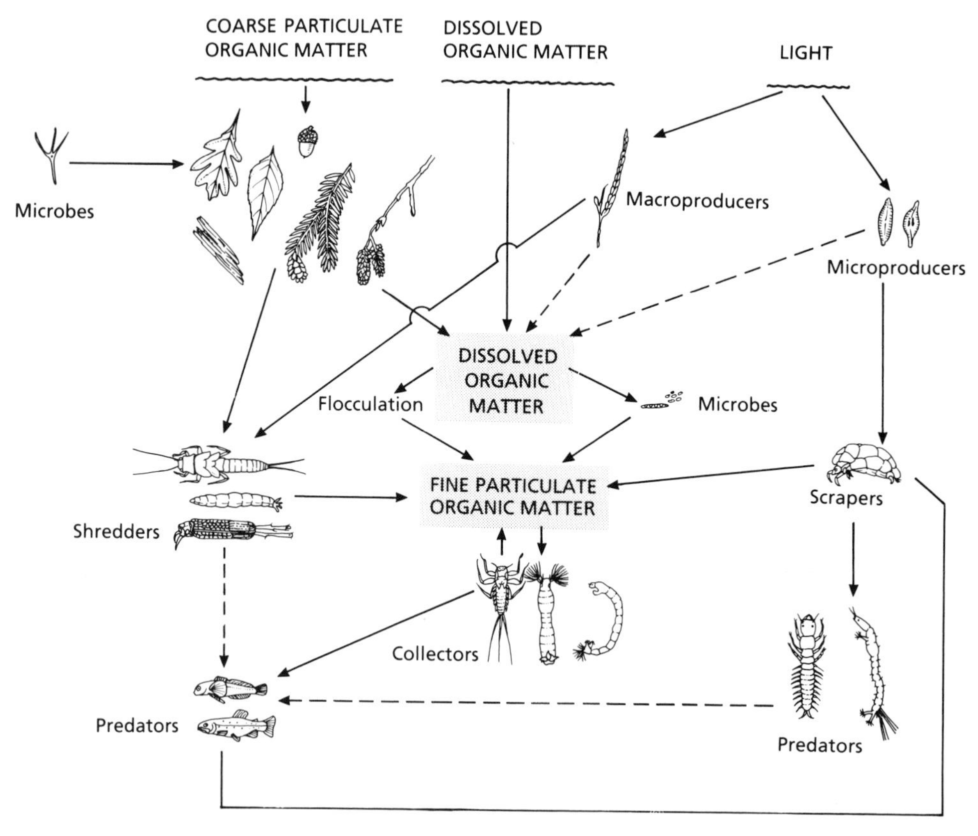

Fig. 5.1 A conceptual model of stream ecosystem structure with emphasis on invertebrate functional groups (shredders, collectors, scrapers and predators) and their food resource pools (coarse and fine particulate organic matter (CPOM and FPOM), periphyton (attached, non-filamentous algae and associated material) and prey) (after Cummins 1974, with permission).

exhaustive. Rather it is focused on selected aspects of running-water invertebrates, bearing on their roles in lotic ecosystem functioning and on their use in monitoring and predicting changing conditions in streams and rivers.

5.2 TAXONOMY, THE OPEN-ENDED AGENDA

The major accomplishments in defining the lotic invertebrate fauna have been in Europe and North America (e.g. Thienemann 1950; Edmondson 1959; Pennak 1978; Merritt & Cummins 1984). The task stands at a more complete state for Europe than North America because holarctic groups had much more restricted refugia in Europe during the last glacial period, although the longer period of anthropogenic environmental disturbance and related extinctions have also been significant (Hynes 1960, 1970). In fact, because of the present global scale of environmental alteration of running waters, many species are being, and certainly will be, lost before they are known to science. Running-water invertebrates are an important component of the general concern over loss of global biodiversity (e.g. Wilson 1988). Taxonomic knowledge is clearly related to accessibility of environments: by the time the lotic invertebrates of a geographical region are reasonably well known, the majority of the habitats have been altered, and undetermined numbers of species lost. For example, in Europe very little 'virgin' habitat remains. In North America, the major ecoregions of the eastern (eastern deciduous forest) and central (tall and short grass prairies) portions have been severely altered, save for a few small virgin preserves, and only 10% of the virgin forests of the Pacific Northwest remain. The tropics, subtropics and major island systems are obviously on the same course, currently representing the areas of most accelerated loss of biodiversity (e.g. Wilson 1988). Therefore, even if we never know what the taxonomic identity of the world's lotic invertebrate fauna *is*, it is certain we shall never know what it *was*.

The level of taxonomic uncertainty for lotic invertebrates, notably the holometabolous insects, is exacerbated by the fact that the aquatic immature stages are often the most difficult to distinguish (a usual feature of neotenous similarity) and, classically, the systematics are based on the reproductive morphology of adult males that are rarely encountered during aquatic sampling. Often, special efforts are required to sample the adult males, because of their brief longevity (usually days to weeks) and reclusive (e.g. night active) or inaccessible (e.g. swarming at great height) terrestrial habits (e.g. Wessenberg-Lund 1943; Merritt & Cummins 1984). The necessary light trap and sweep collections in the riparian zone along running-water habitats need to be adopted as standard procedure in all lotic invertebrate studies concerned with species diversity (Merritt *et al* 1984).

5.3 SAMPLING AND DATA ANALYSIS IN LOTIC HABITATS

These activities have been referred to as the 'four S's' of lotic invertebrate ecology: *S*ampling, *S*orting, *S*ystematics and *S*tatistics (Cummins 1966, 1972, 1975a). The standard reference on sampling design and statistical evaluation of the resulting freshwater benthic samples is Elliott (1977). Since the 1977 revision of this excellent treatment, there is little to add save the debate about 'pseudoreplication' (Hurlbert 1984), that is the failure to adequately replicate observations or experiments in space and/or time (see also Chapter 10). Invertebrate ecologists have long been aware that true replication in running-water studies is often difficult or impossible (e.g. the problem of a replicate for the Amazon River). An example is the problem of comparing the invertebrate standing stocks in 'paired' streams, in which each member of the 'pair' needs to be replicated as well as each time period, such as a season or year. Of course, there is the problem as to whether the assumptions of similarity of the 'pairs' are valid. Related problems arise concerning the need for replicate sampling within a single example of a given habitat type, such as a riffle, that does not serve as appropriate replication to generalize about all riffles in the stream being sampled. As described in detail in Elliott (1977), the number of samples required to arrive at a selected level of precision (defined as variation as a percentage of the mean) can be determined by the analysis of a preliminary set of data (see also Merritt & Cummins 1984; Chapter 10).

Where and how often to sample also remain perennial problems. It is clear from essentially all studies of benthic lotic invertebrates that their distributions are clumped, fitting a negative binomial distribution, and sampling design must account for these habitat-related patchy distributions (e.g. Macan 1958; Cummins 1966; Elliott 1977; Merritt *et al* 1984). Such non-continuous distributions have led many ecologists to employ stratified random sampling, which starts with the assumption that distributions are fundamentally different between major habitats such as riffles or pools, or to sample using a transect method that runs along known gradients from bank to channel thalweg (path of maximum current). The type of sampler is clearly dictated by the nature of the substrate, depth of the water, and current velocity (e.g. Merritt *et al* 1984). For example, larger, deeper rivers have been sampled with scuba gear (e.g. Minshall *et al* 1983a–b, 1985). The depth into the substrate to which the sample is taken should correspond generally to the depth of oxygenation, given the obligate aerobe status, at least in the long term, of the invertebrates. In some systems with good interstitial flow, organisms can be found to significant depth and into the bank as well (e.g. Williams 1989). It seems clear that the extent of this hyporeic zone should be evaluated in each system, and the sampling strategy adjusted accordingly. If half-metre sections of clean wood dowling are driven into the stream bottom in a transect across the channel with the position of the sediment water interface marked, removal of the rods after about a week (50–100 degree-days) will easily reveal the depth of the anoxic zone by the line of discoloration. This type of information can be used to establish approximate sampling depths in a given running-water reach.

Sorting of samples has been a major, time honoured and immensely time-consuming task in the study of running-water invertebrates. A variety of methods, including differential density solutions, various agitation and settling procedures, and specific stains combined with special wavelength lighting, all have been used. Unfortunately, even the best of these is only partially successful in shortening the task and is never universally applicable to all types of lotic samples. Thus, for the forseeable future, there will continue to be armies of sorters. The mesh size used to prepare the sediment and debris samples containing the invertebrates, in the field and/or the laboratory, strongly influences both the dimensions of the task and the accuracy of the estimates. The larger the sample and the finer the mesh the more likely it will be necessary to utilize some subsampling method (e.g. Merritt *et al* 1984). The still prevalent use of mesh sizes in the 0.5–1.0-mm range ensures that the early life stages of most invertebrates, as well as the majority of the meiofauna, will be lost. Historically, this has greatly reduced the number of midge larvae in stream and river benthic samples. Or, as the attitude could be summarized, 'If it's a group you despise, use a coarser mesh size'! Most macroinvertebrate studies now use a mesh size in the range of 250 μm which retains most insect first instars. The use of a 125-μm mesh retains essentially all insect larvae and a 50-μm mesh is reasonably efficient at retaining the meiofauna.

5.4 IN SEARCH OF PRODUCTION ESTIMATES

The methods used to calculate production of invertebrates vary depending on the portion of the community over which the estimate is intended to extend and the resolution of the data available relative to size (age) classes and seasonal patterns (e.g. Waters 1977, 1979a, 1979b; Benke 1984, see also Calow 1994).

The usual objective in studies of lotic invertebrate production is to estimate the elaboration of new biomass per square metre over some specified time span, normally a year. The calculation of total, or gross, production of the aquatic growth stage is independent of the fate of the biomass, for example loss to predation or emergence. Net production of a generation can be measured as the biomass of surviving adults. The field techniques require data on invertebrate size-specific biomass over the growth period. Most frequently the biomass is calculated by using length versus weight regressions (e.g. Meyer 1989) or by volume displacement coupled with the assumption of a specific gravity slightly greater than 1 (e.g. Stanford 1972) to convert size or volume to dry (or ash-free dry) mass; this is the size-frequency method. Although direct mass determinations should be made on dried (or ashed) fresh specimens, as a last resort some workers have weighed preserved material. These results will be highly variable owing to the amount of leaching of soluble fractions from the specimens which is dependent on the type of preservative, length of exposure, and thickness and extent of cuticle cover of the invertebrates.

Many estimates of lotic invertebrate production have taken a more limited taxonomic approach (e.g. Huryn & Wallace 1986; Benke & Parsons 1990; Huryn 1990; Lugthart *et al* 1991) as opposed to entire invertebrate community assessments (e.g. Benke *et al* 1984; Smock *et al* 1985). The size-frequency method of production estimation can be used at the level of a single species population, or some combination of species ranging up to the entire community. The method is very sensitive to differences in lifecycles. For example, including non-growing or diapausing stages in the time interval over which the calculations are applied yields an underestimate of production. Benke (1979) proposed that a cohort production interval (CPI) correction be made. That is, on an annual basis, the production estimate should be multiplied by 365 days/CPI so as to include only the growth period.

Because the growth of most lotic invertebrates

approaches a logarithmic function, individual instantaneous growth rate (IGR) can be calculated as the natural log of the final or maximum biomass minus the initial or minimum biomass divided by the time interval (Cummins & Merritt 1984). The relative growth rate (RGR), which is the change in biomass over a time interval divided by the median mass during that interval, multiplied by 100 gives the expression of growth as a percentage of bodyweight per time interval (e.g. Waldbauer 1968). Over short time spans, the RGR approximates the IGR very closely (e.g. Cummins *et al* 1973). The IGR calculated over an entire lifecycle approximates the ratio of annual or cohort production divided by the mean annual or cohort standing stock biomass. This ratio, P/B, is an expression of turnover; that is, the rate at which biomass is replaced (the units of biomass produced per unit of standing stock biomass per unit area over the time span). A variety of studies have yielded P/B ratios of 3–6, with a mean of 5, for most univoltine species (e.g. Waters & Crawford 1973; Waters & Hokenstom 1980; Benke 1984). Bi- and multivoltine species would be expected to have ratios >6, and for those with very rapid turnover populations the ratio might be as high as 100 (Hauer & Benke 1991). Turnover rates of this magnitude would probably resolve the so-called 'Allen's paradox' which points to the difficulty of measuring an adequate invertebrate prey base to support lotic fish populations (Waters 1988).

The ratio of P/B or the turnover ratio (TR), which is often in the range of 4–5 but may be much higher in polyvoltine species, can be used to estimate the production (P) portion of the P/B relationship. A knowledge of average generation time of a given species allows an estimate of TR. For example, a typical univoltine species in the temperate zone would have a TR of 5. With an estimate of TR, a measurement of average standing stock biomass (B) can in turn yield a production estimate. The other most commonly used method of production estimation is to utilize independent laboratory (e.g. Cummins *et al* 1973) or field enclosure (e.g. Hauer & Benke 1991) measurements of average cohort instantaneous growth. That is, $G = \ln B_f/B_i \times t^{-1}$, where B_i and B_f are average initial and final biomasses of the cohort respectively, and t is the time interval. Multiplying this growth rate (G) by the numerical standing stock provides the production estimate for the species in question. Knowledge of generation times and measurements of G for selected species were used, together with laboratory data on ingestion and respiration rates, to arrive at the system level production values, summed by functional group (Fig. 5.1), that are given in Fig. 5.2 (where production = assimilation − respiration (CO_2) and assimilation = ingestion − egestion).

It is quite useful to express the growth rate of individuals, or the production of the population or community as a whole, in a form normalized for temperature, that is, degree-days (Andewartha & Birch 1954). This summation, which is the daily temperature (usually taken as the median temperature) cumulated over the time interval, allows comparisons or rates measured in lotic systems that differ in temperature. This allows factors regulating growth other than temperature to be exposed, or when the normalized rates are the same indicates the primary control by temperature.

The thermal regulation of growth and reproductive success, the latter being interpreted as 'fitness', in lotic invertebrate populations has been embodied in the Thermal Equilibrium Hypothesis as proposed by Vannote and Sweeney (1980). The basic feature of this hypothesis is that fecundity, which is proportional to female body size, will be maximized at the optimal temperature regime for the species along any gradient (latitude, altitude). This results in reduced fitness of populations living at temperatures above or below the optimum. In a test of the hypothesis by Rader and Ward (1990), mean body size and fecundity were predicted better than population density and biomass. Various sources of mortality and influences on growth other than temperature, such as food quality, are clearly also at work. For example, because of the critical importance of micro-organisms as a detrital component, and the direct relation between microbial growth and temperature, increases in temperature are likely to increase food quality for shredders and collectors that feed on particulate organic matter. This means that food quality would be, on average, greater at temperatures above the optimum for

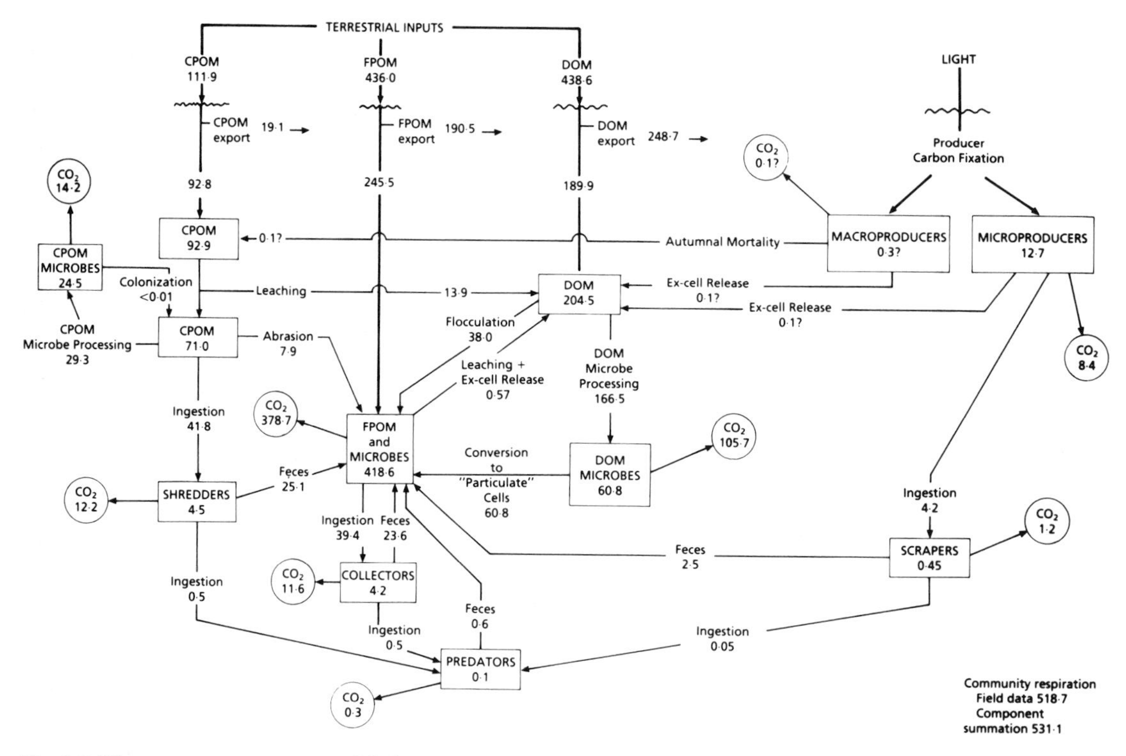

Fig. 5.2 The stream ecosystem model shown in Fig. 5.1 with actual measurements of state variables (invertebrate functional groups and their food resource pools), transfer functions, and carbon dioxide losses. Units are dry mass m^{-2} $year^{-1}$ (after Boling *et al* 1975; Cummins *et al* 1973, 1981, 1983; Petersen *et al* 1989; KW Cummins & RC Petersen unpublished data).

many species. Further, the availability of high quality food has been shown to override the regulation of temperature on body size (e.g. Ward & Cummins 1979).

5.5 FUNCTIONAL ANALYSIS OF SPECIES ASSEMBLAGES

Habitats

There has been little substantive modification of the separation of running-water habitats into erosional and depositional, usually referred to as riffles and pools, since the classic paper of Moon (1939; see also Shelford 1914). The subdivision of these two basic habitat types on geomorphic and hydrological bases has received significant attention (e.g. Davis & Barmuta 1989). Fast-flowing areas have been variously separated into rapids runs and glides, in addition to riffles, which still stands as a general descriptor for all erosional habitats. In addition, pools as a general descriptor, backwaters and side-channels have been identified as specific depositional lotic habitats.

Habits

The adaptations of invertebrates to the unidirectional flow, modified by turbulence, of running-water habitats has served as a major attraction to ecologists for some time (e.g. Shelford 1914; Wesenberg-Lund 1943; Hynes 1970; Macan 1974; Anderson & Wallace 1984). Morphological and behavioural adaptations of invertebrates to the hydraulic characteristics of their running-water environment serve as important examples in the comprehensive and delightful treatments by Vogel (1981, 1988).

The adaptations can be classified into three categories: (1) those involving position of the organism, such as locomotion, attachment (short or long term), and concealment (e.g. burrowing); (2) those associated with feeding (see functional feeding group classification below); and (3) those pertaining to reproduction (e.g. male egg 'protection' by adult male belostomatid hemipterans; e.g. Merritt & Cummins 1984).

Functional feeding groups

The technique of functional analysis of invertebrate feeding was first described 18 years ago (Cummins 1973) and has been modified in some details since then (e.g. Cummins 1974; Cummins & Klug 1979; Merritt & Cummins 1984; Cummins & Wilzbach 1985). The macroinvertebrate functional feeding group method is based on the association between a limited set of feeding adaptations found in freshwater invertebrates and their basic nutritional resource categories (Fig. 5.1; Cummins & Wilzbach 1985). These food resource categories are (Cummins 1974):

1 Detritus: coarse (CPOM) or fine (FPOM) particulate organic matter and the associated microbiota.
2 Periphyton: attached algae and associated entrained material.
3 Live macrophytes.
4 Prey.

The level of morphological and behavioural adaptation of the invertebrates that allows them to exploit these resource categories can be obligate or facultative (Cummins & Klug 1979). The obligate specialist forms are more readily displaced and the facultative generalists are more tolerant under conditions of disturbance. The presence and abundance of the various functional feeding groups, and the dominance of obligate or facultative representatives, is a direct reflection of the availability of the required food resources (both quantity and quality) and the condition of the related environmental parameters. An example here would be a change in the species and timing of the inputs of leaf litter to a headwater stream from the adjacent riparian zone (e.g. Cummins *et al* 1989).

The invertebrate functional groups are (see Fig. 5.1) (Merritt & Cummins 1984; Cummins & Wilzbach 1985):

1 Shredders feeding on CPOM (primarily litter of terrestrial origin from the riparian zone) or live macrophytes.
2 Collectors feeding on FPOM either by filtering from the water column (filtering collectors) or by 'mining' the sediments or browsing surface deposits (gathering collectors).
3 Scrapers feeding on periphyton.
4 Piercers feeding on macroalgae by piercing individual cells.
5 Predators feeding on prey.

The functional group analysis is made on a hierarchical basis of increasing levels of resolution. The first (lowest) level of resolution allows separation of live invertebrate collections in the field at an efficiency of 80–85%. The second level of resolution increases the efficiency by another 5–10%. When comparisons are to be made between sites on a regional basis, the level of resolution must be set so that all workers involved in the assessment can accomplish the task, with levels of greater resolution allowing groups with the appropriate expertise to produce more detailed analyses. Assignments to functional group of most of the North American genera of aquatic insects can be found in the ecological tables in Merritt and Cummins (1984). The feeding activity of shredders, primarily on plant litter of riparian origin, has a significant impact on the overall dynamics of organic matter fluxes in headwater lotic ecosystems. For example, at least 30% of the processing of coarse litter (CPOM) to finer particles (FPOM) can be attributed to shredder activity (e.g. Petersen & Cummins 1974; Cuffney *et al* 1990). In addition, the generation of FPOM can affect the growth of collectors that feed on the fine particles (e.g. Short & Maslin 1977) and shredder feeding can enhance the release of dissolved organic matter (DOM; Meyer & O'Hop 1983).

Invertebrate functional group analysis is sensitive to ecosystem properties of running waters. It is sensitive to both the normal pattern of geomorphic and concomitant biological changes that occur along river systems from headwaters to lower reaches (see Fig. 5.3) as well as to alterations in these patterns resulting from human impact. In headwaters, or extensively braided channels in the mid-reaches of river systems, the influence of riparian vegetation, through shading and litter inputs, is expressed in the general heterotrophic nature of such areas. Litter of terrestrial origin favours shredders, serving as their major food supply once it is appropriately conditioned by aquatic micro-organisms. As the litter is converted to finer organic particles (FPOM), it supports populations of collectors. The exclusion of light by riparian vegetation restricts in-stream primary production and consequently also limits the periphyton-grazing scrapers. The ratio of gross primary production (*P*) to community respiration (*R*) has been shown to be a useful index (*P*/*R*) of the balance between autotrophy and heterotrophy in running-water ecosystems (e.g. Vannote *et al* 1980; Table 5.1). The ratio of scrapers to shredders or to shredders + total collectors (SC/(SH + COL)) can serve as an index of *P*/*R* (Table 5.1).

In addition, the apportionment of coarse particulate (CPOM) to fine particulate (FPOM) organic matter shifts from relatively more to relatively less CPOM as the direct influence of the riparian vegetation is reduced—either as a natural consequence of increasing river size (i.e. reduced width of the riparian zone relative to river width) or because of human alterations (e.g. forest harvest or clearing for agriculture). For this relationship, the ratio of shredders to total collectors (SH/COL) is an index of the degree of the direct riparian influence as a litter source (Table 5.1). A further example can be found in the relative importance of FPOM in suspension (transport) compared with the depositional (benthic storage) component of total FPOM. If the input of FPOM is continuous, as it often is under disturbed conditions (e.g. organic effluents, severe erosion), then the nutritional resource base for filtering collectors (FC) is abundant and continuous. Usually this is in contrast to natural systems in which the input and movement of FPOM tracks the major pattern of the hydrograph, and the major reservoirs of particulate organic matter (POM) are in benthic storage for most of the year. This balance between storage and transport of POM can be reflected in the ratio of filtering to gathering collectors (FC/GC).

Table 5.1 Comparison of stream ecosystem parameters (*P*/*R*, CPOM/FPOM, storage POM/transport POM) and invertebrate functional feeding group ratios (SH/COL, SC/SH, SC/(SH + COL)) for the Kalamazoo River basin in southern Michigan, USA

	Stream order			
Parameter	1	2	3	5
Stream width (m)	1	5	10	45
Trophic status	Heterotrophic	Autotrophic	Heterotrophic	Autotrophic
P/*R*	0.47	1.13	0.90	1.23
Transport, CPOM/FPOM	0.022	0.016	0.019	0.022
Storage, CPOM/FPOM	0.36	0.11	0.15	0.10
POM, storage/transport	0.10	0.16	0.23	0.16
Mean annual				
invertebrates $m^{-2} \times 10^3$	19.6	15.0	63.6	41.7
SH/COL	0.22	0.003	0.002	0.001
FC/GC	0.67	0.42	0.45	1.50
SC/SH	0.18	12.23	3.99	16.91
SC/(SH + COL)	0.08	0.24	0.11	0.05

P, gross annual primary production; *R*, annual community respiration; POM, particulate organic matter; CPOM, coarse POM; FPOM, fine POM; SH, shredders; COL, total collectors; FC, filtering COL; GC, gathering COL. All invertebrate data used were means of fall–winter and spring–summer densities m^{-2} of individuals >0.5 mm (After Cummins *et al* 1981).

In the example shown in Table 5.1 for the Kalamazoo River Basin in southern Michigan, USA, the functional group ratios reflect the autotrophic/heterotrophic nature of the four stream orders that were studied. The sites on orders 1 and 3 were heavily wooded; order 1 was canopy closed and order 3 was nearly so. Order 2 had been altered by timber removal, and the replacement grass and shrub riparian zone provided reduced shading and litter inputs (Cummins *et al* 1981). This is reflected by the ecosystem ratios, *P*/*R* and CPOM/FPOM. The ratio of SC to SH (or SC/(SH + COL)) follows the autotrophic/heterotrophic status as measured by *P*/*R*. The reduced ratio of SC/SH (or SC/(SH+ COL)) at the order 5 site, despite the maximum *P*/*R* value, is an indication of the dominance of vascular hydrophyte and filamentous macroalgal primary producers which are not suitable food resources for scrapers. However, the sloughed tissue and cells from these macrophytes probably yield significant FPOM of high quality (Cummins & Klug 1979) supporting large populations of filtering collectors as indicated by the FC/FG ratio in Table 5.1.

A generalized model of the shifts in the relative abundances of invertebrate functional groups along a river tributary system from headwaters to mouth as predicted by the river continuum concept (RCC; Vannote *et al* 1980; Minshall *et al* 1983b, 1985) is given in Fig. 5.3. The general pattern reflects: (1) the importance of litter inputs, normally maximized in the headwaters, influencing the relative density of shredders; (2) increases in scrapers where light and nutrients favour increased microphyte production, normally in the wide, shallow mid-reaches; (3) the link between the abundance of collectors and FPOM, either in the headwaters related to litter processing, or in the lower reaches as the result of import from upstream tributaries and floodplain capture; and (4) the fairly constant relative abundance (approximately 10%) of predators in all reaches.

Microbial–invertebrate relationships

A major relationship between the invertebrates and the micro-organisms that co-occur with them in running-water environments concerns the general area of feeding biology. For example, the detrital feeding functional groups (shredders and collectors) have resident microflora, primarily in the hind-gut (Klug & Kotarski 1980), able to digest resistant plant compounds such as cellulose and lignin derivatives (e.g. Martin *et al* 1980, 1981a, 1981b; Lawson & Klug 1989). Regular resident microbial components in most shredders and collector hind-guts are large filamentous spore-forming bacteria. These are notably absent in the algal-feeding scrapers. Absorption occurs through the hind-gut wall of organic molecules from resident microbes that are useful to the invertebrate (Lawson & Klug 1989), and is enhanced by the heavy concentration of mitochondria in the area. However, materials in the hind-gut resulting from digestion by resident microbes also can be refluxed forward into the mid-gut, the normal site for absorption.

The invertebrate shredder–hyphomycete fungus relationship in streams is an example of a specific invertebrate–microbial association (see Chapter 8). Feeding by shredders can be shown to be correlated with the presence of the fungi on the leaf substrates, including selection over the presence of bacteria. Further, the key biochemical components involved in the selection have been shown to be specific 16- and 18-carbon 2 and 4 polyunsaturated lipids (Hanson *et al* 1983, 1985; Cargill *et al* 1985a, 1985b).

Drift

Invertebrate drift remains one of the most intriguing phenomena in running waters (e.g. Waters 1972), and the mechanisms involved are still widely debated (e.g. Allan & Russek 1985; Wilzbach *et al* 1988). In particular, the role of directed ('behavioural') versus non-directed ('accidental') drift is a major point of interest (e.g. Allan *et al* 1986; Wilzbach 1990). Wilzbach *et al* (1988) have proposed an integrated behavioural classification for lotic invertebrates that combines drift pattern, habit and functional feeding group characteristics. For example, invertebrate scrapers (functional group) tend to be clingers (habit) and usually show non-directed drift (accidental drift), while many gathering collectors are swimmers or sprawlers and exhibit directed drift behaviour.

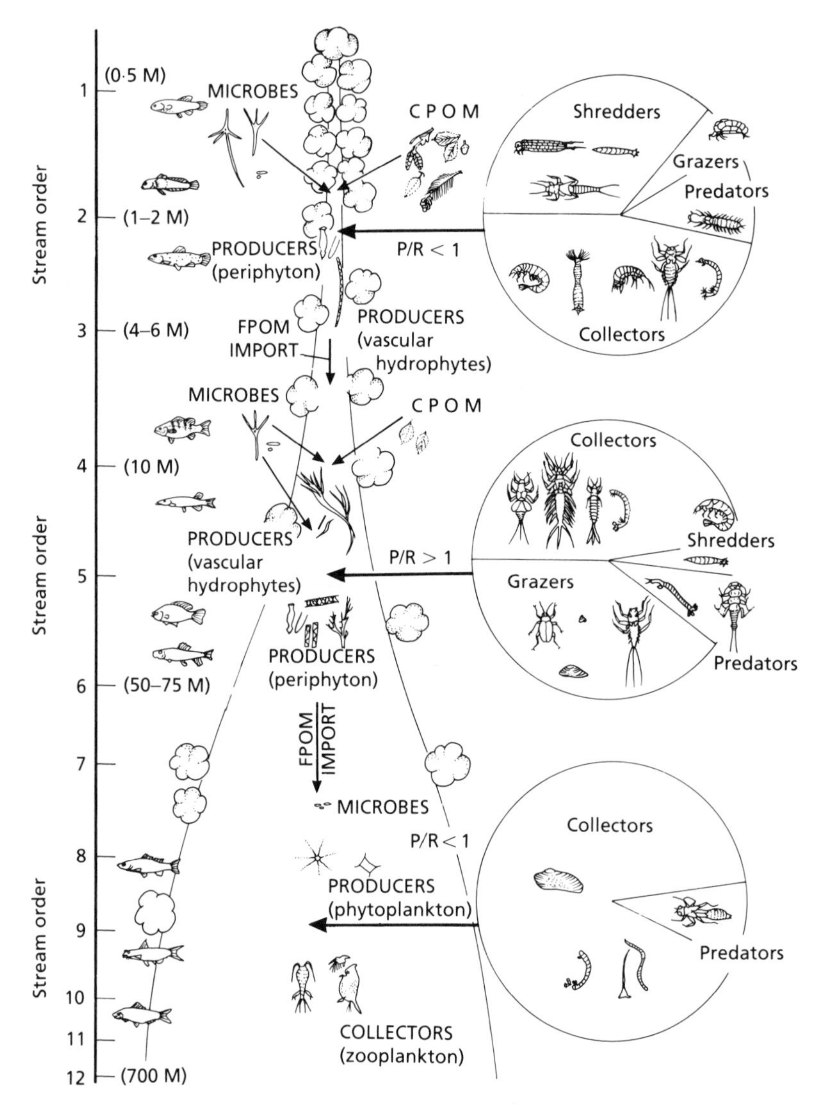

Fig. 5.3 A generalized model of the shifts in the relative abundances of invertebrate functional groups along a river tributary system from headwaters to mouth as predicted by the river continuum concept (RCC; e.g. Vannote *et al* 1980). The river system is shown as a single stem of increasing stream order and width. The headwaters (orders 1–3) are depicted as dominated by riparian shading and litter inputs resulting in a heterotrophic $P/R < 1$. The invertebrates are dominated by shredders which utilize riparian litter as their food resource once it has been appropriately conditioned by aquatic micro-organisms (especially aquatic hyphomycete fungi) and collectors that feed on fine particulate organic matter (FPOM). The mid-reaches (orders 4–6) are less dependent upon direct riparian litter input and with increased width and reduced canopy shading they are autotrophic with a $P/R > 1$. The shredders are reduced and the scrapers are relatively more important as attached microalgae become more abundant. The larger rivers are dominated by FPOM (and therefore collectors), and the increased transport load of this material together with increased depth results in reduced light penetration, and the system is again characterized by a $P/R < 1$ (after Cummins 1975b).

Drift has usually been viewed as the major mechanism of dispersal leading to population expansion through colonization of new habitats. However, the significantly higher mortality rate among drifting individuals relative to their non-drifting benthic counterparts (Cummins & Wilzbach 1988) and the more than adequate compensation for drift by local turnover (e.g. Waters 1961; Wilzbach & Cummins 1989) brings this view into question. The roles of food limitation and risk of predation remain the most often invoked drivers of the drift phenomenon, but the data are conflicting, suggesting that a generalized paradigm applying to all lotic invertebrate species may not be possible (Cummins & Wilzbach 1988; Wilzbach & Cummins 1989).

5.6 HABITAT QUALITY ASSESSMENT AND MANAGEMENT

As the demand for competing water uses continues to grow, so does the need for accurate and rapid biological monitoring of river water quality. Although chemical measurements continue to be critical for water quality assessment (e.g. Averett & McKnight 1987), biological measurements are often preferred because of their time-integrative nature — effects can be detected long after a short-term pulsed chemical input has dispersed. As the analysis of stresses to running-water ecosystems progressed from instantaneous measurements of selected chemical contaminants to biological measurements, such as biochemical oxygen demand (e.g. Calow *et al* 1990), the invertebrates emerged most frequently as the aquatic group of choice for water quality evaluations (Hynes 1970). The invertebrates are universally abundant in running waters, easily collected, large enough to be examined in the field with the unaided eye or a simple hand lens, and most have lifecycle lengths compatible with seasonal or annual sampling regimes. As indicated above, from an ecosystem perspective, macroinvertebrates are ideal integrators between the micro-organisms that dominate nutrient cycling and the fish that are often the product of interest in aquatic systems.

Both the field and laboratory phases of lotic studies, in which invertebrates are used for environmental evaluation, regardless of the specific approach applied, should be conducted on a hierarchical basis. If local (intra- or inter-site), regional, or global comparisons are to be made, a minimum level of resolution must be set so that all groups conducting the analyses can accomplish the task. The hierarchical approach, described above for functional group analysis, allows increased levels of resolution by those sampling who have the appropriate expertise, but allows all studies to be compared at the basic level agreed upon. This was the approach used in the river continuum studies in North America (e.g. Minshall *et al* 1983b, 1985).

The concept of index species (e.g. Hilsenhoff 1987; Lenat 1988) should certainly be modified to some form of index associations. A currently widely used method is the index of biotic integrity (e.g. Karr 1987, 1991). The described functional feeding group approach is another example. The index approach using stream or river invertebrates has been hindered by the incomplete state of taxonomic knowledge referred to above. That is, a methodology that seeks to measure system redundancy, namely many individuals but few species, usually finds redundancy in the taxonomy through an inability to separate species. This is a more significant problem in North America and in 'underdeveloped' countries than in Europe. For example, in one North American third-order woodland stream, the midge (Chironomidae) fauna is composed of more than 175 species on an annual basis (Coffman 1973, 1974). Often this dipteran family is the most diverse one represented in streams, and yet it is most frequently identified and counted only to the family, or possibly subfamily, tribe or (rarely) genus level.

Thus, although structural comparative analysis of running waters with respect to water quality or general environmental conditions (e.g. status of the riparian zone) can be assessed taxonomically, the incomplete state of knowledge about most invertebrate groups in most river systems results in poor levels of resolution unless immense effort is expended. Approaches that focus on functional aspects of invertebrate associations, such as the feeding groups described above, provide insight that is based on combinations of diverse taxa that require less time in categorization. This does not replace the taxonomically based structural analysis, but allows for evaluation of sytem functional status until such time as invertebrate systematics has progressed to the point that reliable alternative approaches are possible.

Large-scale regional inventories of running-water habitats usually include some assessment of invertebrate diversity. An important component of such inventories is the concept of reference sites and evaluations organized by ecological regions, or ecoregions (e.g. Hughes *et al* 1986; Hughes & Larsen 1988). In general, the reference sites are viewed as those with the highest diversity in a given ecoregion or subregion. In the USA, for example, the Environmental Protection Agency (EPA) Environmental Monitoring Assessment Program (EMAP) effort is organized according to ecoregions, and the US Geological Survey's National Water Quality Assessment (NAQWA) Program has established long-term reference sites.

Another area of important new initiatives for management involves the methodology of remote sensing. There has been greatly expanded attention focused on the riparian zone, as a terrestrial system inseparably linked to the biology of the associated running waters, especially the invertebrates (e.g. Cummins 1988). Inclusion of riparian corridors in management strategies for the biota, including the invertebrates, of a particular watercourse is taken as given in many cases. The analysis on a regional scale of the terrestrial biomes through which rivers and their tributary streams flow provides considerable insight into the specific nature of the linkage between invertebrate life histories and these terrestrial vegetation systems (e.g. Cummins 1988; Cummins *et al* 1989). Macro-scale remote sensing from satellite or low-level overflight by helicopter or fixed-wing aircraft can provide the necessary imagery for such analyses using appropriate geographical information system (GIS) software (e.g. White & MacKenzie 1986).

5.7 SOME RESEARCH FRONTIERS

Perhaps the most promising 'frontier' of research on running-water invertebrates involves their relationships with micro-organisms. Among the promising topics are: (1) the coevolution of invertebrates and micro-organisms resulting in controlled habitat (gut) advantages to the microbes and nutrient uptake advantages to the host invertebrates; (2) feeding strategies directed toward specific microbial populations (e.g. the hyphomycete–shredder linkage) or that result in the accumulation by the invertebrate of microbial enzymes for digestion of its gut contents; and (3) microbially mediated pathogenic mortality in natural lotic invertebrate populations (Cummins & Wilzbach 1988).

With regard to the first point, the coevolution of lotic invertebrates and micro-organisms can likely be viewed as a fundamental organizing principle in running-water environments worldwide. Invertebrate shredder feeding on hyphomycete fungi, which was discussed above is a good case in point. The aquatic insect taxa that dominate the shredder functional group evolved from terrestrial stock, as did the hyphomycetes that are probably aquatic vegetative forms of terrestrial Basidiomycetes. The coevolution of these two groups has been the key to terrestrial leaf litter turnover in headwater streams. The hyphomycetes, and the products of their metabolism of leaves in streams, constitute the nutritional base for shredders. In turn, shredder feeding is instrumental in opening up new colonization sites for these obligate aerobic fungi (e.g. Cummins 1974; Cummins & Klug 1979).

The matter of the natural regulation of lotic invertebrate populations by pathogens (e.g. May 1983; Cummins & Wilzbach 1988) will undoubtedly receive increasing attention because of the relationship to basic ecological population theory and questions related to the control of running-water pest and disease species such as the blackflies (Simuliidae).

As the release of genetically engineered organisms becomes a more common reality in the next decade, the relationship between the diversity of species as reproductively isolated entities and the genetic diversity residing within each species will receive increasing attention (e.g. Futuyma & Petersenk 1985; West-Eberhard 1989). In lotic ecosystems it is probable that there is some sort of general inverse relationship between interspecific taxonomic diversity and intraspecific genetic diversity. This means that a lotic habitat with few species, most of which might be generalists, could persist in time at least partly owing to the large genetic diversity embodied in most of the species. The alternative would be diverse invertebrate communities made up largely from genetically more narrow specialists. From the standpoint of lotic ecosystem persistence, the maintenance of functional diversity, or a particular functional balance, would be the critical issue. This functional balance could be achieved through either species or intraspecific genetic diversity.

A high priority in running-water studies should be the integration of the paradigms that are related to the understanding and prediction of the distribution and abundance of invertebrate populations. The integration of such paradigms as the river continuum concept (RCC; e.g. Vannote *et al* 1980) and the concepts related to patch dynamics and disturbance (e.g. Townsend 1989) would constitute a significant contribution to our approach to invertebrate studies and lotic ecology as a whole. Pringle *et al* (1988) concluded that the concepts of patch dynamics could be usefully integrated into the major running-water paradigms, the RCC and nutrient spiralling (Elwood *et al* 1983). Townsend has suggested that the RCC is not generally applicable, but patterns of patch dynamics might serve as a general organizing principle for running waters. Frid and Townsend (1989) concluded that patch dynamics theory, which was developed for terrestrial forest and marine intertidal systems, needed expansion before appropriate applications to other systems, such as lotic communities, could be made. Townsend (1989) has presented a convincing treatment of such an application of the theory to streams and rivers. However, most of the aspects of the RCC were rejected out of hand and no serious attempt at integration was made. Some useful points for studies of lotic invertebrates relative to patch dynamics and disturbance will be the more complete incorporation of island biogeography theory (e.g. Minshall *et al* 1983a) and the significance of the finding that drifting invertebrates have a higher pathogen load and represent likely mortality rather than highly successful colonists, even if settling space is available (e.g. Cummins & Wilzbach 1988).

REFERENCES

Allan JD, Russek E. (1985) The quantification of stream drift. *Canadian Journal of Fisheries and Aquatic Science* **42**: 210–15. [5.5]

Allan JD, Flecker AS, McClintock NL. (1986) Diel epibenthic activity of mayfly nymphs, and its nonconcordance with behavioral drift. *Limnology and Oceanography* **31**: 1057–65. [5.5]

Anderson NH, Wallace JB. (1984) Habitat, life history, and behavioral adaptations of aquatic insects. In: Merritt RW, Cummins KW (eds) *An Introduction to the Aquatic Insects*, 2nd edn, pp 38–58. Kendall/Hunt, Dubuque, Iowa. [5.5]

Andewartha HG, Birch LC. (1954) *The Distribution and Abundance of Animals*. University of Chicago Press, Chicago. [5.4]

Averett RC, McKnight DM (eds) (1987) *Chemical Quality of Water and the Hydrologic Cycle*. Lewis Publishers, Michigan. [5.6]

Benke AC. (1979) A modification of the Hynes method for estimating secondary production with particular significance for multivoltine populations. *Limnology and Oceanography* **24**: 168–71. [5.4]

Benke AC. (1984) Secondary production of aquatic insects. In: Resh VH, Rosenberg DM (eds) *The Ecology of Aquatic Insects*, pp 289–322. Praeger, New York. [5.1, 5.4]

Benke AC, Parsons KA. (1990) Modelling black fly production dynamics in blackwater streams. *Freshwater Biology* **24**: 167–80. [5.4]

Benke AC, van Arsdall TC, Gillespie DH, Parish FK. (1984) Invertebrate productivity in a subtropical blackwater river: the importance of habitat and life history. *Ecological Monographs* **54**: 25–63. [5.4]

Boling RH Jr, Petersen RC, Cummins KW. (1975). Ecosystem Modeling for Small Woodland Streams. In: Patten BH (ed.) *Systems Analysis and Simulation in Ecology*, Vol. 3, pp 183–204. Academic Press, New York. [5.4]

Calow P. (1994) Energy budgets. In: Calow P, Petts GE (eds) *The Rivers Handbook*, vol. 1, pp 370–8. Blackwell Scientific Publications, Oxford. [5.4]

Calow P, Armitage P, Boon P, Chave P, Cox E, Hildrew A. *et al* (1990) River water quality. *Ecology Issues* 1. British Ecology Society, London. [5.6]

Cargill AS II, Cummins KW, Hanson BJ, Lowry RR. (1985a) The role of lipids, fungi, and temperature in the nutrition of a shredder caddisfly, *Clistoronia magnifica*. *Freshwater Invertebrate Biology* **4**: 64–78. [5.5]

Cargill AS II, Cummins KW, Hanson BJ, Lowry RR. (1985b) The role of lipids as feeding stimulants for shredding aquatic insects. *Freshwater Biology* **15**: 455–64. [5.5]

Coffman WP. (1973) Energy flow in a woodland stream

ecosystem: II. The taxonomic composition and phenology of the Chironomidae as determined by the collection of pupal exuviae. *Archives in Hydrobiology* **71**: 281–322. [5.6]

Coffman WP. (1974) Seasonal differences in the diel emergence of a lotic chironomid community. *Ent Tidskr* **95**: 42–8. [5.6]

Corkum LD. (1989) Patterns of benthic invertebrate assemblages in rivers of northwestern North America. *Freshwater Biology* **21**: 191–205. [5.1]

Cuffney TF, Wallace JB, Lugthart GJ. (1990) Experimental evidence quantifying the role of benthic invertebrates in organic matter dynamics of headwater streams. *Freshwater Biology* **23**: 281–99. [5.5]

Cummins KW. (1966) A review of stream ecology with special emphasis on organism–substrate relationships. *Special Publications of the Pymatuning Laboratory of Ecology, University of Pittsburgh* **4**: 2–51. [5.3]

Cummins KW. (1972) What is a river? – Zoological description. In: Oglesby RT, Carlson CA, McCann JA. (eds) *River Ecology and Man*, pp 35–52. Academic Press, New York. [5.3]

Cummins KW. (1973). Trophic relations of aquatic insects. *Annual Review of Entomology* **18**: 183–206. [5.1, 5.5]

Cummins, KW. (1974) Structure and function of stream ecosystems. *BioScience* **24**: 631–41. [5.1, 5.5, 5.7]

Cummins KW. (1975a) Macroinvertebrates. In: Whitton BA (ed.) *River Ecology*, pp 170–98. Blackwell Scientific Publications, New York. [5.1, 5.3]

Cummins KW. (1975b) The ecology of running waters; theory and practice. In: Baker DB, Jackson WB, Prater BL (eds) *Proceedings of the Sandusky River Basin Symposium, International Joint Commission, Great Lakes Pollution*, pp 277–93. Environmental Protection Agency, Washington, DC. [5.5]

Cummins KW. (1988) The study of stream ecosystems: a functional view. In: Pomeroy LR, Alberts JJ (eds) *Concepts of Ecosystem Ecology*, pp 247–62. Springer-Verlag New York. [5.1, 5.6]

Cummins KW, Klug MJ. (1979) Feeding ecology of stream invertebrates. *Annual Review of Ecology and Systematics* **10**: 147–72. [5.5, 5.7]

Cummins KW, Merritt RW. (1984) Ecology and distribution of aquatic insects. In: Merritt RW, Cummins KW (eds) *An Introduction to the Aquatic Insects of North America*, pp 59–65. Kendall/Hunt, Dubuque, Iowa. [5.4]

Cummins KW, Wilzbach MA. (1985) *Field procedures for the analysis of functional feeding groups in stream ecosystems*. Pymatuning Laboratory of Ecology, University of Pittsburgh, Linesville, Pennsylvania. [5.5]

Cummins KW, Wilzbach MA. (1988) Do pathogens regulate stream invertebrate populations? *Verhandlungen Internationale Vereinigung für Theoretische und Angewandt Limnologie* **23**: 1232–43. [5.5, 5.7]

Cummins KW, Petersen RC, Howard FO, Wuycheck JC, Holt VI. (1973) The utilization of leaf litter by stream detritivores. *Ecology* **54**: 336–45. [5.4]

Cummins KW, Klug MJ, Ward GM, Spengler GL, Speaker RW, Ovink RW *et al* (1981) Trends in particulate organic matter fluxes, community processes, and macroinvertebrate functional groups along a Great Lakes Drainage Basin river continuum. *Verhandlungen Internationale Vereinigung für Theoretische und Angewandt Limnologie* **21**: 841–9. [5.4, 5.5]

Cummins KW, Sedell JR, Swanson FJ, Minshall GW, Fisher SG, Cushing CE *et al* (1983) Organic matter budgets for stream ecosystems: problems in their evaluation. In: Barnes JR, Minshall GW (eds) *Stream Ecology Application and Testing of General Ecological Theory*, pp 299–353. Plenum Press, New York. [5.4]

Cummins KW, Minshall GW, Sedell JR, Petersen RC. (1984) Stream ecosystem theory. *Verhandlungen Internationale Vereinigung für Theoretische und Angewandt Limnologie* **22**: 1818–27. [5.1]

Cummins KW, Wilzbach MA. Gates DM, Perry JB, Taliaferro WB. (1989) Shredders and riparian vegetation. *Bioscience* **39**: 24–30. [5.5, 5.6]

Davis JA, Barmuta LA. (1989) An ecologically useful classification of mean and near-bed flows in streams and rivers. *Freshwater Biology* **21**: 271–82. [5.5]

Edmondson WT (ed.) (1959) *Freshwater Biology*, 2nd edn. John Wiley & Sons, New York. [5.2]

Elliott JM. (1977) *Some Methods for the Statistical Analysis of Samples of Benthic Invertebrates*, 2nd edn. Scientific Publication of the Freshwater Biology Association UK, 25. [5.3]

Elwood JW, Newbold JD, O'Neill RV, VanWinkle W. (1983) Resource spiralling: an operational paradigm for analyzing lotic ecosystems. In: Fontaine TD, Bartell SM (eds) *The Dynamics of Lotic Ecosystems*, pp 3–27. Ann Arbor Science. Ann Arbor, Michigan. [5.7]

Frid CLJ, Townsend CR. (1989) An appraisal of the patch dynamics concept in stream and marine benthic communities whose members are highly mobile. *Oikos* **56**: 137–41. [5.7]

Futuyma DJ, Petersenk SC. (1985) Genetic variation in the use of resources by insects. *Annual Review of Entomology* **30**: 217–38. [5.7]

Hanson BJ, Cummins KW, Cargill AS, Lowry RR. (1983) Dietary effects on lipid and fatty acid composition of *Clistoronia magnifica* (Trichoptera: Limnephilidae). *Freshwater Invertebrate Biology* **2**: 2–15. [5.5]

Hanson BJ, Cummins KW, Cargill AS, Lowry RR. (1985) Lipid content, fatty acid composition, and the effect of diet on fats of aquatic insects. *Comparative Bio-*

chemistry and Physiology **80B**: 257–76. [5.5]

Hauer FR, Benke AC. (1991) Rapid growth rates of snag-dwelling chironomids in a black water river: the influence of temperature and discharge. *Journal of the North American Benthological Society* **10**: (in press). [5.4]

Hilsenhoff WL. (1987) An improved biotic index of organic stream pollution. *Great Lakes Entomology* **20**: 31–9. [5.1, 5.6]

Hughes RM, Larsen DP. (1988) Ecoregions: an approach to surface water protection. *Journal of the Water Pollution Control Federation* **60**: 486–93. [5.6]

Hughes RM, Larsen DP, Omernik JM. (1986) Regional reference sites: a method for assessing stream pollution. *Environmental Management* **10**: 629–35. [5.6]

Hurlbert SH. (1984) Pseudoreplication and the design of ecological field experiments. *Ecological Monographs* **54**: 187–211. [5.3]

Huryn AD. (1990) Growth and voltinism of lotic midge larvae: patterns across an Appalachian mountain landscape. *Limnology and Oceanography* **35**: 339–51. [5.4]

Huryn AD, Wallace JB. (1986) A method for obtaining *in situ* growth rates of larval Chironomidae (Diptera) and its application to studies of secondary production. *Limnology and Oceanography* **31**: 216–32. [5.4]

Hynes HBN. (1960) *The Biology of Polluted Waters*. Liverpool University Press, Liverpool.

Hynes HBN. (1970) *The Ecology of Running Waters*. University of Toronto Press, Toronto. [5.1, 5.2, 5.5, 5.6]

Hynes HBN. (1975) The stream and its valley. *Verhandlungen Internationale Vereinigung für Theoretische und Angewandt Limnologie* **19**: 1–15. [5.1]

Iverson TB, Dall P. (1989) The effect of growth pattern, sampling interval and number of size classes on benthic invertebrate production estimated by the size-frequency method. *Freshwater Biology* **22**: 232–331. [5.1]

Karr JR. (1987) Biological monitoring and environmental assessment: a conceptual framework. *Environmental Management* **11**: 249–56. [5.6]

Karr JR. (1991) Biological integrity: a long neglected aspect of water resource management. Ecological Applications **1**: 26–35. [5.1, 5.6]

Klug, MJ, Kotarski S. (1980) Bacteria associated with the gut tract of larval stages of the aquatic cranefly *Tipula abdominalis* (Diptera: Tipulidae). *Applied Environmental Microbiology* **40**: 408–16. [5.5]

Lawson DL, Klug MJ. (1989) Microbial fermentation in the hind guts of two stream detritivores. *Journal of the North American Benthological Society* **8**: 85–91. [5.5]

Lenat DR. (1988) Water quality assessment of streams using a qualitative collection method for benthic macroinvertebrates. *Journal of the North American Benthological Society* **7**: 222–53. [5.6]

Lugthart GJ, Wallace JB, Huryn AD. (1991) Secondary production of chironomid communities in insecticide-treated and untreated headwater streams. *Freshwater Biology* 143–56. [5.4]

Macan TT. (1958) Methods of sampling the bottom fauna in stony streams. *Mitteilungen Internationale Vereinigung für Theoretische und Angewandte Limnologie* **8**: 1–21. [5.3]

Macan TT. (1962) The ecology of aquatic insects. *Annual Review of Entomology* **7**: 261–88. [5.1]

Macan TT. (1977) The influence of predation on the composition of freshwater animal communities. *Biological Reviews of the Cambridge Philosophical Society* **52**: 45–70. [5.1]

Macan TT. (1974) *Freshwater Ecology*, 2nd edn. John Wiley & Sons, New York. [5.1, 5.5]

Macan TT, Worthington EB. (1951) *Life in Lakes and Rivers*. Collins, London.

Martin MM, Martin JS, Kukor JJ, Merritt RW. (1980) The digestion of protein and carbohydrate by the stream detritivore, *Tipula abdominalis* (Diptera, Tipulidae). *Oecologia* **46**: 360–4. [5.5]

Martin MM, Kukor JJ, Martin JS, Merritt RW. (1981a) Digestive enzymes of larvae of three species of caddisflies (Trichoptera). *Insect Biochemistry* **11**: 501–5. [5.5]

Martin MM, Martin JS, Kukor JJ, Merritt RW. (1981b) The digestive enzymes of detritus-feeding stonefly nymphs (Plecoptera: Pteronarcyidae. *Canadian Journal of Zoology* **59**: 1947–51. [5.5]

May RM. (1983) Parasitic infections as regulators of animal populations. *American Scientist* **71**: 36–45. [5.7]

Merritt RW, Cummins KW. (eds) (1984) *An Introduction to the Aquatic Insects of North America*. Kendall/Hunt, Dubuque, Iowa. [5.1, 5.2, 5.3, 5.5]

Merritt RW, Cummins KW, Resh VH. (1984) Collecting, sampling, and rearing methods for aquatic insects. In: Merritt RW, Cummins KW. (eds) *An Introduction to the Aquatic Insects of North America*, pp 11–26. Kendall/Hunt, Dubuque, Iowa. [5.3]

Meyer JL, O'Hop J. (1983) Leaf-shredder insects as a source of dissolved organic carbon in headwater streams. *American Midland Naturalist* **109** 175–83. [5.5]

Minshall GW, Andrews DA, Manuel-Faler CY. (1983a) Application of island biogeographic theory to streams: macroinvertebrate recolonization in the Teton River, Idaho. In: Barnes JR, Minshall GW. (eds) *Stream Ecology*, pp 279–97. Plenum Press, New York. [5.3, 5.7]

Minshall GW, Petersen RC, Cummins KW, Bott TL, Sedell JR, Cushing CE, Vannote RL. (1983b) Interbiome comparison of stream ecosystem dynamics.

Ecological Monographs **53**: 1–25. [5.3, 5.5, 5.6]

Minshall GW, Cummins KW, Petersen RC, Cushing CE, Bruins DA, Sedell JR, Vannote RL. (1985) Developments in stream ecology. *Canadian Journal of Fisheries and Aquatic Sciences* **42**: 1045–55. [5.1, 5.5, 5.6]

Moss D, Furse MT, Wright JF, Armitage PD. (1987) The prediction of the macro-invertebrate fauna of unpolluted running-water sites in Great Britain using environmental data. *Freshwater Biology* **17**: 41–52. [5.1]

Meyer E. (1989) The relationship between body length parameters and dry mass in running water invertebrates. *Archives of Hydrobiology* **117**: 191–203. [5.4]

Meyer JL, O'Hop J. (1983) Leaf-shredding insects as a source of dissolved organic carbon in headwater streams. *American Midland Naturalist* **109**: 175–83. [5.5]

Moon HP. (1939) Aspects of the ecology of aquatic insects. *Transactions of the British Entomological Society* **6**: 39–49. [5.5]

Pennak RW. (1978) *Freshwater Invertebrates of the United States*, 2nd edn. John Wiley & Sons, New York. [5.2]

Petersen RC, Cummins KW. (1974) Leaf processing in a woodland stream. *Freshwater Biology* **4**: 343–68. [5.5]

Petersen RC, Cummins KW, Ward GM. (1989) Microbial and animal processing of detritus in a woodland stream. *Ecological Monographs* **59**: 21–39. [5.4]

Pringle CM, Naiman RJ, Bretschko G, Karr JR, Oswood MW, Welcomme RL, Winterbourn MJ. (1988) Patch dynamics in lotic systems. *Journal of the North American Benthological Society* **7**: 503–24. [5.7]

Rader RB, Ward JV. (1990) Mayfly growth and population density in constant and variable temperature regimes. *Great Basin Naturalist* **50**: 97–106. [5.4]

Resh VH, Rosenberg DM. (eds) (1984) *The Ecology of Aquatic Insects*. Praeger Publishers, New York. [5.1]

Shelford VE. (1914) An experimental study of the behavior agreement among animals of an animal community. *Biological Bulletin* **26**: 294–315. [5.5]

Short RA, Maslin PE. (1977) Processing of leaf litter by a stream detritivore: effect on nutrient availability to collectors. *Ecology* **58**: 935–8. [5.5]

Smock LA, Gilinsky E, Stonebrunner DL. (1985) Macroinvertebrate production in a southeastern United States blackwater stream. *Ecology* **66**: 1491–503. [5.4]

Stanford JA. (1972) A centrifuge method for determining live weights of aquatic insect larvae, with a note on weight loss in preservative. *Ecology* **54**: 499–551. [5.4]

Thienemann A. (1950) *Verbreitungsgeschicte der susswassertierwelt Europas*. Die Binnengewasser, Stuttgart 18. [5.2]

Townsend CR. (1980) *The Ecology of Streams and Rivers*. Edward Arnold, London. [5.1]

Townsend CR. (1989) The patch dynamics concept of stream community ecology *Journal of the North American Benthological Society* **8**: 36–50. [5.7]

Vannote RL, Sweeney BW. (1980) Geographic analysis of thermal equilibria: a conceptual model for evaluating the effect of natural and modified thermal regimes on aquatic insect communities. *American Naturalist* **115**: 667–95. [5.4]

Vannote RL, Minshall GW, Cummins KW, Sedell JR, Cushing CE. (1980) The river continuum concept. *Canadian Journal of Fisheries and Aquatic Sciences* **37**: 130–7. [5.5, 5.7]

Vogel S. (1981) *Life in Moving Fluids*. Princeton University Press, Princeton, New Jersey. [5.5]

Vogell S. (1988) *Life's Devices*. Princeton University Press, Princeton, New Jersey. [5.5]

Waldbauer G. (1968) The consumption and utilization of food by insects. *Advances in Insect Physiology* **5**: 229–88. [5.4]

Ward GM, Cummins KW. (1979) Food quality on growth of a stream detritivore. *Paratendipes albimanus (meigen)* (Diptera: Chironomidae). *Ecology* **60**: 57–64. [5.4]

Ward JV, Voelz NJ. (1990) Gradient analysis of interstitial meiofauna along a longitudinal stream profile. *Stygologia* **5**: 93–9. [5.1]

Waters TF. (1961) Standing crop and drift of stream bottom organisms. *Ecology* **42**: 352–7. [5.5]

Waters TF. (1972) The drift of stream insects. *Annual Review of Entomology* **17**: 253–72. [5.5]

Waters TF. (1977) Secondary production in inland waters. *Advances in Ecological Research* **10**: 91–164. [5.4]

Waters TF. (1979a) Influence of benthos life history upon the estimation of secondary production. *Journal of the Fisheries Research Board of Canada* **36**: 1425–30. [5.1, 5.4]

Waters TF. (1979b) Benthic life histories: summary and future needs. *Journal of the Fisheries Research Board of Canada* **36**: 342–5. [5.1, 5.4]

Waters TF. (1988) Fish production — benthos production relationships in trout streams. *Polskie Archiwum Hydrobiologii* **35**: 548–61. [5.1, 5.4]

Waters TF, Crawford GW. (1973) Annual production of a stream mayfly population: a comparison of methods. *Limnology and Oceanography* **18**: 286–96. [5.4]

Waters TF, Hokenstrom JC. (1980) Annual production and drift of the stream amphipod *Gammarus pseudolimnaeus* in Valley Creek, Minnesota. *Limnology and Oceanography* **25**: 700–10. [5.4]

Wesenberg-Lund C. (1943) *Biologie der Susswasserinsekten*. Springer, Berlin. [5.2, 5.5]

West-Eberhard MJ. (1989) Morphological variation and

width of ecological niche. *Annual Review of Ecology and Systematics* **20**: 249–78. [5.7]

White PS, MacKenzie MD. (1986) Remote sensing and landscape pattern in Great Smokey Mountains biosphere reserve, North Carolina and Tennessee. In: Dyer MI, Crossley DA (eds) *Coupling of ecological studies with remote sensing: potentials at four biosphere reserves in the United States*, pp 52–70. Bureau of Oceans and International Environmental Affairs, US Dept. State Publ. 9504, Washington, DC. [5.6]

Williams DD. (1989) Towards a biological and chemical definition of the hyporeic zone of two Canadian rivers. *Freshwater Biology* **22**: 189–208. [5.3]

Wilson EO. (ed.) (1988) *Biodiversity*. National Academic Press, Washington, DC. [5.2]

Wilzbach MA. (1990) Non-concordance of drift and benthic activity in Baetis. *Limnology and Oceanography* **35**: 945–52. [5.5]

Wilzbach MA, Cummins KW. (1989) An assessment of short-term depletion of stream macroinvertebrate benthos by drift. *Hydrobiologia* **185**: 29–39. [5.5]

Wilzbach MA, Cummins KW, Knapp R. (1988) Towards a functional classification of stream invertebrate drift. *Verhandlungen Internationale Vereinigung für Theoretische und Angewandt Limnologie* **23**: 1244–64. [5.5]

Yount JD, Niemi GJ (eds) (1990) Recovery of lotic communities and ecosystems following disturbance: theory and applications. *Environmental Management* (special issue) **4**: 515–762. [5.1]

6: Riverine Fishes

P.B.BAYLEY AND H.W.LI

6.1 INTRODUCTION

This chapter puts the challenges of understanding and managing riverine fish and fisheries in an ecological and evolutionary context. An appreciation of the adaptations of fish and the reasons for characteristic assemblages will lead to a more strategic view of riverine fisheries management than the tactical ones often employed, such as maximizing the yield of a single species while ignoring environmental variation. We propose that most serious fisheries management problems result from actions that have changed the hydrological regime, habitats and/or fish fauna, thereby disrupting the long-term, dynamic patterns to which the indigenous fishes are adapted. This has resulted in the need for restoration in many systems. We discuss the options available to research and management agencies in the light of current limitations on our ability to sample fish quantitatively and our knowledge of their spatial and temporal dependence on their environment.

Apart from their aesthetic value, riverine fishes are important because they can be harvested for human consumption, caught for recreation, are useful as indicators of the well-being of the environment, or serve as appropriate subjects for testing principles of population or community ecology. Considerable knowledge has resulted from the independent pursuit of these interests. However, this knowledge is dwarfed by information of which we are ignorant, which includes many of the concepts and tools necessary to make inferences about fish populations or communities based on the few systems we can afford to study intensively.

In order that projects pursuing any of these interests are directed towards conservation or recovery of the natural system, have long-term economic viability, and have general application in fish resource management, we propose four requirements. First, the project must recognize (or investigate) the constraints imposed by the evolutionary adaptations (section 6.3) and interactions (section 6.5) of the taxa concerned. Second, the biases and variance of the sampling process must be known with sufficient accuracy (sections 6.4, 6.7; Chapter 10), and the long-term management costs of monitoring the resource with that accuracy must be included. Third, the investigator should be aware of the feasibility and cost of restoring a damaged system (section 6.6). Last, but not least, the project must be designed and reported to enable generalizations across systems on appropriate scales and classifications (sections 6.2, 6.7). Considering that many publications still account for few, if any, of these factors, there is much room for improvement.

6.2 CLASSIFICATIONS AND UNIFYING CONCEPTS FOR RIVERINE FISHES

There are about 8500 freshwater fish species (Lowe-McConnell 1987), most of which occur in rivers or connected floodplains. Current technology and resources (section 6.4) are probably insufficient to complete ecological studies by species and to predict population trends by stock before some of these species become extinct naturally. In view of the unnatural changes occurring in systems since the industrial revolution and the subsequent human population explosion, it is clearly impossible to achieve a moderate level of predictive capability for each stock of each species of interest with respect to their exploitation or conservation.

Therefore, more studies must focus on comparisons among systems or their components, so that information from intensive, localized studies can be used to manage, conserve and restore fish populations and communities across many systems. This requires classifications of ecologically equivalent units that comprise functionally similar species and/or life stages so that generalizations can be tested. Classification of units, gradients of key variables within units, and the scale adopted depend on the problem and the information available. The pluralistic approach (Schoener 1986, 1987) emphasizes differences between ecological communities based on organismic and environmental axes. Elements of these axes, such as body size (section 6.3) and stream discharge values (section 6.2), may jointly indicate appropriate boundaries for working definitions of classification units. Inadequacies in data are common, and generalization can be more limited by appropriate survey information across systems than by results from localized studies.

Classifications at different spatial scales (sections 6.2–6.5) and unifying concepts within and across scales (section 6.6) are presented as heuristic tools to understand how fishes are organized in river systems within the hierarchy of spatial and temporal scales available.

Spatial and temporal scales and hierarchies

The spatial and temporal scales of environmental units available for studying river fishes are correlated (Fig. 6.1(a)). Unifying concepts and classifications of fish assemblages need to recognize this correlation as well as the hierarchical structure of these units, whose physical characteristics persist on scales of 10^5–10^6 to 10^{-1}–10^0 years from landscapes to microhabitats, respectively (Frissell *et al* 1986), and are extended to evolutionary scales at the zoogeographical level in Fig. 6.1(a). Hierarchical scaling promotes the most effective solution of ecological problems (Allen & Starr 1982). The hierarchy implies that the larger, more stable, environment imposes limits on the smaller, more variable, environmental units. Habitats, for example, can be classified within broader units, and thus lend themselves to statistically nested designs for the testing of differences in fish assemblages or other attributes.

Johnson (1980) suggested that resource selection by species follows a hierarchy from the zoogeographical range (first order), through microhabitat scales, with resource selection in each order being conditional on a lower order. Although we question his unidirectional dependence of selection (e.g. the home range can depend on the selection of habitat as well as vice versa) and the separation of feeding and habitat usage into different orders, it is important to understand resource selection in the context of spatial and temporal scales. Resource selection by a fish depends on a series of conditions: (1) ability to disperse among fluvial systems on a zoogeographical scale; (2) seasonal migrations of some species limited by basin extent, geomorphology and habitat availability; (3) home range limited by physicochemical factors (habitat distribution) on the reach or stream scale; and (4) activity under the constraints of biological interactions which include the probability of being killed, availability of prey and reproductive requirements at the microhabitat scale (Fig. 6.1(b)).

The time scales are complex, because those relating to persistence of environmental units (Fig. 6.1(a)) extend to evolutionary scales, and are two to three orders of magnitude greater (at similar spatial scales) than the ecological scales corresponding to the response times of individual fish (Fig. 6.1(b)). The formation of species assemblages depends on zoogeographical limits derived in evolutionary time scales (Fig. 6.1(a)), morphological and physiological preadaptations constraining distributional limits in ecological time scales (Fig. 6.1(b)), and interactions among species which fine-tune assemblage structure in ecological and evolutionary time scales. The following four subsections discuss the usefulness of classifying and predicting properties of fish and environments at decreasing spatial scales.

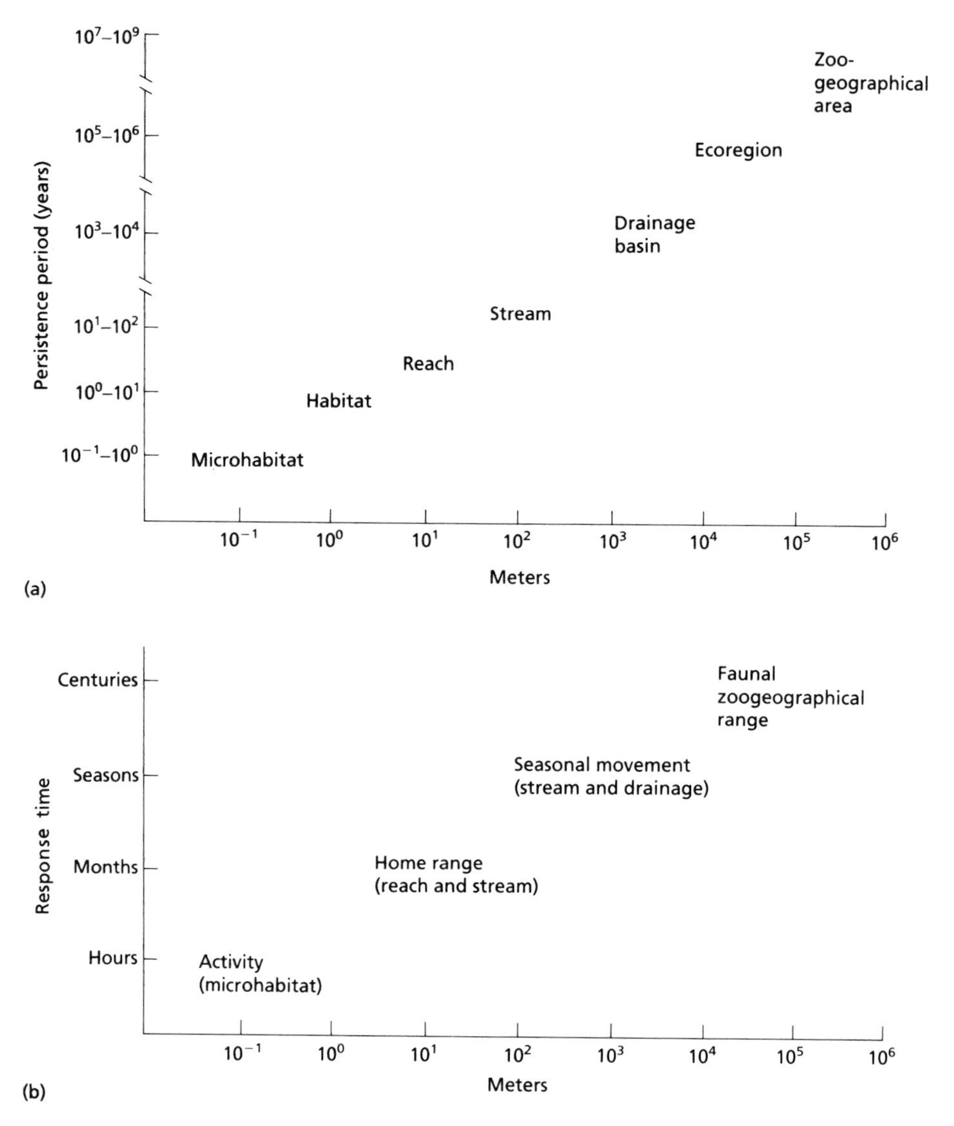

Fig. 6.1 (a) Persistence period versus spatial scale of environmental units (after Frissell *et al* 1986). (b) Response time versus spatial scale of fish movements.

Zoogeographical scales

What can be learned from comparative zoogeography that is useful to the manager or the ecologist who is not concerned with historical reasons for current fish assemblages? Can one generalize among similar systems of different zoogeographical regions? Only with great care; for instance, physically similar drainage systems in Poland and Ontario, Canada differ in species richness and body-size distributions among species (Mahon 1984). In contrast, Moyle and Herbold (1987) found great similarity in assemblage structure among cold headwater streams of Europe and eastern and western North America, and found common structural features among the warm-water fishes of Europe and western North America. These patterns may have resulted from the degree of similarity of conditions during Pleistocene glaciation events. In a comparison of external morphological measurements among stream fish assemblages from North and southern South America, Strauss (1987) found that correlations were high among North American zoogeographical areas and low between the two subcontinents, and reflected the phylogenetic constraints on morphology. To the extent that morphologies reflect ecological attributes, this evidence suggests that it is unwise to infer ecological similarities among assemblages that are distantly related phylogenetically. Likewise, evolutionary convergence of body form is not inevitable because there is not a unique morphological solution to a specific environment (Mayden 1987; Strauss 1987).

This phylogenetic limitation applies to comparisons of ecologies at the species level. At the opposite extreme of comparing gross energy transfers, such as fishery yield from a large proportion of the fish taxocene (section 6.4), similarities across zoogeographical areas reflect similarities in environment rather than in phylogenies.

Regional scales: ecoregions and hydrology

Ecoregional classifications such as those pioneered by Omernik (1987) and his co-workers (e.g. Larsen *et al* 1986; Hughes *et al* 1987; Whittier *et al* 1988; Hughes *et al* 1990) describe fish distributions at the landscape level by relating species presence to geomorphic/land-use patterns. Its use in anticipating where species may become endangered or in formulating broad management policies is not disputed, but in Wisconsin the classification was found to be fairly imprecise and fish assemblages were better classified by general habitat variables (Lyons 1989).

Ecoregions can divide basins and combine adjacent basins, resulting in a departure from the zoogeographical > basin > stream > habitat hierarchy of spatial units. Departing from this hierarchy and devising a network of scales is possible technically, but migratory fish can occupy two or more ecoregions. A classification that predicts and explains fish assemblages should have mechanistic links to finer scales; an ecoregional approach will not successfully predict or explain the presence of a migratory species which also depends on environments in adjoining ecoregions, including ocean habitats for anadromous species.

In addition, there is a cumulative effect of landscape that can also affect non-migratory fishes. A point in a stream downstream of an ecoregional boundary may reflect properties, such as hydrology, temperature and chemistry (e.g. Elwood *et al* 1983), that are controlled by the upstream ecoregion rather than by the ecoregion in which the point lies. Landscape elements from an ecoregional classification may be useful if they reflect environmental features affecting fish populations; the problem lies in assigning those elements within boundaries independently of watersheds and in implying that their effect is independent of their distance to the lotic area of interest (see Osborne & Wiley 1988 for an alternative approach).

Poff and Ward (1989) developed a regional classification of stream communities based upon variation in streamflow patterns of 78 streams across the USA. In Fig. 6.2(a) we have attempted to summarize the essential features of Poff and Ward's nine stream types plus large river-floodplains, in terms of four key hydrological attributes that are independent of spatial scale. Although this condensed representation assumes interdependence among four attributes and cannot represent all hydrological types precisely, these attributes were defined and positioned to account for Poff and Ward's significant correlations. Thus, data from most individual streams would occupy relatively small areas within the corresponding triangle. All types show some overlap with neighbors. Data for snowmelt, snow and rain, and winter rain showed considerable overlap (Poff & Ward 1989) and they, in turn, form a continuum with surrounding perennial types in the right triangle and with intermittent types in the left triangle (Fig. 6.2(a)). The left triangle is drawn incomplete because there is not a complementary relationship between intermittency and low annual flow variability, and examples do not exist along that axis.

Figure 6.2(b) shows some of the expected trends in fish population and community properties across different combinations of these hydrological attributes. In addition to these, Poff and Ward (1989) have suggested other plausible

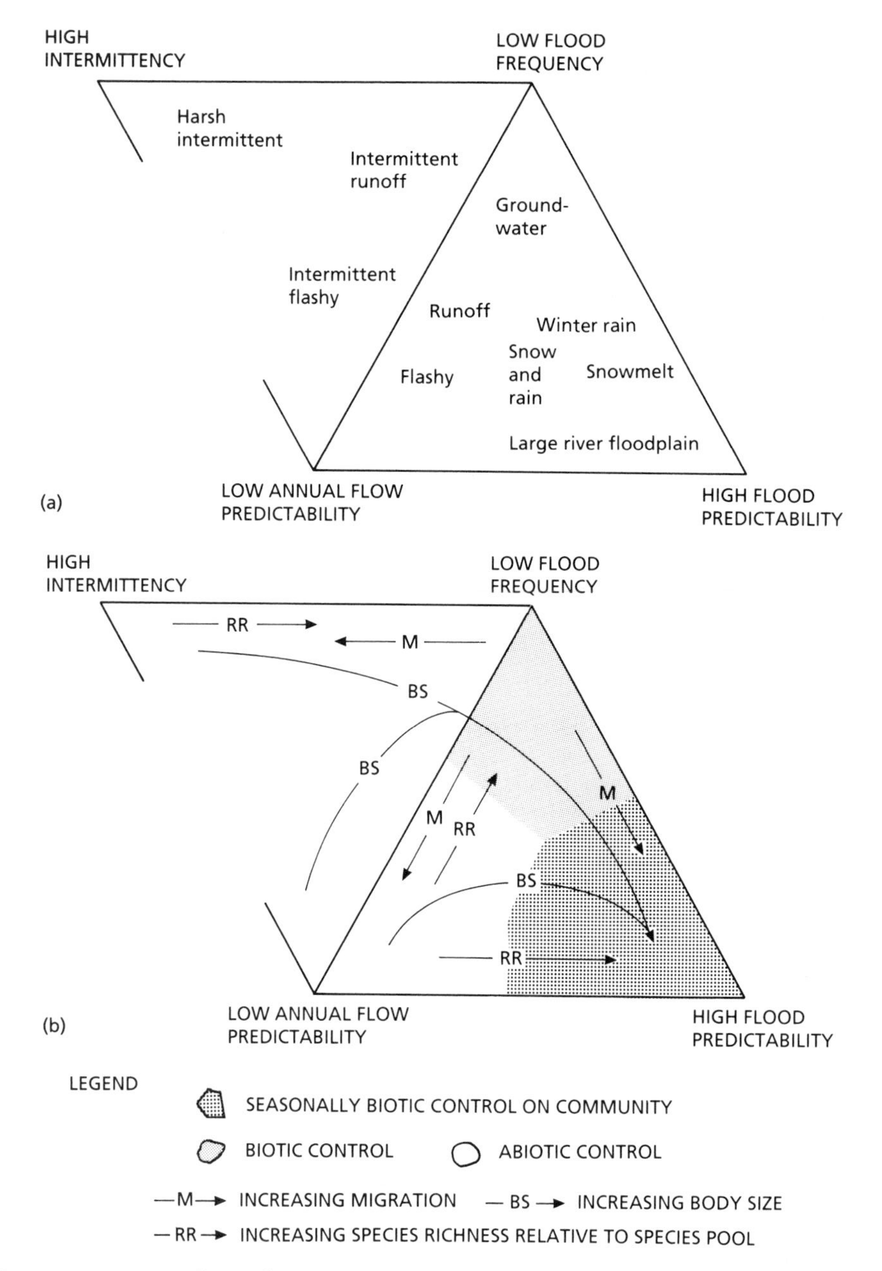

Fig. 6.2 (a) Stream types arranged according to approximate proportions of hydrological attributes (estimated partly from data in Poff & Ward 1989, Table 3). (b) Ecological attributes of fish species and assemblages corresponding to stream types in (a). Right-hand triangle in (a) and (b) includes all perennial stream types.

properties. Most published work has failed to put fish studies in an adequate hydrological framework, but there are exceptions. For instance, fish inhabiting streams subject to regular desiccation and flash floods were found to respond differently to streamflow changes than those restricted to small, clear creeks with permanent flow, resulting in different assemblages (Rohm *et al* 1987; Matthews 1988; Matthews *et al* 1988).

We believe that a hydrological approach, possibly in conjunction with different temperature regimes and landscape properties, may provide a superior regional template for understanding community function, life-history patterns and making inferences than an ecoregional classification, because the hydrology is more directly associated with physical constraints on fish habitats while reflecting some geomorphological features of the basin. Also, one could monitor ecologically significant departures from the natural regime due to anthropomorphic disturbance.

Basin and reach scales: zonation

The basis of zonation is to find large gradients (generic sense) to which the fauna must respond. Zonation schemes have used stream order (Sheldon 1968; Lotrich 1973; Horwitz 1978), hydraulic stress and power (Statzner 1987), temperature (Gard & Flittner 1974), habitat heterogeneity (Gorman & Karr 1978; Gorman 1988) and physicochemical gradients (Echelle & Schnell 1976; Matthews & Styron 1981). Huet's (1947, 1959) longitudinal zonation used a combination of gradient and stream width in European rivers to relate reaches to fish communities characterized by individual species. This approach is more difficult to apply to rich faunal assemblages covering various climatic and geomorphic zones (Allen 1969), such as in North America and the tropics, but it can be useful when empirically derived for a particular area (e.g. Angermeier & Karr 1983; Moyle & Senanayake 1984; Matthews 1986b). However, longitudinal zonation does not explain how stream reaches influence assemblages, does not account for potadromy, and does not explain the distribution of fishes in rivers with significant floodplains.

Habitat and microhabitat scales

It is surprising that the distribution of habitat types within reaches has not received much attention because the measurement of riffle: pool ratios to characterize stream reaches is a cherished tradition. However, Bisson and his associates have refocused awareness on habitats as channel units (e.g. side-scour pools, step pools, riffles, glides), first by creating a typology based on hydrological features (Bisson *et al* 1982), followed by an examination of fish distributions among these habitats (Bisson *et al* 1988). Building on this theme, Hicks (1990) found that the physical characteristics and distribution of habitats and faunal assemblages were different for sandstone versus basaltic drainages within the same ecoregion. Furthermore, habitats in sandstone drainages were more sensitive to changes in low summer discharge following logging than basaltic ones (Hicks *et al* 1991).

The Instream Flow Incremental Methodology (IFIM) is based primarily on microhabitat use patterns of fishes, e.g. current, depth and substrate (Bovee 1982). Some workers have claimed that IFIM shows promise for cold, headwater salmonid stream assemblages (Newcombe 1981; Moyle & Baltz 1985), where conditions and assemblages are quite consistent worldwide (Moyle & Herbold 1987), and for a few obligate warmwater stream fishes (Orth & Maughan 1982). IFIM does not work well when fishes do not have stereotyped behaviours on a limited spatial scale. Many stream fishes have extensive home ranges, select habitats at the channel unit level, or are relatively non-selective in microhabitat choice or are selective only under severe conditions (Angermeier 1987; Felley & Felley 1987; Ross *et al* 1987; Scarnecchia 1988). IFIM depends on fish biomass density being linearly related to the area of each available habitat type, which has been shown to be invalid (Mathur *et al* 1985; Conder & Annear 1987). In addition, IFIM does not work well when fishes are influenced by behavioural trade-offs such as risk avoidance or competitive spatial partitioning (Baltz *et al* 1982) (section 6.5). Finally, IFIM relates ephemeral, short-term behaviours to microhabitats, and such small-scale relationships will not necessarily maintain their predictive quality when extrapolated to larger scales (see Fig. 6.1(b)) that are more appropriate for management. The same general criticisms hold for the related Habitat Suitability Index (HSI) models (US Fish and Wildlife Service 1981; Terrell *et al* 1982).

River continuum and flood pulse concepts

Two unifying concepts, the river continuum (Vannote *et al* 1980) and the flood pulse (Junk *et al* 1989), provide ecological templates which can be used to compare and contrast fish communities or guilds within and among systems (Fig. 6.3). Both concepts have the potential to guide the derivation of sets of hypotheses to identify dominant mechanisms, in particular those operating between adjacent spatiotemporal scales, and either should improve the derivation of classifications at various scales to provide predictions useful for management in appropriate systems.

Although both concepts are designed to work up to a drainage basin scale, they are mutually exclusive in low-gradient potamon reaches, because the flood pulse concept recognizes the periodic nature of the interaction between the flood pulse and the floodplain which influences the adaptations of fish species. In contrast, the phytoplankton-dominated description of low-gradient reaches in the river continuum concept has more in common with heavily regulated rivers that have been denied access to floodplains, such as the Thames (Mann *et al* 1970). Such rivers, which are widespread in the temperate zone, present a dilemma in formulating a unifying ecological concept. Can we produce useful classifications of systems which are manipulated to the extent that adaptive and coevolutionary features of the fish species are no longer relevant, or do we have to undertake detailed investigations in each unique system?

The flood pulse concept has, at most, peripheral importance in the higher gradient rhithron, where many of the longitudinal processes of the continuum concept provide a more appropriate description. However, the original continuum concept needs to be adapted to, or excepted from, the following: differences in upstream riparian vegetation (Barmuta & Lake 1982; Wiley *et al* 1990), discontinuities (Statzner 1987; Naiman *et al* 1988; Pringle *et al* 1988) and upstream transport of nutrients and biomass through migrations of temperate (Hall 1972; Li *et al* 1987) and tropical fishes (Petrere 1985; Welcomme 1985).

In conclusion, the development of unifying con-

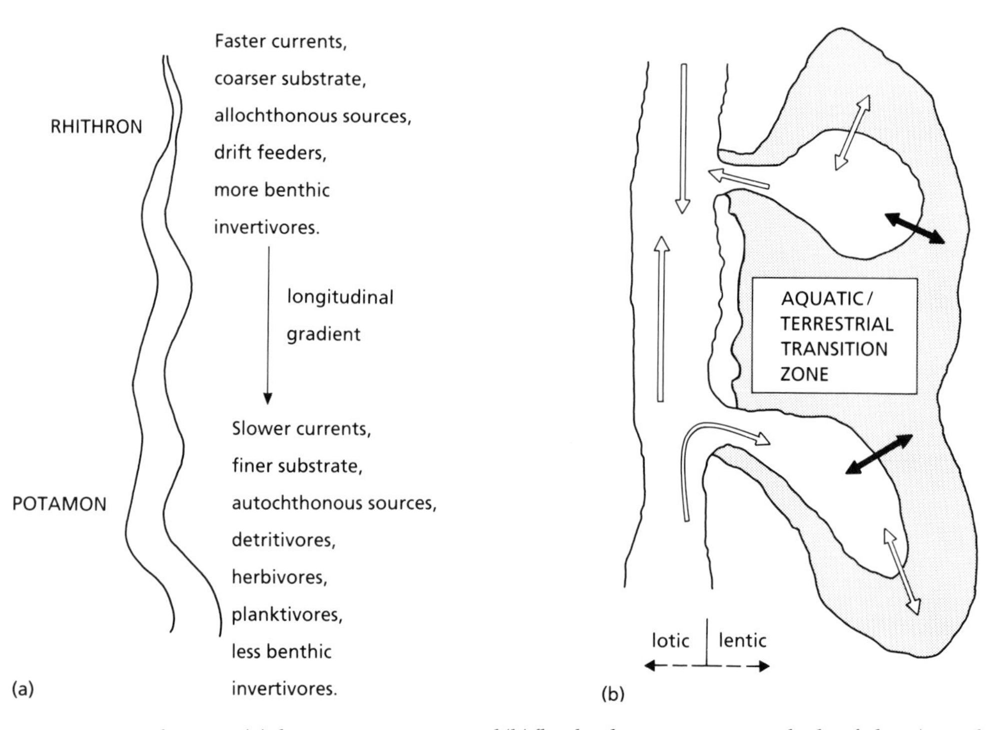

Fig. 6.3 Some contrasts between (a) the river continuum and (b) flood pulse concepts as applied to fishes. (⇨ 'white' fishes; ➡ 'black' fishes; arrows indicate migrations.)

cepts is still in its infancy and will remain so unless more studies and surveys follow guidelines such as those outlined in section 6.7. Nevertheless, the concepts are currently helpful in understanding life-history adaptations and in classifying functional groups to act as templates for testable hypotheses.

6.3 LIFE-HISTORY ADAPTATIONS

A Gleasonian view of fish species adaptations based on physiological and morphological responses to the environment is presented in this section. This is a complementary explanation of longitudinal zonation to that presented by (Horwitz 1978) with lateral effects of floodplains incorporated. We propose that, at least in the headwaters, constraints of habitat size, habitat variety (see also Gorman & Karr 1978) and hydraulic forces limit diversification of body form and limit resource partitioning according to the principle of limiting similarity (MacArthur & Levins 1967). We describe life-history adaptations to different classes of the environment in terms of body morphology and activity, *P/B* ratio and body size, trophic properties, reproductive strategies, and trade-offs and polymorphisms.

The rhithron and the potamon impose different requirements on fish life (Fig. 6.3). The classic, high-gradient rhithron is usually more restrictive: it is smaller, cooler, more highly oxygenated, and comprised primarily of fastwater habitats: rapids, riffles, cascades and step-pools. The low-gradient rhithron typical of surface-drained, low-lying areas is characterized by more variable temperature, oxygen and discharge (Wiley *et al* 1990). The potamon comprises habitats with a greater variety of size, depth and flow: large river channels and pools, braided stream channels, oxbows and sloughs, and habitats of the floodplain. The gradient is usually minimal, temperatures are higher, and some habitats become hypoxic. These features, and local variants, have a profound effect on the adaptations discussed below.

Morphology and activity

The attributes of rhithron fishes are more constrained by habitat size (Schlosser 1982) and hydraulic power and variability (Statzner 1987; Wiley *et al* 1990) than the fishes of the potamon. This is especially noticeable in terms of their size, shape and metabolism (Table 6.1). Smaller body size is not related just to the scale of available in-stream cover (Schlosser 1987); the demands for acceleration and agility in a turbulent environment also favour a smaller body mass (Webb & de Buffrénil 1990). A larger body increases speed, but at the expense of agility and manoeuvrability (Webb & de Buffrénil 1990).

Body shapes of non-benthic rhithron fishes should be closely distributed around an optimal fineness ratio (FR = ratio of body length to depth) of 4.5 (Webb 1975), in contrast to greater variation in the potamon. Scarnecchia (1988) found greater variation in FR for fish assemblages from reaches with complex flow than from those inhabiting fast, channelized reaches.

The greater muscular force needed for agility in fast currents of the rhithron suggests that those fishes should have higher metabolic rates than potamonic fishes of similar size. Clausen's (1936) initial evidence for this argument was flawed because allometric relationships were ignored. However, our survey of the metabolic literature suggests that Clausen's intuition may have been correct (Table 6.2). Haemoglobin of fishes inhabiting fast, cold stream reaches delivers greater amounts of oxygen to active tissues than in those fish found in typical potamonic habitats (Powers 1972; Cech *et al* 1979; Powers *et al* 1979). These generalizations apply to fishes swimming in the water column. Substrate-oriented fishes, including those that inhabit the hyporheos, have metabolic rates and morphologies reflecting low-flow and benthic environments (Facey & Grossman 1990).

P/B ratios and body size versus habitats

Even though metabolic expenditures are higher in the rhithron, fish communities in smaller streams tend to have higher *P/B* ratios (production/biomass or turnover rate on an annual basis; see section 6.4) than in larger streams (Lotrich 1973). We found a significant correlation between body size (as mean weight) and *P/B* ratios for 32 freshwater fish species ($r = -0.47$,

Table 6.1 Typical distributions of fish attributes along the stream gradient

Attribute	Stream gradient		
	Rhithron	Intermediate	Potamon
Temperature guild	Stenotherms (high gradient) Eurytherms (temperate-continental, except springfed)	Mesotherms or eurytherms (temperate)	Eurytherms (temperate)
Metabolism	High standard metabolism		Low standard metabolism (except floodplain habitats)
Fineness ratio*	FR near 4.5 (optimal score)		FR variable
Oxygen-binding affinity or haemoglobin	Low (high gradient)		High
Fish size	Smaller		Variable
Life span	Relatively short		Greater longevity
P/B ratio (index of r_{max})	Relatively high		Relatively low (except floodplain habitats)
Reproductive guild (Balon 1975)	Lithophils and phytophils	Phytolithophils and psammophils and pelagophils	Lithopelagophils

* FR, fineness ratio (body length/depth).

$P < 0.01$), which confirms an earlier analysis of diverse groups (Banse & Mosher 1980) and recent analyses of cohorts within species (Boudreau & Dickie 1989).

Small body size is encountered in the rhithron in the form of small species and young of larger species which move upstream to spawn. Small body size is also dominant in seasonally inundated habitats in floodplains where high growth rates and production occur (Bayley 1983, 1988b; Junk *et al* 1989). Both environments have two significant factors in common: they are shallow and are subjected to flooding and dewatering to a much greater extent than the main channel of the potamon. The P/B ratio is strongly related to the intrinsic rate of increase, r_{max}, of the population. Small species and young stages of larger species with high r_{max} values and other r-traits (*sensu* Pianka 1970) are expected to dominate habitats in the rhithron (Hall 1972) and floodplains (Junk *et al* 1989). An exception to this may be small streams of unusual constancy of flow (groundwater type; see Fig. 6.2(a)). In the lotic component of the potamon, which is normally characterized by lower productivity, larger species with low P/B ratios and r_{max} values, and possibly species with relatively more K-traits (Pianka 1970), are expected to be relatively common, although large r-trait species whose young occupy the rhithron or floodplain may dominate.

Trophic adaptations

One morphological design often affects the rest of the body plan (Thompson 1942). Body size places limits on trophic specialization, life span, and reproductive capacity. Average food particle size is approximately 0.07 of fish length (L) (Kerr 1974), ranges up to 0.33 L for obligate piscivores (Popova 1978), but generally decreases as L increases (Webb & de Buffrénil 1990). Typical rhithron fishes are small and are primarily adapted to

Table 6.2 Standard metabolic rates of stenotherms and eurytherms among temperate fishes

Type	Fresh weight (g)	Temperature (°C)	Standard metabolic rate (mg kg^{-1}h^{-1})	Reference
Stenotherms				
Salmo trutta	100	9.5	73	Elliott (1979)
Oncorhynchus clarki	83–93	10	7	Dwyer & Kramer (1975)
		20	129	
Oncorhynchus mykiss	100	15	11	Rao (1968)
Oncorhynchus nerka	100	15	76	Brett (1965)
Salvelinus fontinalis	100	20	147	Beamish & Mookherjii (1964)
Eurytherms				
Gila atraria	100	18–20	42	Rajagopal & Kramer (1974)
Rhinichthys cataractae	100	18–22	40	Rajagopal & Kramer (1974)
Carassius auratus	87	20	50	Smit (1965)
Cyprinus carpio	100	10	16	Beamish & Mookherjii (1964)
		20	48	
Mylopharodon conocephalus	100	20	51	Alley (1977)
Catostomus catostomus	100	10	35	Beamish (1964)
		20	110	
Lepomis macrochirus	100	10	31	Wohlschlag & Juliano (1959)
		20	116	
Ictalurus nebulosus	100	10	20	Beamish (1964)
		20	66	

consume small aquatic and terrestrial invertebrates, especially terrestrial drift.

More diversification is possible in the potamon because greater habitat and food diversity confers advantages to large and small fishes (Welcomme 1985). A wide range of trophic adaptations results from the enormous productivity and variety of food in the tropical floodplain (Junk *et al* 1989). Seasonal changes result in oppportunistic behaviour. There are many generalists that consume invertebrates and plant matter of aquatic and terrestrial origin (Bayley 1988b).

Detritus and associated microflora appear to be the main repository of organic matter in river-floodplains. Subsequently many fishes have evolved to specialize on fine detritus (e.g. Bowen *et al* 1984). Such fish in the tropics start feeding on detritus when very small (Bayley 1988b) but become large (2–5 kg) compared with the small particle sizes they depend on.

The combination of seasonal food availability and low-water periods has resulted in many species building up large fat reserves to survive a season without feeding and to provide additional energy for migrations (Junk 1985). This applies to tropical and temperate rivers; the season of fasting for most species is the dry season in the tropics and the winter in the temperate zone (Cunjak & Power 1987; Cunjak *et al* 1987). Seasonal fluctuation in food supply demands that most fishes be flexible in food selection and/or energy storage within the limits of their size.

Reproductive strategies

Changes in substrates along the river continuum influence the mode of reproduction. Substrate size is much larger in high-gradient rhithron zones, corresponding to the greater hydraulic power in the steeper gradient (Richards 1982). Therefore, reproductive guilds (*sensu* Balon 1975) tend to be ordered from lithophils in high-gradient rhithron to lithopelagophils and psammophils in the potamon (see Table 6.1). Further generalization is possible by coupling Balon's reproductive guilds with Hokanson's (1977) concepts of temperature guilds. Hokanson (1977) recognized that physiological adaptations to seasonal temperatures affected patterns of growth, gonadal development and reproductive timing of temperate fishes (Table 6.3). In general, temperate fishes in the rhithron spawn primarily from the autumn to the spring, where oxygen conditions permit longer development or dormant periods, whereas most of the fishes remaining in the potamon spawn during the spring, summer or autumn months.

Large floodplains are still connected to rivers in many tropical regions and a few temperate ones, resulting in a full spectrum of lotic to lentic habitats and strong seasonal effects due to the flood pulse (Fig. 6.3) which may be independent of temperature. All spawning strategies seem to point towards giving as many young as possible access to the very productive, newly flooded, shallow areas created during the flood pulse (Junk *et al* 1989). However, the means to obtain this goal vary. The South-East Asian classification of 'black' fish and 'white' fish communities is a useful first-order differentiation of river-floodplain species (Welcomme 1985), and is partly based on reproductive strategy. Black fish prefer lentic habitats and undertake local migrations between floodplain habitats only in response to water level changes (Fig. 6.3). Many black fish species are multiple spawners and practice parental care. White fish undertake longitudinal migrations in the river, tend to be annual spawners influenced by the flood cycle, but also use the floodplain for feeding in nursery areas and, in some species, for spawning (Welcomme 1985; Junk *et al* 1989). Many white species are pelagophils, spawning in the main channel just before, or on the rise of, the flood pulse (de Godoy 1954; Bayley 1973; Schwassmann 1978). This results in the eggs and larvae being dispersed widely over the floodplain during the flood pulse.

There are equivalents to this classification in temperate systems. For example, in North America the bluegill, *Lepomis macrochirus*, and the largemouth bass, *Micropterus salmoides*, are nestbuilding, floodplain lake species that correspond to black fish. The white basses, *Morone* spp., undergo longitudinal migrations when permitted and produce semibuoyant eggs in open water, and correspond to white fish.

Trade-offs in life-history strategies and polymorphism

Trade-offs in life histories and body designs are caused by conflicting selective pressures, particularly with reproductive strategies. Darwinian fitness of fishes is related to survival and fecundity, which are both size related (Calow 1985). The critical energy trade-off is between investment in gametes or in somatic tissues. Fish subjected to physically taxing spawning migrations tend to be semalparous and invest more energy in egg production; iteroparous fish do not exhaust body stores during the migration and tend to allocate energy to ensure post-spawning survival (Schaffer & Elson 1975; Glebe & Leggett 1981a, 1981b).

Iteroparous and semalparous life histories can occur in a single species; e.g. steelhead trout, *Oncorhynchus mykiss* (Ward & Slaney 1988); Atlantic salmon, *Salmo salar* (Schaffer & Elson 1975); and American shad, *Alosa sapidissima* (Glebe & Leggett 1981a, 1981b). A species possessing this polymorphic attribute is buffered against environmental uncertainty through increased specialization without losing fitness from con-

Table 6.3 Hokanson's (1977) temperature guilds for freshwater fishes in temperate zones

Guild	Optimal temperature (°C)	UILT (°C)	Gonadal growth phase	Spawning phenology
Stenotherm	20	<26	Summer (<20°C)	Autumn to spring (<15°C)
Mesotherm	0–28	<28–34	Autumn and winter (<28–34°C)	Spring (3–23°C)
Eurytherm	<28	<34	Increasing photoperiod (12°C)	Spring to autumn (15–32°C)

UILT, upper incipient lethal temperatures.

flicting gene recombinations. The species survives as a metapopulation, some populations of which are in the process of expanding their range and compensating for contractions by others in response to a changing environment.

Polymorphism may reflect adaptive responses to selective forces (Seghers 1974; Suzumoto *et al* 1977; Endler 1980; Taylor & McPhail 1985a, 1985b; Wade 1986), genetic drift (Allendorf & Phelps 1980) or a combination of both processes (Hatch 1990). Polymorphism can reflect fine tuning to local conditions (e.g. Zimmerman & Wooten 1981; Calhoun *et al* 1982; Matthews 1986b; Hulett 1991) or responses to large environmental gradients (Riddell & Leggett 1981; Riddell *et al* 1981; Schreck *et al* 1986; Hatch 1990). Some traits exhibit high degrees of heritability as demonstrated in breeding experiments focusing on the genetics of disease resistance (Wade 1986; Withler & Evelyn 1990). Polymorphism has been attributed to phenotypic plasticity responding to environmental change (Gee 1972; Stearns 1980; Metcalfe 1989).

Sedentary and anadromous phenotypes among salmonids have been attributed to phenotypic plasticity with arctic charr (Nordeng 1983) and to genetic differentiation with rainbow trout (Currens *et al* 1990). Various stocks exhibit polymorphic life-history traits, including migratory/sedentary tendencies among some white fishes of the river-floodplains (Welcomme 1985) and the diverse potamodromous life histories exhibited by Yellowstone cut-throat trout, *Oncorhynchus clarki bouvieri* (Varley & Gresswell 1988).

How should these stocks be managed? The most conservative, but recommended, approach is to assume that polymorphic traits are under genetic control and to adopt a policy that conserves the diversity of the metapopulation. This relates to problems with hatchery-reared fish.

Augmenting stocks from hatcheries

Hatchery fish account for the production of approximately 80% of the salmonids in the Columbia River. However, mitigation using current hatchery techniques will, at best, result in a pyrrhic victory. Although it is theoretically possible to manage hatcheries to conserve genetic resources (Nyman & Ring 1989) and create conditions for heterozygous populations (Schreck *et al* 1986), much genetic diversity is lost through genetic drift and founder's effect (Allendorf & Phelps 1980; Ryman & Stahl 1980; Quattro & Vrijenhoek 1989) and a recent study has shown differences in selection between local hatchery and wild populations of chinook salmon (Hulett 1991). Hatchery-raised fishes tend to be more aggressive and dominate wild cohorts (Nickelson *et al* 1986; Noble 1991). The degree to which aggressive dominance is influenced by early exposure to hatchery conditions compared with genetic influences is still unknown. In any case, if hatcheries are to be used, genetic and environmental factors must be accounted for.

6.4 QUANTITATIVE MEASURES OF FISH POPULATIONS AND YIELD PREDICTION

Measures of the quantities by number or weight of fish in populations, guilds or communities is important for the management of fish for human exploitation and for the testing of effects of environmental change. Such changes can be caused by natural events (section 6.5) or human impacts (section 6.6). Details of the biases and variance associated with specific sampling methods are discussed in Chapter 10.

Rational exploitation for food or recreation requires prediction of the yield or the proportion of production that is available for human use on a sustained basis. In principle, this can be estimated through traditional dynamic pool models, from production estimates, or from comparative approaches. There are distinctions between the approaches to problems of exploitation and environmental change, but there are common practical limitations that will become apparent below.

Fish population dynamics

Fish population models (Schaefer 1954; Beverton & Holt 1957; Gulland 1969; Ricker 1975) have evolved to predict trends in intensive fisheries of significant value as measured by market economists, such as those exploiting offshore marine and anadromous populations. They explicitly or implicitly incorporate birth, death and growth rates on a given number of individuals. The simplest, the logistic or surplus-yield model (Schaefer 1954), requires the definition of a stock or viable population, yield estimates, and reliable indices of fishing mortality rate (fishing effort) and of population abundance (catch per unit effort) for a series of years. Typically, the most critical assumptions are that yield estimates be at equilibrium with the current exploitation rate, catch per unit effort be proportional to abundance density, and that the environment limiting the population is constant.

The yield-per-recruit model (Beverton & Holt 1957) requires knowledge of the age structure and natural mortality rate of the population in addition to the logistic model requirements. It provides an estimate of the fishing mortality rate that maximizes yield per fish recruited to the fishery. So far no models described have required population size estimates, but to estimate yield from the yield-per-recruit model the number of recruits obviously needs to be estimated each year (e.g. Pope 1972). Predicting recruits from abundance of parents has a good theoretical foundation (Ricker 1975), but with rare exceptions (Elliott 1985) these models are very noisy and have no predictive capability.

Can these models help us with river fisheries? They are sound theoretically, but their application is limited by the quality of data, the cost of obtaining data, and changes in the environment. Elliott's experience suggests that stream fish are more appropriate for the testing of population regulation mechanisms. More accurate estimates are possible in streams, although with more effort than most biologists realize (Bayley 1985), and fisheries in large rivers are generally more difficult to sample than benthic fish in many continental shelf environments. Fishery studies have tended to underrate environmental effects in all systems, but streams and smaller rivers are typically the most physically variable of them all.

A principal physical variable is discharge. Because fish recruitment and production are generally considered to be most affected by events in the first year of life, the variability of abundance of 0+ and 1+ fish was estimated from the literature (Table 6.4). Coefficients of variation of annual fish population measures were compared with those of annual discharge measures (Fig. 6.4). Despite the variety of streams, species, methods and lengths of stream sampled there is a clear dependence between year-to-year variability of population size and discharge. There are probably various mechanisms that connect discharge to population size, and critical periods of flow within systems may provide clues. Considering that for a given accuracy and time scale discharge is cheaper to measure than population size. It is disappointing that discharge is so often inadequately monitored.

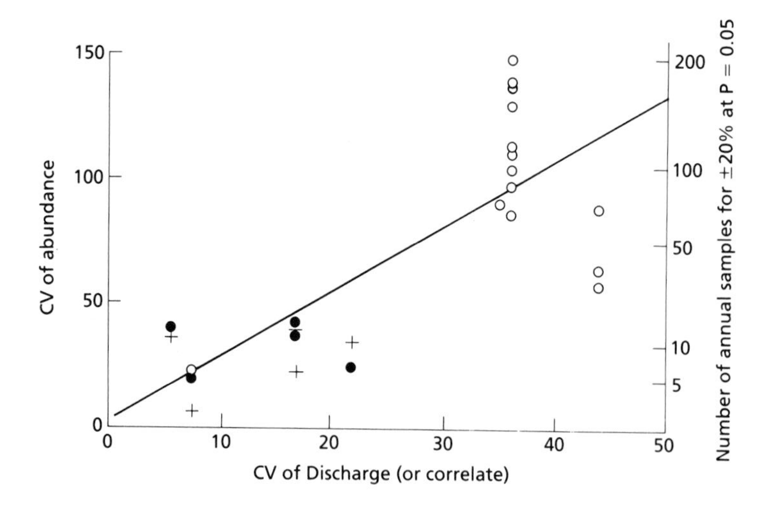

Fig. 6.4 Coefficient of variation (CV) of autumn fish abundance versus CV of annual discharge or correlate (see Table 6.4 for further explanation of data and sources). Number of samples (=years) required for a precision of ±20% at $P = 0.05$, based on central limit theory applied to CV of abundance predicted by regression, is shown at right. (•, 0+ fish; +, 1+ fish; ○, mostly 0+ and 1+ fish.)

It is sobering to consider the number of samples, which in this case is years, required to obtain estimates of cohort sizes within ±20% of the mean with 95% confidence (see right ordinate, Fig. 6.4). Although in some cases the methods could be improved to reduce bias, reduction in measurement variance usually requires more sampling, which may itself influence the population in a small stream. The streams with lowest discharge variability (Table 6.4), situated in northern Michigan and Wisconsin, are fed by groundwater. The other examples of salmonid streams, although they are surface run-off, are unusually stable. In contrast, the examples of higher discharge variability, which happen to be surface run-off midwestern streams, show what to some would be an alarming variability, although the examples from Missouri include two drought years. While much of this variability might be explained by discharge or other factors within streams, this does not help prediction if next year's weather has to be predicted first. Thus, there is an important distinction between a *post-hoc* investigation, which for example tests the effect of an environmental change given a discharge pattern, and one that needs to predict the future trend of a population for management of a fishery. Problems of assessing effects of environmental change on salmonid populations have been assessed by Hall and Knight (1981).

Although high recruitment variability is not restricted to stream fish (Dennis *et al* 1985), the resources available to adequately sample stream fish, define each stock, and account for environmental change are quite inadequate to consider using traditional models on the majority of stream or river populations. An exception is the valuable sockeye salmon mixed-stock fishery of the River Fraser, Canada, but even with 34 years of spawner-recruit data, extreme harvest rate reductions on some stocks over two generations (8 years) were deemed necessary to give >50% chance of detecting an increase in spawner abundance (Collie *et al* 1990).

Fish production

Production is defined as the total addition of biomass or equivalent potential energy to a population, including that from individuals not surviving the time period concerned. Therefore, production can represent resources available for other trophic levels. The International Biological Program (IBP) prompted many biologists to undertake fish production studies since the 1960s. These efforts have been useful in providing first-order estimates of trophic efficiency and energy flow, but have limited utility for estimating the proportion of production available to humans or to predict changes in the community due to an environmental change. These limitations result as much from the lack of accuracy in methods as from the lack of appropriate comparisons among systems.

Unfortunately, many biologists have pursued fish production estimates, which are expensive endeavours, without a clear purpose. In addition, some bemoan the lack of production estimates for the first year of life, implying that only estimates for the whole population are useful. Sometimes the contrary is true. For instance, the production of exploited year-classes furnishes an upper limit to the yield potential. The production of a particular size-range of fish may better reflect the influence of a critical habitat type. The implication that the production of a fish population represents a single trophic level is usually false. Young fish can be trophically more similar to (or even consumed by) some invertebrates than older conspecifics. Designs for studies requiring knowledge of annual production of young fish should be aware that such estimates in streams may be as variable as those for 0+ abundance levels (Fig. 6.4) and be subject to similar biases, because both estimates depend on the same limited options for estimating abundance density (Bayley 1985). A similar argument applies to the use of bioenergetics to estimate production or yield.

It is useful to estimate the production of the whole fish taxocene to compare the broad, upper-trophic-level productivity of different systems whose fauna are dominated by fish (e.g. Table 6.14 in Welcomme 1985) or to estimate the primary production sources required to support the taxocene (Bayley 1989). This approach could be refined by stratifying the production along criteria of individual body size and trophic group (or guild within zoogeographical areas), but should prefer-

Table 6.4 Coefficients of variation for fish abundances estimated near the end of the growing season and for discharges

Stream	Country	Years of study	Species	Mean dimensions of stream segment (m)		Age class	No. of years	% Coefficient of variation of annual	
				Length	Width			Abundance	Discharge
Black Brows Beck	UK	1967–83[1]	*Salmo trutta*	75	0.8	0+	17	24	26
			Salmo trutta	75	0.8	1+	16	34	26
Hunt Creek	USA	1949–62[2]	*Salvelinus fontinalis*	2816	5.6	0+	14	20	7.4
			Salvelinus fontinalis	2816	5.6	1+	14	6.3	7.4
Hunt Creek	USA	1968–86[3]	*Salvelinus fontinalis*	1610	4.0	0/1+	20	22	7.4
Laurence Creek	USA	1953–70[4]	*Salvelinus fontinalis*	5630	—	0+	18	39	5.5
			Salvelinus fontinalis	5630	—	1+	18	36	5.5
Shelligan Burn	UK	1966–75[5]	*Salmo trutta*	167	3.6	0+	10	41	17
			Salmo trutta	167	3.6	1+	10	22	17
			Salmo salar	167	3.6	0+	10	37	17
			Salmo salar	167	3.6	1+	10	39	17
Maquoketa River	USA	1978–82[6]	*Micropterus dolomieui*	3540	27.0	0/1+	5	90	35
Middle Fork Salt River	USA	1985–89[7]	*Lepomis cyanellus*	250	1.1	0/1+	5	86	36
			Pimephales promelas	250	1.1	0/1+	5	104	36
			Lepomis humilis	250	1.1	0/1+	5	113	36
			Notemigonus crysoleucas	250	1.1	0/1+	5	97	36
			Cyprinella lutrensis	250	1.1	0/1+	5	129	36

Titus Creek	USA	1985–89[7]	*Hybopsis dorsalis*	250	2.9	0/1+	5	148	36
			Lepomis cyanellus	250	2.9	0/1+	5	137	36
			Etheostoma nigrum	250	2.9	0/1+	5	139	36
			Semotilus atromaculatus	250	2.9	0/1+	5	138	36
			Pimephales promelas	250	2.9	0/1+	5	133	36
Jordan Creek	USA	1950–53[8]	*Pimephales notatus*	1450	6.7	0/1+	4	88	44
			Campostoma anomalum	1450	6.7	0/1+	4	63	44
			Ericymba buccata	1450	6.7	0/1+	4	57	44

[1] August/September abundance data from Elliott (1985); discharge estimate uses annual rainfall variation from June to August (Elliott 1984).
[2] September abundance data from McFadden *et al* (1967); CV of flow estimated from range of 21–27 cubic feet per second.
[3] Autumn abundance data of 2–4.9-inch fish in control stretch in Alexander and Hansen (1988), CV discharge assumed same as during previous period on Hunt Creek.
[4] September abundance data from R. Hunt, Wisconsin Department of Natural Resources, personal communication. Discharge CV estimated from 1961–6 data only.
[5] Autumn abundance data from Egglishaw and Shackley (1977); discharge estimate uses annual rainfall variation as a proxy.
[6] Autumn abundance data of fish <20 cm (mostly 0+ fish) from Paragamian (1987); discharge data May–September from Paragamian and Wiley (1987).
[7] Autumn abundance data of Smale and Rabeni (1990); the five most abundant species from each stream are shown; discharge estimate uses annual rainfall variation as a proxy.
[8] August abundance data from Larimore (1955); discharge estimate uses US Geological Survey data from River Vermillion downstream.

ably include all significant fauna and not just fish. Conversely, comparing the production of fish populations at the species level or between species from different environments has no utility, unless differences in competitive or predatory environments and in population structure are accounted for. (For more discussion, see Calow 1994.)

Multispecies yields

The development and application of dynamic models, such as the surplus-yield model, to multispecies fisheries is in its infancy, and is severely limited by empirical data (Pauly & Murphy 1982). However, there is surprising consistency among yields of tropical river-floodplain fisheries (Welcomme 1985) that can largely be explained by fishing effort using a function related to the surplus-yield model (Bayley 1988a). This consistency across fisheries may be a result of common socioeconomic development of fishing communities in response to high species diversity (yields of species groups tend to be more stable than individual species), the dominance of native, coevolved fish fauna in tropical systems, and/or the relatively natural flood pulses and floodplain habitats in the tropics.

However, residual variability is significant, and the explanation of 74% of the variance in yield per unit floodplain area by fishing intensity (Bayley 1988a) should not hide the fact that the prediction of an individual system is still too imprecise for most management needs in individual river fisheries. Other factors at the comparative level, such as indices of system productivity or, at the individual system level, the hydrological regime in previous years (Welcomme 1985) should be explored to refine these 'top-down' approaches. However, the current models are adequate to alert politicians and economists of the current or potential animal protein yields that may be lost, and are currently being lost, by many basin development projects.

Summary

Three principal problems limit our ability to measure and predict fish populations: deriving sampling methods of known accuracy (bias and precision), accounting for temporal and spatial variability, and accounting for environmental effects, particularly discharge (see Chapter 10). We consider age determination to be a secondary issue with most populations. Available models are not considered to be as limiting as good quality empirical data. Simulation models (e.g. DeAngelis *et al* 1991) can suggest plausible sets of mechanisms and driving forces, and reasons for the high variance found empirically. Such models indicate probable areas requiring better empirical data, which can be obtained to generate another cycle of improved simulations. In the absence of a large injection of empirical data, however, it is unlikely that simulation modelling will increase the accuracy of annual prediction that managers often require, even if such predictions are conditional on future weather patterns. More empirical data based on parallel time series of independent systems is one approach to developing more efficient syntheses and predictions (section 6.7).

6.5 COMMUNITY ECOLOGY AND MANAGEMENT

Fish resources have not traditionally been managed at the assemblage or ecosystem level. At best, we have monitored the welfare of a small fraction of the assemblage and concentrated management activities within the confines of the river channel. The difficulties of estimating populations (section 6.4), the need to account for environmental constraints and adaptations (section 6.3), and the effects of changes in single components on the assemblage (this section) all indicate the necessity to understand community- or assemblage-level processes at the habitat, stream and ecosystem levels, and to develop fisheries management approaches at those levels. But do we know enough to provide viable alternatives to current management methods?

Abiotic or biotic control in response to disturbance

Fisheries managers need to know the extent to which a change in one species affects the responses of the other species in the assemblage. If biotic interactions such as predation or competition are strong, one presumes that the loss or addition of a single species would change the structure of the remaining species of the assemblage. If a physical change, such as an environmental disturbance or intense exploitation, principally affected one species, would this result in more changes in the assemblage than would be expected from only the autecologies of the remaining species? If physical forces are highly dynamic, perhaps biotic factors do not have sufficiently short response times to intervene (Hutchinson 1961; Wiens 1984). These regulatory issues are at the heart of the controversy over whether abiotic or biotic effects govern assemblage structure of stream fishes (Moyle & Li 1979; Grossman *et al* 1982; Herbold 1984; Rahel *et al* 1984; Yant *et al* 1984; Grossman *et al* 1985).

Considerations of scale and generally acceptable definitions of terms (Schoener 1987) can increase the odds of solving this difficult problem. First, regional influences on hydrology (section 6.2; Fig. 6.2) should serve as the largest scale in which to assess the role of abiotic factors in structuring fish assemblages. Second, it is important to distinguish between contingency and magnitude as factors of disturbance (Colwell 1974). High flushing flows during normal periods of peak discharge, within typical bounds, do not disturb fauna coadapted to seasonal contingencies (Resh *et al* 1988). However, such flows during the base flow period can severely impact fauna. In systems with variable flood frequency and poor predictability, such as the prairie streams described by Matthews (1987), disturbances such as drought and flash floods are frequent events and when measured at small scales (habitat units and days) result in apparently randomly varying assemblages; but at larger scales (reaches and years) assemblages appear to be relatively stable (Ross *et al* 1985; Matthews 1986a; Matthews *et al* 1988). (Comparisons between scales should ensure that a difference is not an artefact of different measurement precision; for instance, a smaller standard error can result from a larger statistical unit associated with the larger scale.)

Intermittent disturbances such as floods of low predictability can act as resetting mechanisms (Starrett 1951; Fisher 1983; Matthews 1986a) or mediate competitive and predatory interactions (Meffe 1984). Conversely, the lack of a flood in a river-floodplain where predictability is high (see Fig. 6.2(a)) comprises a disturbance because the fauna are adapted to a regular flood pulse (Junk *et al* 1989). Variable and stable assemblage components were observed following a severe drought over a 40-month period (Freeman *et al* 1988). Some fishes remained persistent, characterized by relatively stable populations, whereas others were less persistent and highly variable. The effect on the assemblage depended on timing, especially in relation to recruitment dynamics, frequency and magnitude of the event, and regional adaptation (Freeman *et al* 1988).

Predation and competition

Biological interactions are apparent at smaller spatial scales, such as at the stream reach or channel unit level. Studies indicating competition or predation suggest that assemblage structure may be controlled by non-random factors because observations of behavioural patterns, especially in experiments, appear to be predictable and repeatable.

In general, examples of competitive displacement from large sections of stream result from species introductions (Fausch & White 1981, 1986; Cunjak & Green 1983, 1984; Meffe 1984; Castleberry & Cech 1986; Fausch 1988). Competition among native assemblages generally evokes resource partitioning through selective and interactive segregation (Baker & Ross 1981; Baltz & Moyle 1984; Moyle & Senanayake 1984; Baltz *et al* 1987; Gorman 1987; Ross *et al* 1987; Wikramanayake & Moyle 1989; Wikramanayake 1990). For instance, the innate hierarchy of similar-sized salmonids in Cascadian streams is coho salmon > steelhead trout > cutthroat trout > chinook salmon in order of dominance (Li *et al* 1987). This social hierarchy has a direct effect on selection and use of microhabitats and habitat units among sympatric salmonids. Coho salmon force other salmonids to less desirable locations.

A comparison of stream sites with and without piscivores by Bowlby and Roff (1986) indicated that the biomass of non-piscivorous fish was lower in the presence of piscivores. Piscivores can shape species composition, morphology, behaviour and size structure of their prey, as demonstrated when guppies of Trinidad were exposed to the piscivores *Rivulus harti, Crenicichla alta* and *Hoplias malabaricus* (Seghers 1974; Endler 1980). Predators can influence patterns of habitat use, activity budgets and foraging patterns of the prey through intimidation (Fraser & Cerri 1982; Gilliam & Fraser 1987).

Fish diseases and parasites

The effects of fish diseases on fish assemblage structure have been underestimated. Parasites can have more varied direct and indirect impacts on communities than predators (Holmes 1982; Holmes & Price 1986). Although spectacular epizootics have killed millions of fishes (Rohovec & Fryer 1979; Wurtsbaugh & Tapia 1988), the impacts of disease can be subtle. Patterns of abundance and distribution may appear to be governed by random physical processes when in fact a subset of species may be susceptible to a microparasite while others are not. As stated by Price *et al* (1986, p. 499) '... many interactions that appear to be between two species actually involve a third. Unless this is recognized, models will either fail to match field reality or will match it spuriously.'

Microparasitic infection has probably been a major selective force on fish assemblages of the Pacific Northwest (Li *et al* 1987). Subpopulations of steelhead trout, coho salmon and chinook salmon differ in genetic resistance to various microparasites (Suzumoto *et al* 1977; Winter *et al* 1979; Wade 1986). The degree of immunity reflects the historical distributions in space and time of the microparasite and host (Buchanan *et al* 1983). Microparasites, such as *Ceratomyxa shasta* and *Flexibacter columnaris*, can consequently act as zoogeographical barriers to specific populations. For instance, the epizootic of *F. columnaris* in the Columbia River during 1973 was responsible for the loss of approximately 80% of the spawning run of Snake River steelhead in eastern Oregon and Idaho (Becker & Fugihara 1978). Steelhead stocks that ascend early in the year in typically cold water, when columnaris disease is neither prevalent nor virulent, were especially vulnerable. Columnaris disease became rampant when the Columbia River warmed early in the drought year of 1973, but not all size classes and species of fish were susceptible. In general, larger cypriniform fishes, a group which is tolerant of higher temperatures, were more immune (Becker & Fugihara 1978).

Monitoring changes in fish assemblages

It is well known that sufficient stress or 'press disturbance' on a community will reduce the number of species (species richness). We know that the number of species is related to the size of the basin (Welcomme 1985) and that this relationship changes among regions (Fausch *et al* 1984; Moyle & Herbold 1987) and continents (Welcomme 1985; Moyle & Herbold 1987). To the degree that species richness is predictable in natural systems, departures from the norm at a comparable scale can serve as an indicator of stress upon stream communities. However, even if non-native species are excluded and natural conditions are well defined, relative species richness is a crude measure of stress because it does not reflect different stress tolerances among species, is prone to measurement error because some species are always rare (Sheldon 1987) and suffers from measurement bias when gear selectivity cannot be corrected (Bayley *et al* 1989).

The Index of Biological Integrity (IBI) is an attempt to provide a more robust measure of stress or stream degradation than relative richness, diversity or methods using an indicator species (Karr 1981; Fausch *et al* 1984, 1990; Karr *et al* 1986, 1987), and has received wide publicity and use in the USA. The IBI is a sum of 12 metrics, each scored as odd numbers up to 5, that are estimated to represent the degree to which a particular stream locality is degraded from its natural state. The metrics, which are applied to fish that are not young-of-the-year, include species richness, proportions of trophic groups, total abundance, proportions of stress-tolerant species specific to the region concerned, abundance of all species, extent of hybridization and parasitism. Subsets of these metrics have varying degrees of correlation. Like relative species richness or diversity, the IBI is only as good as the information on the natural fish community for the reach being assessed; it is sensitive (to an unknown degree) to sampling bias; it does not relate directly to known ecological relationships among the species and with their environment; and it does not identify types of stress or disturbance. We are not suggesting that any single number could represent all these factors (multivariate approaches are called for) or that the pertinent ecological information is available (it usually is not). However, much information already exists that quantifies species/habitat relationships (sections 6.2 and 6.3). Although the IBI does not link species composition to physical characteristics, it could be modified to detect disruptions of longitudinal zonation patterns (Fausch *et al* 1984). If physical factors influence fish species distributions, then one should be able to infer 'habitat integrity' from species composition. Conversely, the IBI was insensitive to the environmental impacts of massive military exercises on a small prairie watershed (Bramblett & Faush 1991), where hardy fishes are naturally subjected to flash floods and droughts, and whose presence and structure depend on their rate of colonization rather than habitat changes.

The IBI has definite limits and its assumptions should be carefully examined, as there are other more quantitative alternatives (Fausch *et al* 1990; Bramblett & Fausch 1991). However, the index is a laudable attempt to provide a yardstick for stream managers who have a good knowledge of local natural history (Karr *et al* 1986), but who lack cause-specific ecological information to identify the mechanism(s) and source(s) of degradation. Therefore, in common with any composite index composed of arbitrary elements, the IBI should be regarded as a useful management tool for preliminary diagnosis rather than as an index of ecological or heuristic value.

6.6 ANTHROPOGENIC DISTURBANCES, MITIGATION, AND RESTORATION

We have presented evidence that different species and life stages of fishes are adapted to particular components of the riverscape, and are sensitive to the spatial scales and dynamics in each system. Disruption of riverscape processes causes great damage. River fisheries management operates on the river channel, yet economic and social activities outside of the lotic habitat have profound cumulative impacts on fishes.

The effects of and recovery from disturbance are scale dependent with respect to fishes (see Fig. 6.1(b)) and environmental units (see Fig. 6.1(a)). Microhabitat units are more sensitive to anthropogenic disturbances but are also the quickest to recover. In contrast, damages to the river from the surrounding landscape may be measured in hundreds of years, if one takes into account geomorphic processes and key structural organisms. For example, it is estimated that at least 500 years are required to restore drainages of the Pacific Northwest where ancient forests of Douglas firs have been clearcut (Li *et al* 1987).

Major impacts on regional scales have occurred in the plains streams of Kansas and Missouri (Cross & Moss 1987; Pflieger & Grace 1987). The combination of dams, revetments and jetties created deeper, clearer, more channelized streams of high velocity. The native fishes are morphologically and behaviourally adapted to shallow, silty, highly braided rivers. This native fauna is disappearing and introduced species are becoming dominant. Further west, two-thirds of the fish species in the Colorado basin are introduced and 17 of 54 native species are threatened, endangered or extinct (Carlson & Muth 1989) owing to similar effects of mainstream dams. Dams impede the distribution of material and energy transfer through the drainage basin, obstruct spawning migrations (Bonetto *et al* 1989; Barthem *et al* 1991), inhibit reproduction by altering thermal regimes (Baxter 1977), alter faunal structure through habitat change (Bain *et al* 1988), are centers for disease transmission (Becker & Fugihara 1978) and create unstable fish assemblages (Gelwick & Matthews 1990). Impoundments in tropical river systems, where extant floodplains have demonstrated high fish yields (Welcomme 1985; Bayley 1991), can devastate fish production, in particular downstream where the flood pulse is affected (Bonetto *et al* 1989; Junk *et al* 1989).

The temptation is to mitigate for damage using

complicated, expensive, engineering solutions because these are politically attractive; but this has unpleasant consequences that are expensive in the long run. Gore (1985) correctly stated that river restoration should be equated to 'recovery enhancement' because it should accelerate the natural process of recovery. In general, the most cost-effective mitigation procedures mimic natural processes. Such procedures will also enable relatively painless restoration in the long term.

Mitigation to restore native fishes in the intermountain western USA will require burning and logging stands of newly invaded juniper (J. Sedell, personal communication, Forest Sciences Laboratory, Oregon State University). Native bunch grasses have virtually disappeared in just over a century. They did not coevolve with large grazing ungulates and declined when livestock was introduced. Exotic plants soon became established and artificial suppression of the natural fire cycle led to poorer water infiltration, greater siltation and runoff, and the massive invasion of junipers. Each mature juniper evapotranspirates 15–35 gallons (68–160 litres) of water daily, which resulted in many streams becoming ephemeral. Restoration of natural fire cycles will provide short-term mitigation, but full restoration of the watershed will require decades.

Habitat restoration has been very successful at reach, channel unit and microhabitat levels in the central USA (White & Brynildson 1967; Hunt 1976). However, the success of enhancement projects from Alaska and the Pacific Northwest has, with few exceptions (House & Boehne 1985, 1986), not been determined (Hall & Baker 1982). Blatant failures occurred when mitigation was attempted without a regional context; what succeeds in a low gradient stream in Wisconsin will not necessarily apply to high gradient streams along the Pacific coast. Recovery times of successful projects are only in terms of years at the reach scale. Hunt (1976) found maximum response to improvements of channel structures in 5 years.

Bayley (1991) has argued that the best short-term approach to restore the function of river-floodplains impacted by navigation impoundments is to control discharge to mimic the pattern of the natural hydrograph, providing that artificial levees are also removed. However, any extensive restoration of large rivers will require watershed restoration which reduces the drainage rate, and dam removal which will restore the natural downstream transport of sediment and permit the natural flood pulse and associated vegetation to return. This is not a restoration that is just in the interest of fish and other native biota, but will provide the most cost-effective flood control (see Belt 1977).

Mitigation using exotic fishes is a great temptation. However, it is one of the poorest tools because exotic fishes can cause more problems than solutions, and the treatment is irreversible. Exotic species have contributed to the defaunation of many areas, especially areas that are disturbed (e.g. Moyle *et al* 1986; Carlson & Muth 1989; Moyle & Williams 1990). Predation and competition have been implicated in many instances (Moyle *et al* 1986), but the role of fish diseases can be insidious. Transfers of exotic fish diseases are occurring at unprecedented rates (Ganzhorn *et al* in press) and coadapted (micro) parasite–host populations are more stable than novel ones. As in germ warfare, parasites carried by resistant hosts can enhance their invasion by infecting native competitors (Holmes 1982; Freeland 1983).

6.7 CONCLUSIONS AND FUTURE RESEARCH DIRECTIONS

We argue that in most parts of the world it is now more important to spend resources on the restoration and conservation of riverine environments than to promote and maximize exploitation of species of current interest. Exceptions are relatively pristine river-floodplains in some developing countries, where sustained food fisheries are a human necessity and are advisable to provide a defence against destructive river-basin development schemes (Bayley & Petrere 1989).

Given an environment to which they are adapted, riverine fish are remarkably persistent because of the intensity and frequency of natural disturbances in evolutionary time, such as extreme floods and droughts. Therefore, in the long term the cost of correcting mistakes in management of the exploitation process is negligible compared to that of correcting or compensating for the effects of permanent changes to the environment, such as dams, water extraction, floodplain habitat removal, chronic pollution or introduction of exotics.

Research on riverine fish is controlled to a large extent by cost which is governed by sampling limitations (Fig. 6.5). There is a high variance in population size (see Fig. 6.4) which may be accounted for only partly by measurement variance, so increased sampling intensity is not necessarily a solution. As is typical with all ecosystems, the population level is the noisiest, with many measures of individual organisms and communities increasing in stability on a comparable temporal scale (Fig. 6.5). Therefore, how can we obtain the information to understand the environmental requirements to which the fish have adapted, estimate an acceptable management protocol, and apply such findings to problems of conservation, restoration, or exploitation in other systems?

We believe that a co-ordinated approach that combines extensive, protracted, empirical data collection with experiments and modelling is required. The co-operation of fishery managers is important so that some experimental control over the exploitation process is achieved. However, fisheries managers have only limited control over most aspects of the environment, and the co-operation of agencies and the private sector that are associated with the hydrology, siltation, chemical loads and in-stream structures is essential. This degree of integration is rare because of the different philosophies and individuals involved, which results in educational and institutional divisions.

Long time-series are advocated as a partial solution to improved empirical data needs. Such series are important in assessing responses by fish to unpredictable disturbances or long-term effects such as global warming. However, when seeking 'natural experiments' from the data, confounding

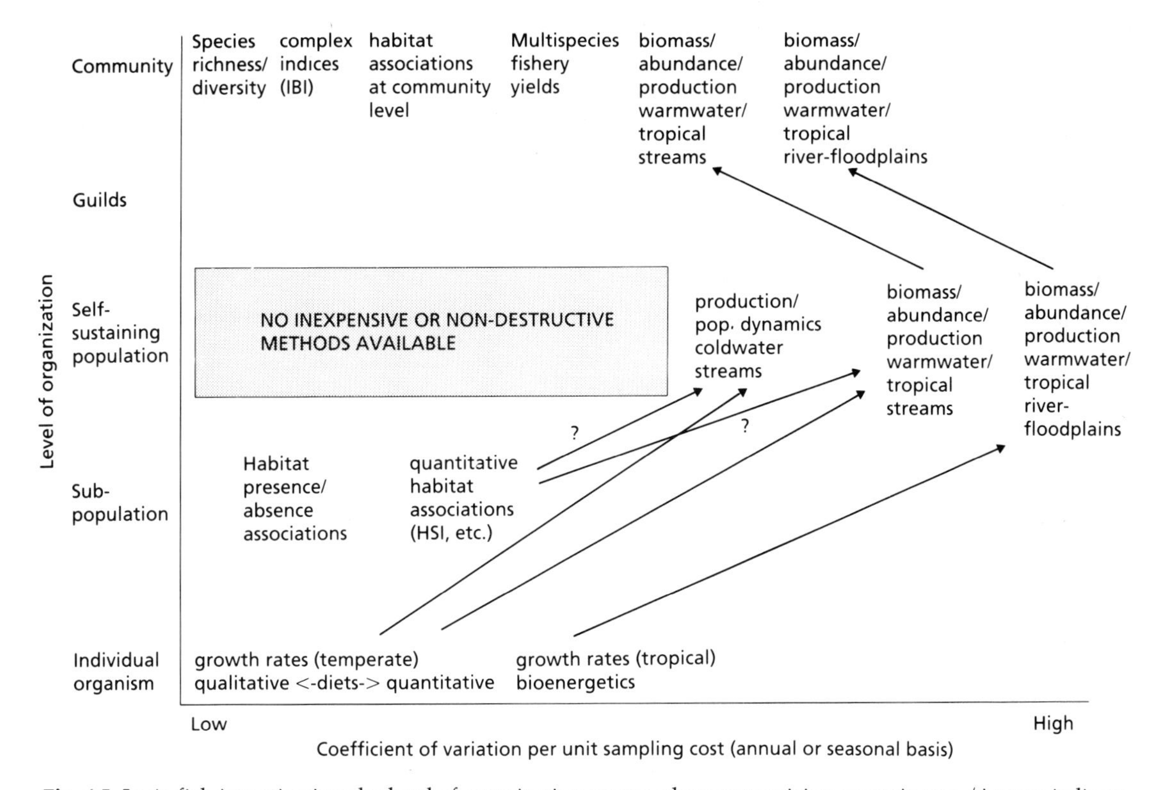

Fig. 6.5 Lotic fish investigations by level of organization versus adequate precision per unit cost. (Arrows indicate which higher level studies typically utilize information from a lower level.)

among time-correlated variables can produce too many alternative explanations. Concurrent time-series among 'ecologically comparable' systems (section 6.2) increases our ability to unravel time-correlated effects, such as when different climatic events affect fish in subsets of the systems being monitored. By ecologically comparable, we mean that the systems have comparable habitats and communities and have environments that are not drastically altered from important pristine conditions, particularly with respect to the hydrological regime. Comparisons among stressed systems are an alternative and are sometimes unavoidable, but the variety of stresses among systems may result in too many variables to control. In some river systems of developed countries this line of investigation will not be useful until some degree of restoration is attained.

Concurrent time-series of sufficient observation frequency and density will facilitate scale-dependent analyses of fish and habitat interactions that will permit more reliable inferences in other systems. One could obtain time-series by spending money on population estimates for a limited number of systems, or by spending a similar amount on community/guild/habitat/organism measures for a greater number of systems. In the majority of situations we advocate the latter approach, provided that every attempt is made to estimate fishery yield and effort. Such analyses should at least indicate the most economical paths towards the goal of restoration in many rivers and provide estimates of yield trends across sets of rivers.

Although extensive empirical data and natural experiments will narrow down the number of possible factors needed to improve prediction and understanding, the generality of results will be limited to those systems or very similar ones. Determination of dominant sets of mechanisms through experimentation and modelling within the context of a sound empirical database will broaden the application to more systems and reduce the future costs of empirical data collection. Reductionist approaches in the absence of preliminary analyses of empirical field data can result in the proposal of biologically plausible but ecologically insignificant mechanisms.

The cost and difficulty of analysing extensive field and experimental data have been reduced by improved computational facilities and methods. However, the cost of obtaining data has at least kept pace with inflation, and inadequate data have often been collected and published. If we are going to overcome the challenges of managing riverine fishes and their environment, researchers must strive for better quality data by investing more time to understand the sampling properties of their methods and to obtain the co-operation of managers and even the public to provide more extensive data. The ideal of cost-effective fishery management in rivers will not be achieved by a dependence on, or subjugation to, engineering approaches or byproducts, such as the development of unnatural hydrological regimes and attempts to utilize artificial habitats with hatchery-reared or exotic fishes.

REFERENCES

Alexander GR, Hansen EA. (1988) *Decline and recovery of a brook trout stream following an experimental addition of sand sediment*. Fisheries Research Report, July 11 1943. Michigan Department of Natural Resources. [6.4]

Allen KR. (1969) Distinctive aspects of the ecology of stream fishes: a review. *Journal of the Fisheries Research Board of Canada* **26**: 1429–38. [6.2]

Allen TFH, Starr TB. (1982) *Hierarchy: Perspectives for Ecological Complexity*. University of Chicago Press, Chicago. [6.2]

Allendorf FW, Phelps SR. (1980) Loss of genetic variation in a hatchery stock of cutthroat trout. *Transactions of the American Fisheries Society* **109**: 537–43. [6.3]

Alley DW Jr. (1977) *The energetic significance of microhabitat selection by fishes in the foothill Sierra stream*. M.S. Dissertation, University of California, Davis, California. [6.3]

Angermeier PL. (1987) Spatiotemporal variation in habitat selection by fishes in small Illinois streams. In: Matthews WJ, Heins DC (eds) *Community and Evolutionary Ecology of North American Stream Fishes*, pp 52–60. University of Oklahoma Press, Norman, Oklahoma. [6.2]

Angermeier PL, Karr JR. (1983) Fish communities along environmental gradients in a system of tropical streams. *Environmental Biology of Fishes* **9**(2): 117–35. [6.2]

Bain MB, Finn JT, Booke HE. (1988) Streamflow regulation and fish community structure. *Ecology* **69**: 382–92. [6.6]

Baker JA, Ross ST. (1981) Spatial and temporal resource utilization by southeastern cyprinids. *Copeia* **1981**: 178–89. [6.5]

Balon EK. (1975) Reproductive guilds of fishes: a proposal and definition. *Journal of the Fisheries Research Board of Canada* **32**(6): 821–64. [6.3]

Baltz DM, Moyle PB. (1984) Segregation by species and size classes of rainbow trout, *Salmo gairdneri*, and Sacramento sucker, *Catostomus occidentalis*, in three California streams. *Environmental Biology of Fishes* **10**: 101–10. [6.5]

Baltz DM, Moyle PB, Knight NJ. (1982) Competitive interactions between benthic stream fishes, riffle sculpin, *Cottus gulosus*, and speckled dace, *Rhinichthys osculus*. *Canadian Journal of Fisheries and Aquatic Sciences* **39**: 1502–11. [6.2]

Baltz DM, Vondracek B, Brown LR, Moyle PB. (1987) Influence of temperature on microhabitat choice by fishes in a California stream. *Transactions of the American Fisheries Society* **116**: 12–20. [6.5]

Banse K, Mosher S. (1980) Adult body mass and annual production biomass relationships of field populations. *Ecological Monographs* **50**: 355–79. [6.3]

Barmuta LA, Lake PS. (1982) On the value of the river continuum concept. *New Zealand Journal of Marine and Freshwater Research* **16**: 227–31. [6.2]

Barthem RB, Ribeiro MLB, Petrere M Jr. (1991) Life strategies of some long-distance migratory catfish in relation to hydroelectric dams in the Amazon Basin. *Biological Conservation* **55**: 339–45. [6.6]

Baxter RM. (1977) Environmental effects of dams and impoundments. *Annual Review of Ecology and Systematics* **8**: 255–84. [6.6]

Bayley PB. (1973) Studies on the migratory characin, *Prochilodus platensis* Holmberg 1889 (Pisces, Characoidei) in the Rio Pilcomayo, South America. *Journal of Fish Biology* **5**: 25–40. [6.3]

Bayley PB. (1983) *Central Amazon fish populations: biomass, production and some dynamic characteristics*. PhD Dissertation, Dalhousie University, Halifax, Nova Scotia. [6.3]

Bayley PB. (1985) Sampling problems in freshwater fisheries. In: O'Hara K, Aprahamian C, Leah RT (eds) *Proceedings of the Fourth British Freshwater Fisheries Conference, 1–3 April*, pp 3–11. University of Liverpool, Liverpool. [6.4]

Bayley PB. (1988a) Accounting for effort when comparing tropical fisheries in lakes, river-floodplains, and lagoons. *Limnology and Oceanography* **33**: 963–72. [6.4]

Bayley PB. (1988b) Factors affecting growth rates of young tropical fishes: seasonality and density-dependence. *Environmental Biology of Fishes* **21**: 127–42. [6.3]

Bayley PB. (1989) Aquatic environments in the Amazon Basin, with an analysis of carbon sources, fish production, and yield. *Special Publication of the Canadian Journal of Fisheries and Aquatic Sciences* **106**: 399–408. [6.4]

Bayley PB. (1991) The flood pulse advantage and the restoration of river-floodplain systems. *Regulated Rivers: Research and Management* **6**: 75–86. [6.4, 6.6]

Bayley PB, Petrere M Jr. (1989) Amazon fisheries: assessment methods, current status, and management options. *Special Publication of the Canadian Journal of Fisheries and Aquatic Sciences* **106**: 385–98. [6.7]

Bayley PB, Larimore RW, Dowling DC. (1989) Electric seine as a fish-sampling gear in streams. *Transactions of the American Fisheries Society* **118**: 447–53.

Beamish FWH. (1964) Respiration of fishes with special emphasis on standard oxygen consumption. II. Influence of weight and temperature on respiration of several species. *Canadian Journal of Zoology* **42**: 178–88. [6.3]

Beamish FWH, Mookherjii PS. (1964) Respiration of fishes with special emphasis on standard oxygen consumption. I. Influence of weight and temperature on respiration of goldfish, *Carassius auratus* L. *Canadian Journal of Zoology* **42**: 161–75. [6.3]

Becker CD, Fugihara MP. (1978) *The bacterial pathogen,* Flexibacter columnaris, *and its epizootiology among Columbia River fish: a review and synthesis*. American Fisheries Society Monographs No. 2, Bethesda, Maryland. [6.5, 6.6]

Belt CB Jr. (1977) The 1973 flood and man's constriction of the Mississippi River. *Science* **189**: 681–4. [6.6]

Beverton RJH, Holt SJ. (1957) *On the Dynamics of Exploited Fish Populations*. Fisheries Investigations Series 2, Vol. 19. UK Ministry of Agriculture and Fisheries, London. [6.4]

Bisson PA, Nielsen JL, Palmason RA, Grove LE. (1982) A system of naming habitat types in small streams, with examples of habitat utilization by salmonids during low stream flow. In: Armandtrout NB (ed.) *Acquisition and Utilization of Aquatic Habitat Information*, pp 62–73. Western Division of the American Fisheries Society, Portland, Oregon. [6.2]

Bisson PA, Sullivan K, Nielsen JL. (1988) Channel hydraulics, habitat use and body form of juvenile coho salmon, steelhead, and cutthroat trout in streams. *Transactions of the American Fisheries Society* **117**: 262–73. [6.2]

Bonetto AA, Wais JR, Castello HP. (1989) The increasing damming of the Paraná Basin and its effects on the lower reaches. *Regulated Rivers: Research and Management* **4**: 333–46. [6.6]

Boudreau PR, Dickie LM. (1989) Biological model of fisheries production based on physiological and ecological scalings of body size. *Canadian Journal of Fisheries and Aquatic Sciences* **46**: 614–23. [6.3]

Bovee KD. (1982) *A guide to stream habitat analysis*

using the Instream Flow Incremental Methodology. Instream Flow Information Paper No. 12 FWS/OBS-82–26. US Fish and Wildlife Service Biological Services Program, Cooperative Instream Flow Service Group. Fort Collins, Co. [6.2]

Bowen SH, Bonetto AA, Ahgren MO. (1984) Microorganisms and detritus in the diet of a typical neotropical riverine detritivore, *Prochilodus platensis* (Piscies: Prochilodontidae). *Limnology and Oceanography* **29**(5): 1120–2. [6.3]

Bowlby JN, Roff JC. (1986) Trophic structure in southern Ontario streams. *Ecology* **67**: 1670–9. [6.5]

Bramblett RG, Fausch KD. (1991) Variable fish communities and the Index of Biological Integrity in a western Great Plains river. *Transactions of the American Fisheries Society* **120**: 752–69. [6.5]

Brett JR. (1965) The relation of size to rate of oxygen consumption and sustained swimming speed of sockeye salmon (*Oncorhynchus nerka*). *Journal of the Fisheries Research Board of Canada* **22**: 1491–501. [6.3]

Buchanan DV, Sanders JE, Zinn JL, Fryer JL. (1983) Relative susceptibility of four strains of summer steelhead to infection by *Ceratomyxa shasta. Transactions of the American Fisheries Society* **112**: 541–3. [6.3]

Calhoun SW, Zimmerman EG, Beitinger TL. (1982) Stream regulation alters acute temperature preferenda of red shiners, *Notropis lutrensis. Canadian Journal of Fisheries and Aquatic Sciences* **39**: 360–3. [6.3]

Calow P. (1985) Adaptive aspects of energy allocation. In: Tytler P, Calow P (eds) *Fish Energetics: New Perspectives*, pp 13–31. The Johns Hopkins University Press, Baltimore, Maryland. [6.3]

Calow P. (1994) Energy budgets. In: Calow P, Petts GE (eds) *The Rivers Handbook*, vol. 1, pp 370–78. Blackwell Scientific Publications, Oxford. [6.4]

Carlson CA, Muth RT. (1989) The Colorado River: lifeline of the American Southwest. *Special Publication of the Canadian Journal of Fisheries and Aquatic Sciences* **106**: 220–39. [6.6]

Castleberry DT, Cech JJ Jr. (1986) Physiological responses of a native and introduced desert fish to environmental stressors. *Ecology* **67**: 912–18. [6.5]

Cech JJ Jr, Mitchell SJ, Massingill MJ. (1979) Respiratory adaptations of Sacramento blackfish, *Orthodon microlepidotus* (Ayers), for hypoxia. *Comparative Biochemistry and Physiology* **63A**: 411–15. [6.3]

Clausen RG. (1936) Oxygen consumption in fresh water fishes. *Ecology* **17**: 216–26. [6.3]

Collie JS, Peterman RM, Walters CJ. (1990) Experimental harvest policies for a mixed-stock fishery: Fraser River sockeye salmon, *Oncorhynchus nerka. Canadian Journal of Fisheries and Aquatic Sciences* **47**: 145–55. [6.4]

Colwell RK. (1974) Predictability, constancy, and contingency of periodic phenomena. *Ecology* **55**: 1148–53. [6.5]

Conder AL, Annear TC. (1987) Test of weighted usable area estimates derived from a PHABSIM model for instream flow studies on trout streams. *North American Journal of Fisheries Management* **7**: 339–50. [6.2]

Cross FB, Moss RE. (1987) Historic changes in fish communities and aquatic habitats in plains streams of Kansas. In: Matthews WJ, Heins DC (eds) *Community and Evolutionary Ecology of North American Stream Fishes*, pp 155–65. University of Oklahoma Press, Norman, Oklahoma. [6.6]

Cunjak RA, Green JM. (1983) Habitat utilization by brook char (*Salvelinus fontinalis*) and rainbow trout (*Salmo gairdneri*) in Newfoundland streams. *Canadian Journal of Zoology* **61**: 1214–19. [6.5]

Cunjak RA, Green JM. (1984) Species dominance by brook trout and rainbow trout in a simulated stream environment. *Transactions of the American Fisheries Society* **113**: 737–43. [6.5]

Cunjak RA, Power G. (1987) The feeding and energetics of stream-resident trout in winter. *Journal of Fish Biology* **31**: 493–511.

Cunjak RA, Curry RA, Power G. (1987) Seasonal energy budget of brook trout in streams: implications of a possible deficit in early winter. *Transactions of the American Fisheries Society* **116**: 817–28. [6.3]

Currens KP, Schreck CB, Li HW. (1990) Allozyme and morphological divergence of rainbow trout (*Oncorhynchus mykiss*) above and below waterfalls in the Deschutes River, Oregon. *Copeia* **1990**: 730–46. [6.3]

de Godoy MP. (1954) Locais de desovas de peixes num trecho do Rio Mogi Guaçu, Estado de São Paulo, Brasil. *Revista Brasileira de Biologia* **14**(4): 375–96. [6.3]

DeAngelis DL, Godbout L, Shuter BJ. (1991) An individual-based approach to predicting density-dependent dynamics in smallmouth bass populations. *Ecological Modelling* **57**: 91–116. [6.4]

Dennis B, Brown BE, Stage AR, Burkhart HE, Clark S. (1985) Problems of modeling growth and yield of renewable resources. *The American Statistician* **39**(4): 374–83. [6.4]

Dwyer WP, Kramer RH (1975) The influence of temperature on scope for activity in cutthroat trout, *Salmo clarki. Transactions of the American Fisheries Society* **104**: 552–4. [6.3]

Echelle AA, Schnell GD. (1976) Factor analysis of species association among fishes of the Kiamichi River, Oklahoma. *Transactions of the American Fisheries Society* **105**: 17–31. [6.2]

Egglishaw HJ, Shackley PE. (1977) Growth, survival and production of juvenile salmon and trout in a Scottish stream, 1966–75. *Journal of Fish Biology* **11**: 647–72. [6.4]

Elliott JM. (1979) Energetics of freshwater teleosts. *Symposia of the Zoological Society of London* **44**: 29–61. [6.3]

Elliott JM. (1984) Numerical changes and population regulation in young migratory trout *Salmo trutta* in a Lake District stream, 1966–83. *Journal of Animal Ecology* **53**: 327–50. [6.4]

Elliott JM. (1985) Population regulation for different life-stages of migratory trout *Salmo trutta* in a Lake District stream, 1966–83. *Journal of Animal Ecology* **54**: 617–38. [6.4]

Elwood JW, Newbold JD, O'Neill RV, van Winkle W. (1983) Resource spiraling: an operational paradigm for analyzing lotic ecosystems. In: Fontaine TD, Bartell SM (eds) *Dynamics of Lotic Ecosystems*, pp 3–28. Ann Arbor Science, Ann Arbor, Michigan. [6.2]

Endler J. (1980) Natural selection on color patterns in *Poecilia reticulata. Evolution* **34**: 76–91. [6.3, 6.5]

Facey DE, Grossman GD. (1990) The metabolic cost of maintaining position for four North American stream fishes. *Physiological Zoology* **63**: 757–76. [6.3]

Fausch KD. (1988) Tests of competition between native and introduced salmonids in streams: what have we learned? *Canadian Journal of Fisheries and Aquatic Sciences* **45**: 2238–46. [6.5]

Fausch KD, White RJ. (1981) Competition between brook trout (*Salvelinus fontinalis*) and brown trout (*Salmo trutta*) for positions in a Michigan stream. *Canadian Journal of Fisheries and Aquatic Sciences* **38**: 1220–7. [6.5]

Fausch KD, White RJ. (1986) Competition among juveniles of coho salmon, brook trout, and brown trout in a laboratory stream, and implications for Great Lakes tributaries. *Transactions of the American Fisheries Society* **115**: 363–81. [6.5]

Fausch KD, Karr JR, Yant PR. (1984) Regional application of an index of biotic integrity based on stream fish communities. *Transactions of the American Fisheries Society* **113**: 39–55. [6.5]

Fausch KD, Lyons J, Karr JR, Angermeier PL. (1990) Fish communities as indicators of environmental degradation. *American Fisheries Society Symposium* **8**: 123–44. [6.5]

Felley JD, Felley SM. (1987) Relationship between habitat selection by individuals of a species and patterns of habitat segregation among species: fishes of the Calcasieu drainage. In: Matthews WJ, Heins DC. (eds) *Community and Evolutionary Ecology of North American Stream Fishes*, pp 61–8. University of Oklahoma Press, Norman, Oklahoma. [6.2]

Fisher SG. (1983) *Succession in Streams*. Plenum Press, New York. [6.5]

Fraser DF, Cerri RD. (1982) Experimental evaluation of predator–prey relationships in a patchy environment: consequences for habitat use patterns in minnows. *Ecology* **63**: 307–13. [6.5]

Freeland WJ. (1983) Parasites and the coexistence of animal host species. *American Naturalist* **121**: 223–36. [6.6]

Freeman MC, Crawford M, Barrett J, Facey DE, Flood M, Stouder D, Grossman GD. (1988) Fish assemblage stability in a southern Appalachian stream. *Canadian Journal of Fisheries and Aquatic Sciences* **45**: 1949–58. [6.5]

Frissell CA, Liss WJ, Warren CE, Hurley MD. (1986) A hierarchical framework for stream habitat classification: viewing streams in a watershed context. *Environmental Management* **10**: 199–214. [6.2]

Ganzhorn J, Rohovec JS, Fryer JL. (in press) Dissemination of microbial pathogens through introductions and transfers of finfish. In: Rosenfield A, Mann R (eds) *Dispersal of Living Organisms into Aquatic Ecosystems*, Maryland Sea Grant Program, College Park, Maryland. [6.6]

Gard R, Flittner GA. (1974) Distribution and abundance of fishes in Sagehen Creek, California. *Journal of Wildlife Management* **38**: 347–58.

Gee JH. (1972) Adaptive variation in swimbladder length and volume in dace, genus *Rhinichthys. Journal of the Fisheries Research Board of Canada* **29**: 119–27. [6.3]

Gelwick FP, Matthews WJ. (1990) Temporal and spatial patterns in littoral-zone fish assemblages of a reservoir (Lake Texoma, Oklahoma-Texas, U.S.A.). *Environmental Biology of Fishes* **27**: 107–20. [6.6]

Gilliam JF, Fraser DF. (1987) Habitat selection under predation hazard: test of a model with foraging minnows. *Ecology* **68**: 1856–62. [6.5]

Glebe BD, Leggett WC. (1981a) Latitudinal differences in energy allocation and use during the freshwater migrations of American shad (*Alosa sapidissima*) and their life history consequences. *Canadian Journal of Fisheries and Aquatic Sciences* **38**: 806–20. [6.3]

Glebe BD, Leggett WC. (1981b) Temporal, intra-population differences in energy allocation and use by American Shad (*Alosa sapidissima*) during the spawning migration. *Canadian Journal of Fisheries and Aquatic Sciences* **38**: 795–805. [6.3]

Gore JA. (1985) Introduction. In: Gore JA (ed.) *The Restoration of Rivers and Streams*, pp vii–xii. Butterworth, Boston. [6.6]

Gorman OT. (1987) Habitat segregation in an assemblage of minnows in an Ozark stream. In: Matthews WJ, Heins DC (eds) *Community and Evolutionary Ecology of North American Stream Fishes*, pp 33–41. University of Oklahoma Press, Norman, Oklahoma. [6.5]

Gorman OT. (1988) The dynamics of habitat use in a guild of Ozark minnows. *Ecological Monographs* **58**: 1–18. [6.2]

Gorman OT, Karr JR. (1978) Habitat structure and stream fish communities. *Ecology* **59**: 507–15. [6.2, 6.3]

Grossman GD, Moyle PB, Whitaker JO Jr. (1982) Stochasticity in structural and functional characteristics of an Indiana stream fish assemblage: a test of community theory *American Naturalist* **120**: 423–54. [6.5]

Grossman GD, Freeman MC, Moyle PB, Whitaker JO Jr. (1985) Stochasticity and assemblage organization in an Indiana stream fish assemblage. *American Naturalist* **126**: 275–85. [6.5]

Gulland JA. (1969) *Manual of methods for fish stock assessment: Part 1. Fish population analysis.* FAO Manuals in Fisheries Science, No. 4. Food and Agriculture Organization, Rome. [6.4]

Hall CAS. (1972) Migration and metabolism in a temperate stream ecosystem. *Ecology* **53**: 586–604. [6.2, 6.3]

Hall JD, Baker CD. (1982) Rehabilitating and enhancing stream habitat: 1. Review and evaluation. In: Meehan WR (ed.) *Influences of forest and rangeland management on anadromous fish habitat in western North America.* USDAS Forest Service, General Technical Report PNW-138, Portland, Oregon. p. 29. [6.6]

Hall JD, Knight NJ. (1981) *Natural variation in abundance of salmonid populations in streams and its implications for design of impact studies.* EPA-600/S3-81–021. July 1981. US Environmental Protection Agency report. Corvallis, Oregon. [6.4]

Hatch KM. (1990) *Phenotypic comparison of thirty-eight steelhead* (Oncorhynchus mykiss) *populations from coastal Oregon.* M.S. Dissertation, Oregon State University, Corvallis, Oregon. [6.3]

Herbold B. (1984) Structure of an Indiana stream fish association: choosing an appropriate model. *American Naturalist* **124**: 561–72. [6.5]

Hicks BJ. (1990) *The Influence of geology and timber harvest on channel morphology and salmonid populations in Oregon Coast Range streams.* PhD Dissertation, Oregon State University, Corvallis, Oregon. [6.2]

Hicks BJ, Hall JD, Bisson PA, Sedell JR. (1991) Response of salmonid populations to habitat changes caused by timber harvest. In: Meehan WR (ed.) *The Influence of Forest and Rangeland Management on Salmonids and their Habitat,* pp 484–518. Special Publication of the American Fisheries Society, Bethesda, Maryland. [6.2]

Hokanson KEF. (1977) Temperature requirements of some percids and adaptations to the seasonal temperature cycle. *Journal of the Fisheries Board of Canada* **34**: 1524–50. [6.3]

Holmes JC. (1982) Impact of infectious disease agents on the population growth and geographical distribution of animals. In: Anderson RM, May RM (eds) *Population Biology of Infectious Diseases,* pp 37–51. Springer-Verlag, Berlin. [6.5, 6.6]

Holmes JC, Price PW. (1986) Communities of parasites. In: Kikkikawa J, Anderson DJ (eds) *Community Ecology: Pattern and Process,* pp 187–213. Blackwell Scientific Publications, Melbourne. [6.5]

Horwitz RJ. (1978) Temporal variability patterns and the distributional patterns of stream fishes. *Ecological Monographs* **48**: 307–321. [6.2, 6.3]

House RA, Boehne PL. (1985) Evaluation of instream structures for salmonid spawning and rearing in a coastal Oregon stream. *North American Journal of Fisheries Management* **5**: 283–95. [6.6]

House RA, Boehne PL. (1986) *Effects of instream structure on salmonid habitat and populations in Tobe Creek, Oregon. North American Journal of Fisheries Management* **6**: 38–46. [6.6]

Huet M. (1947) Aperçu des relations entre la pente et les populations piscicoles des eaux courantes. *Schweizerische Zeitschrift für Hydrologie* **11**: 333–51. [6.2]

Huet M. (1959) Profiles and biology of Western European streams as related to fish management. *Transactions of the American Fisheries Society* **88**: 155–63. [6.2]

Hughes RM, Rexstad E, Bond CE. (1987) The relationship of aquatic ecoregions, river basins and physiographic provinces to the ichthyogeographic regions of Oregon. *Copeia* **1987**: 423–52. [6.2]

Hughes RM, Whittier TR, Rohm CM, Larsen DP. (1990) A regional framework for establishing recovery criteria. *Environmental Management* **14**: 673–83. [6.2]

Hulett PL. (1991) *Patterns of genetic inheritance and variation through ontogeny for hatchery and wild stocks of chinook salmon.* M.S. Dissertation, Oregon State University, Corvallis, Oregon. [6.3]

Hunt RL. (1976) A long-term evaluation of trout habitat development and its relation to improving management-related research. *Transactions of the American Fisheries Society* **105**: 361–4. [6.6]

Hutchinson GE. (1961) Paradox of the plankton. *American Naturalist* **95**: 137–46. [6.5]

Johnson DH. (1980) The comparison of usage and availability measurements for evaluating resource preference. *Ecology* **61**: 65–71. [6.2]

Junk WJ. (1985) Temporary fat storage, an adaptation of some fish species to the waterlevel fluctuations and related environmental changes of the Amazon system. *Amazoniana* **9**: 315–51. [6.3]

Junk WJ, Bayley PB, Sparks RE. (1989) The flood pulse concept in river-floodplain systems. *Special publication of the Canadian Journal of Fisheries and Aquatic Sciences* **106**: 110–27. [6.2, 6.3, 6.5, 6.6]

Karr JR. (1981) Assessment of biotic integrity using fish communities. *Fisheries (Bethesda)* **6**(6): 21–7. [6.5]

Karr JR, Fausch KD, Angermeier PL, Yant PR, Schlosser IJ. (1986) *Assessing biological integrity in running waters. A method and its rationale* Special publication No. 5, September 5. Illinois Natural History Survey. Champaign, Illinois. [6.5]

Karr JR, Yant PR, Fausch KD, Schlosser IJ. (1987) Spatial and temporal variability of the index of biotic integrity in three midwestern streams. *Transactions of the American Fisheries Society* **116**: 1–11. [6.5]

Kerr SR. (1974) Theory of size distribution in ecological communities. *Journal of the Fisheries Research Board of Canada* **31**: 1859–62. [6.3]

Larimore RW. (1955) Minnow productivity in a small Illinois stream. *Transactions of the American Fisheries Society* **84**: 110–16. [6.4]

Larsen DP, Omernik JM, Hughes RM, Rohm CM, Whittier TR, Kinney AJ *et al* (1986) Correspondence between spatial patterns in fish assemblages in Ohio, USA, streams and aquatic ecoregions. *Environmental Management* **10**: 815–28. [6.2]

Li HW, Schreck CB, Bond CE, Rexstad E. (1987) Factors influencing changes in fish assemblages of Pacific Northwest streams. In: Matthews WJ, Heins DC (eds) *Community and Evolutionary Ecology of North American Stream Fishes*, pp 193–202. University of Oklahoma Press, Norman, Oklahoma. [6.2, 6.5, 6.6]

Lotrich VA. (1973) Growth, production, and community composition of fishes inhabiting a first-, second-, and third-order stream of eastern Kentucky. *Ecological Monographs* **43**: 377–97. [6.2, 6.3]

Lowe-McConnell RH. (1987) *Ecological Studies in Tropical Fish Communities*. Cambridge Tropical Biology Series. Cambridge University Press, Cambridge. [6.2]

Lyons J. (1989) Correspondence between the distribution of fish assemblages in Wisconsin streams and Omernik's ecoregions. *American Midland Naturalist* **122**: 163–82. [6.2]

MacArthur RH, Levins R. (1967) The limiting similarity, convergence and divergence of coexisting species. *American Naturalist* **101**: 377–85. [6.3]

McFadden JT, Alexander GR, Shetter DS. (1967) Numerical changes and population regulation in brook trout *Salvelinus fontinalis*. *Journal of the Fisheries Research Board of Canada* **24**: 1425–59. [6.4]

Mahon R. (1984) Divergent structure in fish taxocenes of north temperate streams. *Canadian Journal of Fisheries and Aquatic Sciences* **41**: 330–50. [6.2]

Mann KH, Britton RH, Kowalczewski A, Lack TJ, Mathews CP, McDonald I. (1970) Productivity and energy flow at all trophic levels in the River Thames, England. In: Kajak Z, Hillbricht-Ilkowska A (eds) *Productivity Problems of Freshwaters*, pp 579–96. Proceedings of the IBP-UNESCO Symposium on Productivity Problems of Freshwaters, Kazimierz Dolny, Poland. [6.2]

Mathur D, Bason WH, Purdy EJ Jr, Silver CA. (1985) A critique of the Instream Flow Incremental Methodology. *Canadian Journal of Fisheries and Aquatic Sciences* **42**: 825–31. [6.2]

Matthews WJ. (1986a) Fish faunal structure in an Ozark stream: stability, persistence, and a catastrophic flood. *Copeia* **1986**: 388–97. [6.5]

Matthews WJ. (1986b) Geographic variation in thermal tolerance of a widespread minnow (*Notropis lutrensis*) of the North American mid-west. *Journal of Fish Biology* **28**: 407–17. [6.2, 6.3]

Matthews WJ. (1988) North American prairie streams as systems for ecological study. *Journal of the North American Benthological Society* 7: 387–409. [6.2]

Matthews WJ, Styron JT Jr. (1981) Tolerance of headwater vs. mainstream fishes for abrupt physicochemical changes. *American Midland Naturalist* **105**: 149–58. [6.2]

Matthews WJ, Cashner RC, Gelwick FP. (1988) Stability and persistence of fish faunas and assemblages in three midwestern streams. *Copeia* **1988**: 945–55. [6.2, 6.5]

Matthews WM. (1987) Physicochemical tolerance and selectivity of stream fishes as related to their geographic ranges and local distributions. In: Matthews WJ, Heins DC (eds) *Community and Evolutionary Ecology of North American Stream Fishes*, pp 111–20. University of Oklahoma Press, Norman, Oklahoma. [6.5]

Mayden RL. (1987) Historical ecology and North American highland fishes: a research program in community ecology. In: Matthews WJ, Heins DC (eds) *Community and Evolutionary Ecology of North American Stream Fishes*, pp 210–22. University of Oklahoma Press, Norman, Oklahoma. [6.2]

Meffe GK. (1984) Effects of abiotic disturbance on coexistence of predatory–prey fish species. *Ecology* **65**: 1525–34. [6.5]

Metcalfe NB. (1989) Differential response to a competitor by Atlantic salmon adopting alternative life-history strategies. *Proceedings of the Royal Society of London B* **236**: 21–7. [6.3]

Moyle PB, Baltz DM. (1985) Microhabitat use by an assemblage of California stream fishes: developing criteria for instream flow determinations. *Transactions of the American Fisheries Society* **114**: 695–704. [6.2]

Moyle PB, Herbold B. (1987) Life-history patterns and community structure in stream fishes of western North America: comparisons with eastern North America and Europe. In: Matthews WJ, Heins CD (eds) *Community and Evolutionary Ecology of North American Stream Fishes*, pp 25–32. University of Oklahoma Press, Norman, Oklahoma. [6.2, 6.5]

Moyle PB, Li HW. (1979) Community ecology and predator-prey systems in warmwater streams. In: Stroud RE, Clepper HE (eds) *Predator–Prey Systems in Fisheries Management*, pp 171–81. Sports Fisheries Institute, Washington, DC. [6.5]

Moyle PB, Senanayake FR. (1984) Resource partitioning among the fishes of rainforest streams in Sri Lanka. *Journal of Zoology, London* **202**: 195–223. [6.2, 6.5]

Moyle PB, Williams JE. (1990) Biodiversity loss in the temperate zone: decline of the native fish fauna of

California. *Conservation Biology* **4**: 275–84. [6.6]
Moyle PB, Daniels RA, Herbold BL, Baltz DM. (1986) Patterns in distribution and abundance of a noncoevolved assemblage of estuarine fishes in California. *Fishery Bulletin (NOAA)* **84**: 105–17. [6.6]
Naiman RJ, Décamps H, Pastor J, Johnston CA. (1988) The potential importance of boundaries to fluvial ecosystems. *Journal of the North American Benthological Society* 7(4): 289–306. [6.2]
Newcombe C. (1981) A procedure to estimate changes in fish populations caused by changes in stream discharge. *Transactions of the American Fisheries Society* **110**: 382–90. [6.2]
Nickelson TE, Solazzi MF, Johnson SL. (1986) Use of hatchery coho salmon (*Oncorhynchus kisutch*) presmolts to rebuild wild populations in Oregon coastal streams. *Canadian Journal of Fisheries and Aquatic Sciences* **43**: 2443–9. [6.3]
Noble SL. (1991) *Impacts of earlier emerging steelhead fry of hatchery origin on the social structure, distribution, and growth of wild steelhead fry.* M.S. Dissertation, Oregon State University, Corvallis, Oregon. [6.3]
Nordeng H. (1983) Solution to the 'char problem' based arctic char (*Salvelinus alpinus*) in Norway. *Canadian Journal of Fisheries and Aquatic Sciences* **40**: 1372–87. [6.3]
Nyman L, Ring O. (1989) Effects of hatchery environment on three polymorphic loci in arctic char (*Salvelinus alpinus* species complex). *Nordic Journal of Freshwater Research* **65**: 34–43. [6.3]
Omernik JM. (1987) Ecoregions of the conterminous United States (with map supplement). *Annals of the Association of American Geographers* **77**: 118–25. [6.2]
Orth DJ, Maughan OE. (1982) Evaluation of the incremental methodology for recommending instream flows for fishes. *Transactions of the American Fisheries Society* **110**: 1–13. [6.2]
Osborne LL, Wiley MJ. (1988) Empirical relationships between land use/cover and stream quality in an agricultural watershed. *Journal of Environmental Management* **26**: 9–27. [6.2]
Paragamian VL. (1987) Production of smallmouth bass in the Maquoketa River, Iowa. *Journal of Freshwater Ecology* **4**: 141–8. [6.4]
Paragamian VL, Wiley MJ. (1987) Effects of variable streamflows on growth of smallmouth bass in the Maquoketa River, Iowa. *North American Journal of Fisheries Management* **7**: 357–62. [6.4]
Pauly D, Murphy GI. (eds) (1982) *Theory and Management of Tropical Fisheries.* International Center for Living Aquatic Resources Management, Manila, Conference Proceedings 9. Cronulla, Australia. [6.4]
Petrere M Jr. (1985) *Migraciones de peces de agua dulce en America Latina: algunos comentarios.* COPESCAL Doc. Ocas. No. 1. FAO Organizacion de las Naciones Unidas para la Agricultura y la Alimentacion. COPESCAL/OP1. p. 17. Rome, Italy. [6.2]
Pflieger WL, Grace TB. (1987) Changes in the fish fauna of the lower Missouri River, 1940–1983. In: Matthews WJ, Heins DC (eds) *Community and Evolutionary Ecology of North American Stream Fishes,* pp 166–77. University of Oklahoma Press, Norman, Oklahoma. [6.6]
Pianka ER. (1970) On *r*- and *K*-selection. *American Naturalist* **104**: 592–7. [6.3]
Poff LN, Ward JV. (1989) Implications of streamflow variability and predictability for lotic community structure: a regional analysis of streamflow patterns. *Canadian Journal of Fisheries and Aquatic Sciences* **46**: 1805–18. [6.2]
Pope JG. (1972) An investigation of the accuracy of virtual population analysis using cohort analysis. *International Commission for the Northwest Atlantic Fisheries Research Bulletin* **9**: 65–74. [6.4]
Popova OA. (1978) The role of predacious fish in ecosystems. In: Gerking SD (ed.) *Ecology of Freshwater Fish Production,* pp 215–49. Blackwell Scientific Publications, Oxford. [6.3]
Powers DA. (1972) Hemoglobin adaptation for fast and slow water habitats in sympatric catostomid fishes. *Science* **177**: 360–2. [6.3]
Powers DA, Fyhn HJ, Fyhn UEH, Marin JP, Garlick RL, Wood SC. (1979) A comparative study of the oxygen equilibria of blood from 40 genera of Amazonian fishes. *Comparative Biochemistry and Physiology* **62A**: 67–85. [6.3]
Price PW, Westoby M, Rice B, Atsatt PR, Fritz RS, Thompson JN, Mobley K. (1986) Parasite mediations in ecological interactions. *Annual Review of Ecology and Systematics* **17**: 487–506. [6.5]
Pringle CM, Naiman RJ, Bretschko G, Karr JR, Oswood MW, Webster JR *et al.* (1988) Patch dynamics in lotic systems: the stream as a mosaic. *Journal of the North American Benthological Society* 7(4): 503–24. [6.2]
Quattro JM, Vrijenhoek RC. (1989) Fitness differences among remnant populations of the endangered Sonoran topminnow. *Science* **245**:976–8. [6.3]
Rahel FJ, Lyons JD, Cochran PA. (1984) Stochastic or deterministic regulation of assemblage structure? It may depend on how the assemblage is defined. *American Naturalist* **124**: 583–9. [6.5]
Rajagopal PK, Kramer RH. (1974) Respiratory metabolism of Utah chub, Gila atraria (Girard) and speckled dace, *Rhinichthys osculus* (Girard). *Journal of Fish Biology* **6**: 215–22. [6.3]
Rao GMM. (1968) Oxygen consumption of rainbow trout (*Salmo gairdneri*) in relation to activity and salinity. *Canadian Journal of Zoology* **62**: 866–87. [6.3]
Resh VH, Brown AV, Covich AP, Gurtz ME, Li HW,

Minshall GW *et al.* (1988) The role of disturbance in stream ecology. *Journal of the North American Benthological Society* **7**: 433–55. [6.5]

Richards K. (1982) *Rivers: Form and Process in Alluvial Channels.* Methuen, London. [6.3]

Ricker WE. (1975) Computation and interpretation of biological statistics of fish populations. *Bulletin of the Fisheries Research Board of Canada* **191**: 382. [6.4]

Riddell BE, Leggett WC. (1981) Evidence of an adaptive basis for geographic variation in body morphology and time of downstream migration of juvenile Atlantic salmon (*Salmo salar*). *Canadian Journal of Fisheries and Aquatic Sciences* **38**: 308–20. [6.3]

Riddell BE, Leggett WC, Saunders RL. (1981) Evidence of adaptive polygenic variation between two populations of Atlantic salmon (*Salmo salar*) native to tributaries of the S.W. Miramichi River, N.B. *Canadian Journal of Fisheries and Aquatic Sciences* **38**: 321–33. [6.3]

Rohm CM, Giese JW, Bennett CC. (1987) Evaluation of an aquatic ecoregion classification of streams in Arkansas. *Journal of Freshwater Ecology* **4**: 127–40. [6.2]

Rohovec JS, Fryer JL. (1979) Fish health management in aquaculture. In: Klingeman PC (ed.) *Aquaculture: A Modern Fish Tale*, pp 15–36. Water Resource Research Institute, Oregon State University, Seminar Series, SEMIN WR 026–79, Corvallis, Oregon. [6.5]

Ross ST, Matthews WJ, Echelle AA. (1985) Persistence of stream fish assemblages: effects of environmental change. *American Naturalist* **126**: 24–40.

Ross ST, Baker JA, Clark KE. (1987) Microhabitat partitioning of southeastern stream fishes: temporal and spatial predictability. In: Matthews WJ, Heins DC (eds) *Community and Evolutionary Ecology of North American Stream Fishes*, pp 42–51. University of Oklahoma Press, Norman, Oklahoma. [6.2, 6.5]

Ryman N, Stahl G. (1980) Genetic changes in hatchery stocks of brown trout (*Salmo trutta*). *Canadian Journal of Fisheries and Aquatic Sciences* **37**: 82–7. [6.3]

Scarnecchia DL. (1988) The importance of streamlining in influencing fish community structure in channelized and unchannelized reaches of a prairie stream. *Regulated Rivers: Research and Management* **2**: 155–66. [6.2, 6.3]

Schaefer MB. (1954) Some aspects of the dynamics of populations important to the management of the commercial marine fisheries. *Bulletin of the Inter-American Tropical Tuna Commission* **1**: 27–56. [6.4]

Schaffer WM, Elson PF. (1975) The adaptive significance of variation in life history among local populations of Atlantic salmon in North America. *Ecology* **56**: 577–90. [6.3]

Schlosser IJ. (1982) Fish community structure and function along two habitat gradients in a headwater stream. *Ecological Monographs* **52**(4): 395–414. [6.3]

Schlosser IJ. (1987) A conceptual framework for fish communities in small warmwater streams. In: Matthews WJ, Heins DC (eds) *Community and Evolutionary Ecology of North American Stream Fishes*, pp 17–24. University of Oklahoma Press, Norman, Oklahoma. [6.3]

Schoener TW. (1986) Overview: kinds of ecological communities—ecology becomes pluralistic. In: Diamond J, Case TJ (eds) *Community Ecology*, pp 467–79. Harper and Row, New York. [6.2]

Schoener TW. (1987) Axes of controversy in community ecology. In: Matthews WJ, Heins DC (eds) *Community and Evolutionary Ecology of North American Stream Fishes*, pp 8–16. University of Oklahoma Press, Norman, Oklahoma. [6.2, 6.5]

Schreck CB, Li HW, Hjort RC, Sharpe CS, Currens KP, Hulett PL *et al.* (1986) *Stock identification of the Columbia River chinook salmon and steelhead trout.* Bonneville Power Administration Final Report, Portland, Oregon. [6.3]

Schwassmann HO. (1978) Times of annual spawning and reproductive strategies in Amazonian fishes. In: Thorp JE (ed.) *Rhythmic Activity of Fish*, pp 186–200. Academic Press, London. [6.3]

Seghers BH. (1974) Schooling behavior in the guppy (*Poecilia reticulata*): an evolutionary response to predation. *Evolution* **28**: 486–9. [6.3, 6.5]

Sheldon AL. (1968) Species diversity and longitudinal succession in stream fishes. *Ecology* **49**: 193–8. [6.2]

Sheldon AL. (1987) Rarity: patterns and consequences for stream fishes. In: Matthews WJ, Heins DC (eds) *Community and Evolutionary Ecology of North American Stream Fishes*, pp 203–9. University of Oklahoma Press, Norman, Oklahoma. [6.5]

Smale MA, Rabeni CF. (1990) *Stream fish communities and stream health in headwater streams draining agricultural watersheads in Missouri.* Progress Report to Missouri Department of Natural Resources, Water Quality Division, July 1990. Cooperative Fish and Wildlife Research Unit, University of Missouri, Missouri. [6.4]

Smith H. (1965) Some experiments on the oxygen consumption of goldfish (*Carassius auratus* L.) in relation to swimming speed. *Canadian Journal of Zoology* **43**: 623–33. [6.3]

Starrett WC. (1951) Some factors affecting the abundance of minnows in the Des Moines River, Iowa. *Ecology* **32**(1): 13–27. [6.5]

Statzner B. (1987) Characteristics of lotic ecosystems and consequences for future research directions. *Ecological Studies* **61**: 365–90. [6.2, 6.3]

Stearns SC. (1980) A new view of life-history evolution. *Oikos* **35**: 266–81. [6.3]

Strauss RE. (1987) The importance of phylogenetic constraints in comparisons of morphological structure among fish assemblages. In: Matthews WJ, Heins DC (eds) *Community and Evolutionary Ecology of North American Stream Fishes*, pp 136–43. University of Oklahoma Press, Norman, Oklahoma. [6.2]

Suzumoto BK, Schreck CB, McIntyre JD. (1977) Relative resistances of three transferrin genotypes of coho salmon (*Oncorhynchus kisutch*) and their hematological responses to bacterial kidney disease. *Canadian Journal of Fisheries and Aquatic Sciences* **34**: 1–8. [6.3, 6.5]

Taylor EB, McPhail JD. (1985a) Variation in body morphology among British Columbia populations of coho salmon (*Oncorhynchus kisutch*). *Canadian Journal of Fisheries and Aquatic Sciences* **42**: 2020–8. [6.3]

Taylor EB, McPhail JD. (1985b) Variation in burst and prolonged swimming performance among British Columbia populations of coho salmon, *Oncorhynchus kisutch. Canadian Journal of Fisheries and Aquatic Sciences* **42**: 2029–33. [6.3]

Terrell JW, McMahon TE, Inskip PD, Raleigh RF, Williamson KL. (1982) *Habitat suitability index models: Appendix A. Guidelines for riverine and lacustrine applications of fish HSI models with the habitat evaluation procedures.* FWS/OBS-82/10A. US Fish and Wildlife Service. p. 54. Washington, DC. [6.2]

Thompson DW. (1942) *On Growth and Form*. Mcmillan, New York. [6.3]

US Fish and Wildlife Service. (1981) *Standards for the development of habitat suitability index models.* 103 ESM. US Fish and Wildlife Service, Division of Ecological Services. Washington, DC. [6.2]

Vannote RL, Minshall GW, Cummins KW, Sedell JR, Cushing CE. (1980) The river continuum concept. *Canadian Journal of Fisheries and Aquatic Sciences* **37**: 130–7. [6.2]

Varley JD, Gresswell RE. (1988) Ecology, status and management of the Yellowstone cutthroat trout. *American Fisheries Society Symposium* **4**: 13–24. [6.3]

Wade MG. (1986) *The relative effects of* Ceratomyxa shasta *on crosses of resistant and susceptible stocks of summer steelhead.* M.S. Dissertation, Oregon State University, Corvallis, Oregon. [6.3, 6.5]

Ward BR, Slaney PA. (1988) Life history and smolt-to-adult survival of Keogh River steelhead trout (*Salmo gairdneri*) and the relationship to smolt size. *Canadian Journal of Fisheries and Aquatic Sciences* **45**: 1110–22. [6.3]

Webb PW. (1975) Hydrodynamics and energetics of fish propulsion. *Canadian Bulletin of Fisheries and Aquatic Sciences* **190**: 1–159. [6.3]

Webb PW, de Buffrénil V. (1990) Locomotion in the biology of large aquatic vertebrates. *Transactions of the American Fisheries Society* **119**: 629–41. [6.3]

Welcomme RL. (1985) *River Fisheries*. FAO Fisheries Technical Paper No. 262. Food and Agriculture Organization, Rome, Italy. 330. [6.2, 6.3, 6.4, 6.5, 6.6]

White RJ, Brynildson OM. (1967) *Guidelines for management of trout stream habitat in Wisconsin*. Technical Bulletin 39. Department of Natural Resources, Wisconsin. [6.6]

Whittier TR, Hughes RM, Larsen DP. (1988) Correspondence between ecoregions and spatial patterns in stream ecosystems in Oregon, USA. *Canadian Journal of Fisheries and Aquatic Sciences* **45**: 1264–78. [6.2]

Wiens JA. (1984) On understanding a non-equilibrium world: myth and reality in community patterns and processes. In: Strong DR Jr, Simberloff D, Abele LG, Thistle AB (eds) *Ecological Communities: Conceptual Issues and the Evidence*, pp 439–57. Princeton University Press, Princeton, New Jersey. [6.5]

Wikramanayake ED. (1990) Ecomorphology and biogeography of a tropical stream fish assemblage. *Ecology* **71**: 1756–64. [6.5]

Wikramanayake ED, Moyle PB. (1989) Ecological structure of tropical fish assemblages in wet-zone streams of Sri Lanka. *Journal of Zoology (London)* **281**: 503–26. [6.5]

Wiley MJ, Osborne LL, Larimore RW. (1990) Longitudinal structure of an agricultural prairie river system and its relationship to current stream ecosystem theory. *Canadian Journal of Fisheries and Aquatic Sciences* **47**: 373–84. [6.2, 6.3]

Winter GW, Schreck CB, McIntyre JD. (1979) Resistance of different stocks and transferrin genotyes of coho salmon and steelhead trout to bacterial kidney disease and vibriosis. *Fishery Bulletin (NOAA)* **77**: 795–802. [6.5]

Withler RE, Evelyn TPT. (1990) Genetic variation in resistance to bacterial kidney disease within and between two strains of coho salmon from British columbia. *Transactions of the American Fisheries Society* **119**: 1003–9. [6.3]

Wohlschlag DE, Juliano RO. (1959) Seasonal changes in bluegill metabolism. *Limnology and Oceanography* **4**: 195–205. [6.3]

Wurtsbaugh WA, Tapia RA. (1988) Mass mortality of fishes in Lake Titicaca (Peru–Bolivia) associated with the protozoan parasite *Ichthyophthirius multifiliis. Transactions of the American Fisheries Society* **117**: 213–17. [6.5]

Yant PR, Karr JR, Angermeier PL. (1984) Stochasticity in stream fish communities: an alternative interpretation. *American Naturalist* **124**: 573–82. [6.5]

Zimmerman EG, Wooten MC. (1981) Increased heterozygosity at the MDH-B locus in fish inhabiting a rapidly fluctuating thermal environment. *Transactions of the American Fisheries Society* **110**: 410–16. [6.3]

7: Food Webs and Species Interactions

A.G.HILDREW

7.1 PATTERNS IN FOOD WEBS

The idea that the nutrition of living things links them together into a network of trophic interactions — known as food webs — is perhaps the most familiar in ecology. It is also deceptively simple. The study of food webs involves investigations of both their structure and function. Structure may be revealed when the species in a community are written down systematically and an audit made of the feeding links between them — 'joining up the dots'. As we shall see, there appear to be patterns in the structure of such webs that emerge from the examination of data from many communities. The most important question then is whether these patterns in food-web structure result from the nature of the biological interactions, embodied in the web, for example predation and competition, or from some extrinsic feature(s) of the environment. This is the motivation for juxtaposing in this section food web 'function', the processes by which species interact, with observed patterns in food web structure.

While much of this section may dwell on quite basic aspects of river and stream ecology, a better understanding of food webs is rich in promise for the applied science of river management. Progress in this area could provide a real understanding of river communities, which are important both as targets for environmental management and as tools in environmental assessment. The really remarkable feature of this field is how little we know of food webs in general, and how little effort is going into the study of this unifying feature of natural ecosystems.

Apparent patterns in food webs have emerged by examination of patchy and sometimes inadequate published information, for instance the 113 webs compiled by Briand and Cohen (1987). Some of the best-known patterns follow:

1 Food chains are short, usually five species or less.

2 Connectance (the proportion of possible feeding links between species which are actually realized) declines with increasing species richness.

3 The proportions of 'top', 'intermediate' and 'basal' species in food webs are constant.

4 Omnivory (species feeding at more than one level in the web) is rare.

5 Food webs from 'constant' environments have relatively more omnivory and higher connectance than those from 'fluctuating' environments.

6 Food webs vary with environmental structure and food chains are longer in three- than in two-dimensional habitats.

The reader should refer to Lawton (1989) for a fuller list, references and explanation. Briefly, some of these patterns may be real, some artefacts, but several of them are amenable to explanation by a number of competing hypotheses. The textbook explanation that food chains are short because of the low efficiency of energy transfer along them is now controversial, for food chain length may not vary along gradients from very high to very low productivity (Briand & Cohen 1987). Most of the food-web patterns can be predicted by models based on Lotka–Volterra dynamical interactions. Upper limits to species richness and connectance are then constrained by stability and, further, food-web structure will relate to environmental variability (because complex webs are dynamically 'fragile' and cannot persist in more variable environments). A non-dynamic hypothesis, the cascade model, offers an alternative explanation (Cohen & Newman 1985). It assumes a constant number of trophic links per species and that there is a hierarchy such that species feed only on those below them in this trophic 'cascade'. Warren and Lawton (1987) add the idea that the hierarchy could be based on body size, predators feeding only on species smaller than themselves.

But what of the specifics of freshwater food-webs? Only 21 of Briand and Cohen's (1987) compilation are from freshwaters and of these only nine are from rivers or streams. This is unsurprising because there are very few studies of running waters indeed that provide food-web descriptions sufficiently detailed to compare with the observed patterns. Jeffries and Lawton (1985) showed that there is a constant ratio of predator to prey species in a broad range of freshwaters, including some data from streams and rivers. This constant ratio is widespread in a variety of habitats (Pimm *et al* 1991). Briand (1985) compared food webs in a variety of freshwater habitats with others. He found the shortest food webs in streams. They were also complex (highly connected and wide) compared with lake and river webs which were thinner, longer and exhibited fewer connections. In Table 7.1, data have been added from the food web of Broadstone Stream (Hildrew *et al* 1985), which is far more complete than any in Briand's (1985) compilation. In only one respect does Broadstone Stream conform markedly to Briand's stream characteristics: it has high linkage complexity (SC_{max}, 10.54, Table 7.1).

Connectance is a parameter of central importance in both static and dynamic theories of community structure (Warren 1990). Whereas dynamic theories predict that connectance should decline with increasing species richness, the static cascade model simply assumes an equivalent proposition: that the number of feeding links per species is constant. The latter theory thus needs an independent explanation for patterns of connectance. Pimm (1982) proposed that species may simply be restricted in the range of others they can feed upon because of limitations of behaviour or morphology. Warren (1990), however, suggested that as total species richness increases, prey would become increasingly closely packed in the predator's niche space and the number of feeding links per species would thus also increase. He showed that the number of links will in fact increase in proportion to S^2, and that the proportionality constant will be that fraction of total

Table 7.1 Food-web variables for Broadstone Stream (Hildrew *et al* 1985) and group means from a compilation of webs (consisting of 'trophic' species, see text) by Briand (1985)

	Broadstone Stream		Briand (1985)		
	Taxonomic species	Trophic species	Streams ($n = 7$)	Rivers ($n = 2$)	Others (non-freshwater) ($n = 20$)
S	24	17	11–19	?	?
C_{min}	0.35	0.29			
C_{max}	0.76	0.62			
SC_{max}	18.2	10.54	9.28	5.86	6.38
Max chain	5	5	4.00	5.00	5.25
Mean chain	4.1	3.7	2.29	2.91	2.99
Fraction sp.					
Basal	0.17	0.23	0.19	0.26	0.20
Intermediate	0.71	0.59	0.47	0.61	0.57
Top	0.12	0.18	0.33	0.13	0.23

S, species richness; C, connectance (fraction of all possible links realized; min excludes competitive interactions, max includes them); max and mean chain, longest and average chain lengths in the web; a basal species has no prey, intermediate species have both prey and predators, top species have no predators.

niche space exploited by predators (large for webs of generalists, small for webs of specialists) (Fig. 7.1 (a)).

Data from the benthic invertebrates of two fishless, acidic freshwaters, Skipwith pond (Warren 1990) and Broadstone stream (Hildrew *et al* 1985), place them clearly among the highly connected webs of generalists (Fig. 7.1(b)). It may be premature, however, to categorize all freshwater food webs in this fashion; there are simply too few good data. Indeed, Paine (1988) argues that patterns of connectance in food webs in general may be artefacts, generated by the increasing difficulty of tallying feeding interactions in species-rich communities. Thus, the decline in connectance with species richness in Fig. 7.1(b) may be more apparent than real.

The poverty of the stream data in the Briand (1985) compilation (Table 7.1) is illustrated by the fact that the species richness of the nine stream communities ranged from 11 to 19. Broadstone stream has an extremely impoverished fauna, yet even that has a list of nearly 30 macroinvertebrate taxonomic species and about 20 'trophic' species (those taxonomic species with identical predators and prey are lumped to form a trophic species). To this number must now be added the ten or so species of microarthropods (harpacticoid and cyclopoid copepods, benthic Cladocera and mites) found by Rundle and Hildrew (1990). These are still aggregated in the Broadstone web because their trophic interactions are not known. Many other groups remain unstudied in this 'simple' stream community.

We have so far dwelt mainly on patterns in the structure of food webs, based on whether interactions are present or absent. But these interactions also vary in strength, and some at least may be dynamically trivial. If we can identify strong and weak interactions between 'elements' (often aggregates of species appearing to act in the food web in a similar way) perhaps this will reveal the main features of real webs and the most important restrictions on their structure (Menge & Sutherland 1987). In addition to better descriptions of food webs, we therefore need

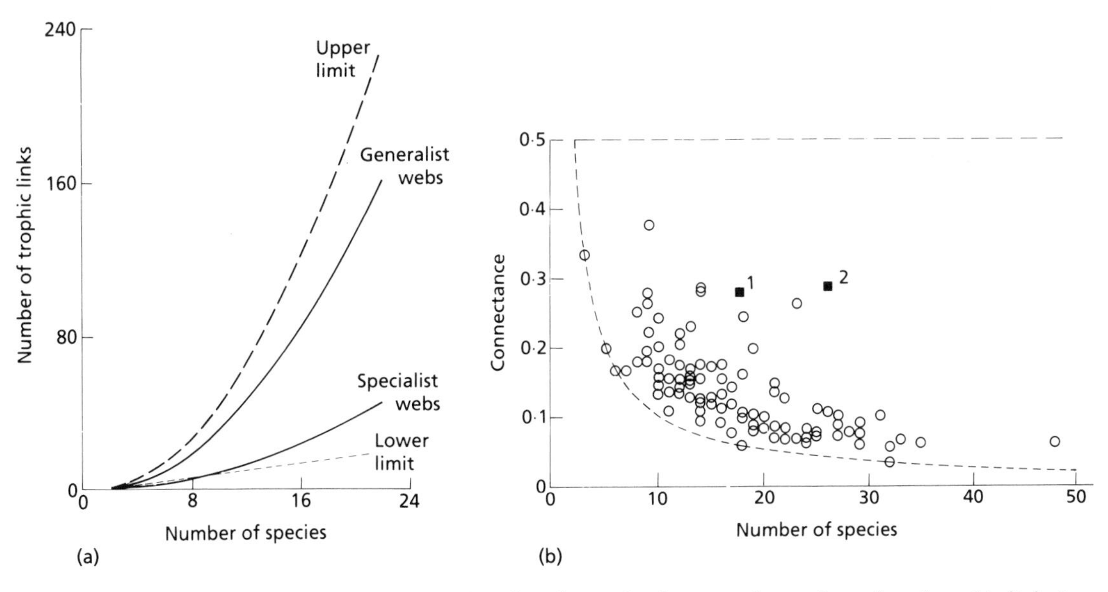

Fig. 7.1 (a) Upper and lower limits (broken lines) to the relationship between the total number of trophic links in a food web and species richness. Webs of generalists and specialists fall close to these upper and lower limits respectively. (b) Connectance declines with species richness in the 113-web compilation of Briand and Cohen (1987) (the broken lines are theoretical upper and lower limits to connectance derived from the limits to trophic links shown in (a)). The solid square symbols add data from a fishless, acidic stream (1, Hildrew *et al* 1985) and a similar pond (2, Warren 1990). (a) and (b) are both after Warren (1990).

experimental evidence on the role and strength of biotic interactions before we can distinguish between the theoretical options.

Lawton (1989) has seen two major difficulties with models of food-web structure based on Lotka–Voltera dynamics. The first is the assumption that trophic links involve closely coupled enemy–victim interactions, and the second is that all interactions take place in a spatially homogeneous world. Both will raise eyebrows among many empirical ecologists, for modern emphasis in the ecology of running waters has been on disturbance and on patchy, stochastic environments (Hildrew & Townsend 1987; Pringle *et al* 1988; Resh *et al* 1988). However, Pimm *et al* (1991) have pointed out that many of the predictions of dynamic models are actually quite robust to these detailed assumptions.

7.2 BIOTIC INTERACTIONS IN RUNNING-WATER FOOD WEBS

The best-known categories of interactions involve either positive/negative effects on the two interactors or negative/negative effects. In running waters the first category is commonly exemplified by herbivory, consumption by detritivores of microbes attached to dead organic matter and by conventional predation. There are then cases of competitive interactions, which are not strictly direct trophic links but may be important in the dynamics of food webs through indirect effects. There are also some indications that mutualisms (positive/positive interactions) may be widespread in freshwater communities (see below).

Herbivory

There are four main sources of autochthonous primary production in streams and rivers: rooted aquatic angiosperms and bryophytes, their algal epiphytes, and the algal elements of biofilms on mineral particles. The reasons for the apparently negligible grazing of living angiosperm tissue in freshwaters are essentially unknown (but see Newman 1991). By far the best known herbivore systems are the invertebrate and fish grazers of algae in stony streams.

There has been a flurry of experimental work on invertebrate/algal interactions in streams in the last 10 years (reviews in Gregory 1983; Lamberti & Moore 1984: recent examples by Feminella *et al* 1989; Lamberti *et al* 1989). This field has undoubtedly been one of the big success stories of stream ecology and has revealed widespread, although not ubiquitous, strong interactions between grazers and grazed. Features of the algal community often altered by grazing include algal biomass, physical structure, species composition, productivity per unit biomass and total primary production. The determinants of the apparent strength of the interaction are: (1) the productive capacity of the algae (usually set by nutrient and/or light supply); (2) the type of grazer tested (caddis and snails are often particularly effective, mayflies sometimes not); (3) the onset of grazing in relation to the development of algal growth (algae may 'escape' control if the onset of grazing is too late); and (4) hydraulic disturbance in the stream (which may overwhelm biotic interactions).

Hill and Knight (1987) established a gradient of mayfly (*Ameletus validus*) densities in a number of *in-stream* perspex channels in a northern California stream. Periphyton biomass declined with increasing grazer density in the channels (Fig. 7.2(a)). They also found increased chlorophyll *a* per unit biomass and a decreased contribution of the loosely attached upper layer of periphyton to total algal biomass. Feminella *et al* (1989) compared the algal biomass accrued on tiles placed either on the streambed or on raised platforms (this treatment excluding crawling grazers, mainly cased caddis and snails) (Fig. 7.2(b)). Raised tiles accumulated more algal biomass than controls in two of the three study streams. Crawling grazers also maintained low algal biomass across a wide range of tree canopy cover in one of the streams (Fig. 7.2(c)).

In a search for indirect trophic effects in river food webs, Hill and Harvey (1991) recently manipulated insectivorous fish and grazing snails in a series of channels in a shaded headwater stream. Snail (*Elimia clavaeformes*) grazing substantially reduced the biomass of loosely attached periphyton but had no significant effect on the tightly attached layer. Snail grazing had opposing effects

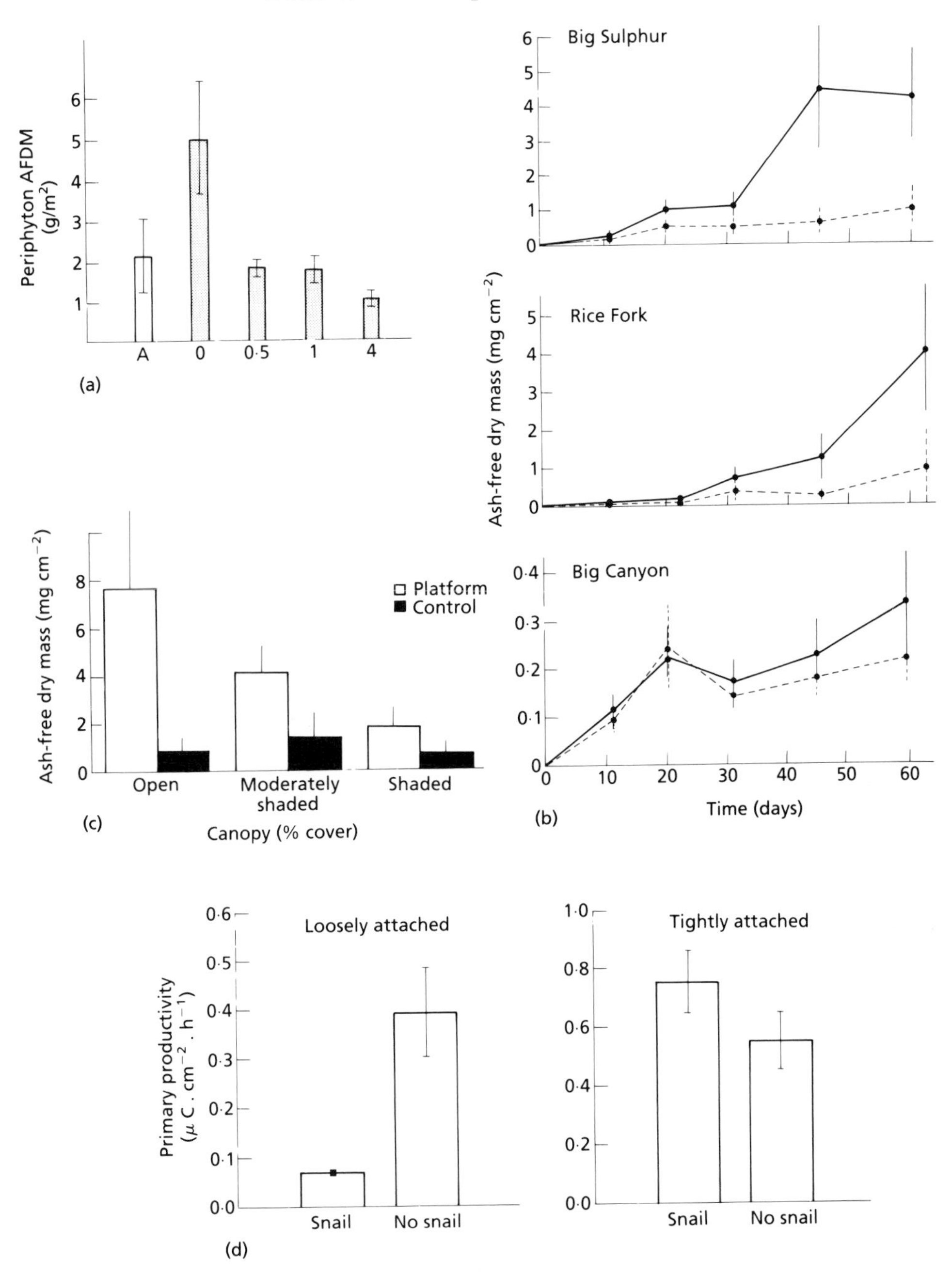

Fig. 7.2 Grazer/algal interactions. (a) The effects of grazing mayflies on mean periphyton biomass (±SE); A, ambient grazer density; 0, 0.5, 1, 4 are multiples of ambient (Hill & Knight 1987). (b) Feminella *et al* (1989) measured the accrual of mean periphyton biomass (±95% CL) on raised tiles (solid lines), from which crawling grazers were excluded, and control tiles (broken lines) in three Californian streams. (c) They found that periphyton biomass responded to light on raised (platform, □) but not on control (■) tiles. (d) Snail grazing in a shaded, headwater US stream reduced mean (±SE) primary production in the loosely attached periphytic layer but increased it in the tightly attached layer (Hill & Harvey 1991).

on primary production in the loosely and tightly attached layers, apparently stimulating it in the latter (Fig. 7.2(d)). Light, however, also varied generally among the channels and it was light, rather than the presence of snails or fish, which accounted for most of the variation in total primary production.

Although grazing clearly affects the species composition of periphyton, its influence on species diversity is less clear. Slight, intermediate and intense grazing by *Baetis* mayflies, the snail *Juga* and the caddis *Dicosmoecus* produced no effect, an increase and a strong decrease in algal diversity, respectively (DeNicola *et al* 1990). Vulnerable species, such as the large, stalked diatoms of the genera *Gomphonema* and *Cymbella*, are often enormously reduced in abundance even by low densities of grazers (McCormick & Stevenson 1989). In terms of these true *species* interactions, however, the analysis of food webs in these grazing systems is in its infancy. Periphytic communities in streams are species rich and no published empirical web offers anything like a complete description.

Community level, structural features of periphyton are thus clearly affected by grazing in streams. The loss of loosely attached, large, or otherwise vulnerable species often leads to a reduction in biomass. However, at least biomass-specific primary production may actually be stimulated by grazers (Fig. 7.2(d)). The mechanism is presumably by enhancing a resource, such as light or nutrients, through the removal of senescent cells, detritus or silt. Overall primary production is commonly then limited by abiotic factors (light in shaded streams, nutrients in open streams) or by disturbance. But what of the reciprocal half of the interaction?

Are the grazers food limited?

There has been much less work on the 'bottom-up' effects of algal food on grazers in streams but several significant results, mainly with caddis larvae, have been reported. In McAuliffe's (1984a) experiments *Glossosoma* reduced periphyton abundance on experimental bricks. He then exposed bricks with differing resource abundance and *Glossosoma* densities to short-term colonization by other mobile grazers. The experiments seem to show quite clearly the effect of exploitative, interspecific competition on colonization (Fig. 7.3(a)). Other experimenters have demonstrated food limitation and intraspecific competition for food. For example, Lamberti *et al* (1987) manipulated densities of *Helicopsyche* in enclosures. The final weight of larvae declined with

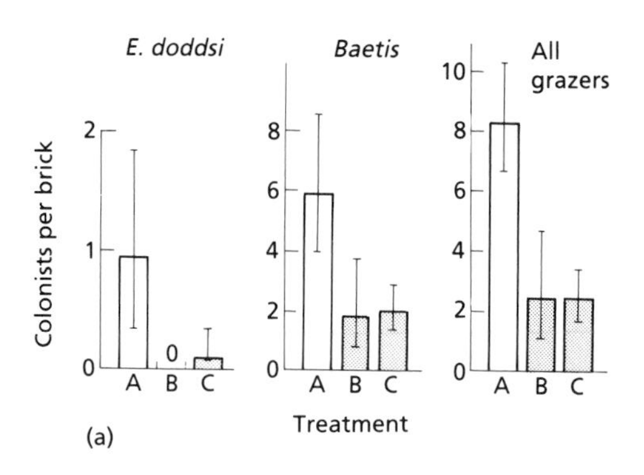

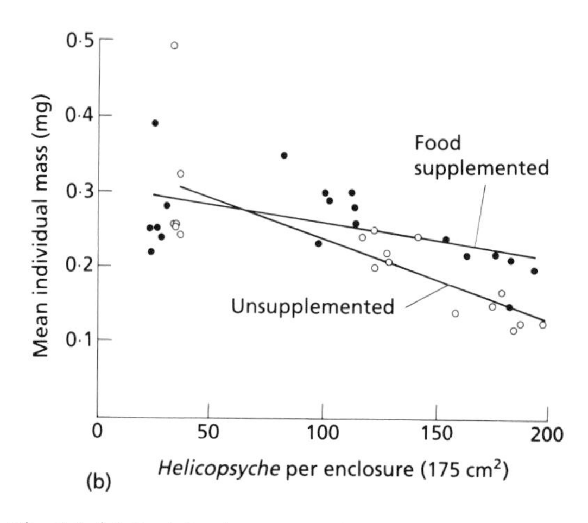

Fig. 7.3 (a) Exploitative competition among grazers. Colonization by the mayflies *Ephemerella doddsi* and *Baetis*, and by total grazers, of experimental bricks in treatments A (high algal biomass and no interference competition from the grazing caddis *Glossosoma*), B (low algal density and no interference competition) or C (low algal density and *Glossosoma* present) (McAuliffe 1984a). (b) Weight attained by the grazing caddis *Helicopsyche* kept for 60 days at various densities in enclosures in a Californian stream. •, Food supplementation; ○, no supplementation (Lamberti *et al* 1987). Lines are least square regressions.

increasing larval density, but when food was supplemented this effect was slight (Fig. 7.3(b)). Essentially similar results were obtained by Hart (1987) and Lamberti *et al* (1989).

In summary, there is evidence both of strong top-down regulation of algal community structure and of bottom-up resource limitation of grazers. In such systems supplementation of limiting resources to primary producers might lead not to an increased biomass of algae but to that of their grazers or both. We must temper this by remembering that most experimental approaches so far have been at small spatial and temporal scales. Also, experimenters tend to emphasize positive results. Frequent resetting of algal succession by spates or droughts may weaken, or render intermittent, this 'enemy–victim' interaction. Biggs and Close (1989) thus found that the hydrological regime in nine gravel-bed rivers in New Zealand explained more of the variation in algal biomass and chlorophyll *a* than did nutrient concentrations. In particular, percentage time in flood was an important variable, and flow disturbances may determine the extent to which grazing interactions are important.

The mysteries of detritus

Most streams, with the particular exception of desert systems, are heterotrophic, with total respiration exceeding gross primary production (Busch & Fisher 1981; Winterbourn 1986). The detrital subsidy comes from the catchment in the form of particulate plant litter and dissolved organic matter. Even autochthonous primary production, particularly that due to aquatic vascular macrophytes, is often incorporated in stream food webs as detritus, rather than being grazed as living material. Detritus is, therefore, quantitatively the most important fuel for running-water food webs.

There is a good deal of circumstantial evidence that detritivorous animals feeding on tree leaves are food limited (e.g. Gee 1988). Recent experiments have tested this contention. Smock *et al* (1989) manipulated the number of debris dams in coastal plain streams of the south eastern USA and found that shredder abundance increased with dam abundance. Richardson (1991) imposed three rates of detritus supply to replicated, stream-side channels. In treatments with extra detritus there were increases in the density of several common detritivorous insects and in the biomass of detritivores overall (Fig. 7.4). It is the retention characteristics of streams which often seems to determine the importance of quantitative food limitation to detritus feeders (Hildrew *et al* 1991). However, there may also be important limitations imposed on detritus feeders by the quality of their food (Groom & Hildrew 1989).

Detritus is a complex mixture of non-living and microbial components, and the precise relationship between detritivores and these two fractions of their food is still a matter of controversy (see Chapter 8). Microbially unconditioned leaf litter may be more or less indigestible by animals, and colonization by microbes and partial decomposition yields a more palatable food source which may support higher individual growth (Groom & Hildrew 1989; Fig. 7.5). In terms of species interactions and food webs there are potentially five separate but not exclusive modes of interaction between decomposer microbes and detritivorous animals (Fig. 7.6):

1 A simple 'predator/prey' interaction in which animals assimilate microbial carbon or nutrients but little of the detrital material itself (Fig. 7.6(a)). Animals certainly do feed on the microbial element, although controversy abounds over whether microbial biomass is sufficient for their

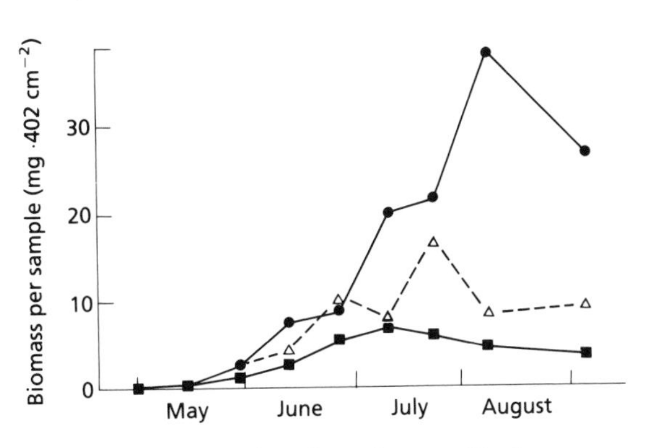

Fig. 7.4 Richardson (1991) supplemented leafy detrital food at two levels above ambient in streamside channels on a Canadian stream. This shows seasonal change in the biomass of a total of nine detritivorous taxa. Each point is a mean of two replicate channels. ●, High; △, intermediate; ■, 'natural'.

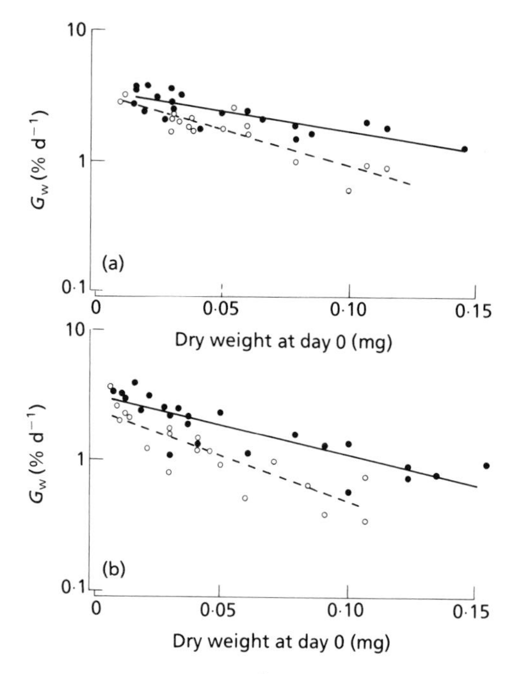

Fig. 7.5 Growth (G_w, % d^{-1}) of the common detritivorous stonefly *Nemurella pictetii* as a function of initial weight when fed on (a) alder and (b) beech leaves which have been conditioned in acid (○) or neutral (•) stream water. Growth is significantly faster on alder than on beech and on neutral- than on acid-conditioned litter. Both indicate an enhancement of growth on microbially well-conditioned litter (Groom & Hildrew 1989).

needs (Findlay *et al* 1986; Arsuffi & Suberkropp 1988).

2 A competitive interaction between microbiota and animals for dead organic matter—microbes are essentially a sink for detrital carbon, which is mainly mineralized and not passed on to animals (Fig. 7.6(b)).

3 A commensal interaction in which animals benefit from the presence of microbial populations, although not principally by the ingestion of microbial biomass *per se* (Fig. 7.6(c)). Microbes soften and partially digest leaf tissue, rendering it more nutritious for animals (Bärlocher & Kendrick 1975).

4 A combination between predation and commensal interactions (Fig. 7.6(d)) is possible if, as well as gaining nutrition from microbial biomass, microbial enzymes can be sequestered, remain

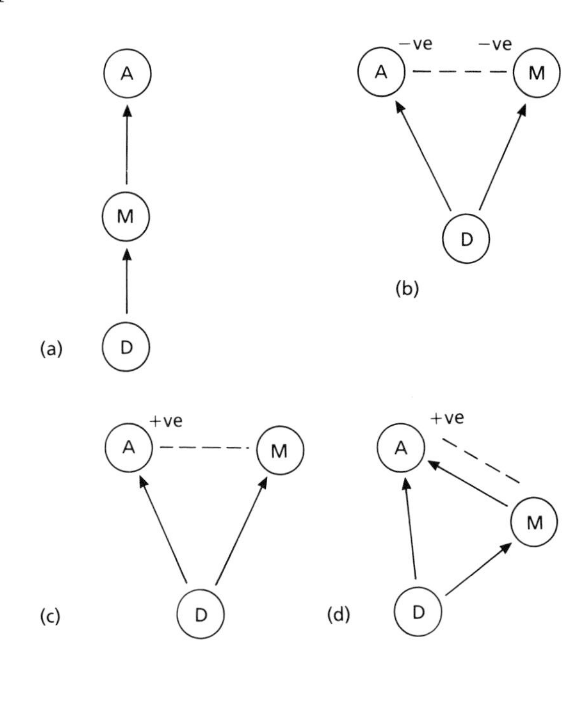

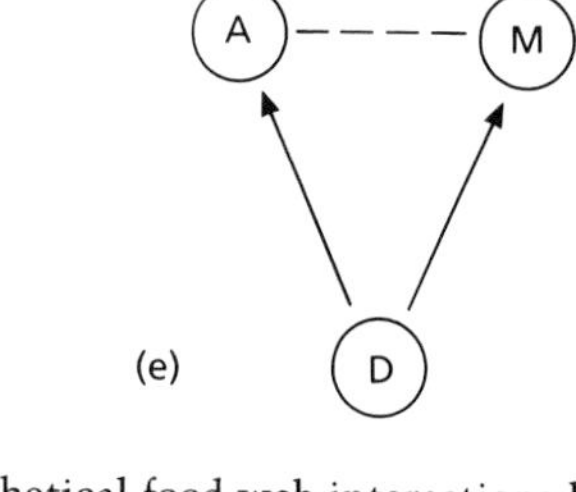

Fig. 7.6 Hypothetical food-web interactions between the dead fraction of detritus (D), micro-organisms (M) and detritivorous animals (A). Trophic interactions (solid lines) show arrows from 'prey' to 'predator'. Non-trophic interactions (broken lines) have positive or negative signs to indicate possible competition, commensalism or mutualism. (a) Simple predator/prey interaction. (b) Animals and micro-organisms as competitors for detritus. (c) Commensal interaction in which animals benefit from partial digestion and softening of leaves by micro-organisms. (d) Commensal/predator–prey interaction in which animals benefit from sequestered enzymes. (e) Mutualism in which animals benefit from microbial conditioning and micro-organisms benefit, for example, from comminution of particles.

active in animal guts and enhance the digestion by the animal of the dead fraction of the detritus (Sinsabaugh *et al* 1985).

5 A mutualistic interaction in which both microbes and animals benefit (Fig. 7.6(e)), for instance the animals by the partial digestion of detritus by microbes, and the microbes by the increased fragmentation of the substrate by animal feeding.

These subtle interactions lie at the very base of running-water food webs, yet we know little of their quantitative relative importance in any whole web. Although good progress has been made in several respects, we need to expand our efforts in microbial ecology perhaps more urgently than in any other area. The incorporation of dissolved organic matter into river food webs, centred on the formation of particles and the growth of slimy biofilms on stones, is understood mainly in 'process terms' (see Chapter 8) and not as part of a taxonomically discriminated food web. It seems inevitable that this component of the river community will remain as an aggregated 'black box' for the forseeable future.

In conclusion, there is evidence of bottom-up control of populations of detritivorous animals, both through food quantity and quality. Whether their grazing on micro-organisms decomposing detritus ever limits microbial biomass or production remains a matter largely of conjecture. It may be that detritus-based communities are examples of 'donor control', in which the consumed influences the consumer but the effect is not reciprocated. At the limit it is clear that running-water animals have little influence on the gross supply of allochthonous organic matter, but they might affect the rate at which it is made available by microbes. Donor-controlled food webs can theoretically have a different structure from those dominated by closely coupled interactions (Pimm 1982; Lawton 1989), with greater connectance and omnivory.

Predation

Predators are numerous in streams and rivers and include a variety of invertebrate taxa, most fish, amphibia such as salamander larvae, and streamside birds and mammals. The role of predation by fish and invertebrates in the population dynamics of their prey and in structuring benthic communities, particularly in stony trout streams, is a matter of enduring controversy (see Cooper *et al* 1990). Many features of the morphology, physiology and behaviour of prey animals seem related to defence against fish and invertebrate predators, indicating that predation is a powerful selective force in their evolution (e.g. Peckarsky 1984; Peckarsky & Penton 1988; Soluk 1990). Estimates of food consumption, both by fish and invertebrate predators, continue to yield estimates greater than the quantities thought to be produced in the benthos (Allan 1983; Waters 1988). These indirect lines of evidence indicate a heavy exploitation of prey populations and possibly a case of 'closely coupled enemy–victim interactions'.

Many ecologists have been tempted to test this idea in field experiments, often using various kinds of enclosures or exclosures in which prey density attained is compared between replicates with and without predators. In sharp contrast to the obvious effects of similar manipulations in many grazer/herbivore systems, the results have rarely been spectacular and often insignificant (Cooper *et al* 1990). The inference is that either predation really is dynamically trivial in running waters or that the experiments are unable to detect the effects through some details of scale, timing or design.

A number of authors have speculated on possible confounding processes in the detection of predator effects (e.g. Gilliam *et al* 1989). Perhaps the most important is the extremely patchy distribution of prey which masks the impact of predators in enclosures. One obvious implication is that a large number of replicates is needed to detect differences between treatments. More interesting is the possibility that prey exchange with the surrounding benthos rapidly replaces prey consumed. One determinant of prey exchange is the mesh size of the enclosure walls, often chosen as that which will just retain predators. In a recent analysis of 52 published experiments, Cooper and colleagues (1990) found a significant relationship between predator impact and mesh size of enclosures. They also found a tendency in their own field experiments for predator impacts to be more easily detected when using fine mesh enclosures. The effects of trout in California stream pools were also greater when prey turnover (drift out of each pool per day divided by the number of invertebrates in the pool) was low. Experiments by Lancaster *et al* (1991) are consistent with the notion that a high rate of exchange of prey with benthos may mask effects; when they restricted exchange by reducing enclosure mesh sizes, predator impact was significant. Both mathematical (Cooper *et al* 1990) and graphical (Lancaster *et al* 1991) models indicate the clear effect of exchange rate on the detection of predator impact at the patch (=enclosure) scale.

Intriguingly, predation on the stream benthos could therefore be a case of a process which is often not apparent at the small scale but which is important at broader spatial scales. Predation on mobile and patchily distributed prey can be intense overall whilst remaining elusive in patch scale experiments. The comparison with benthic algae and their herbivores is well made; algae are sessile and local depredations on them easy to detect. What has slowly been learned of predation in streams has made field manipulations more sophisticated and consistently rewarding. Recent experiments by Lancaster (1991) with predatory stoneflies involved replicate streamside channels rather than mesh enclosures and enabled careful accounting of arrival and departure of prey by drift. The results are rather clearcut and indicate a strong effect of a stonefly predator on the accumulation of two prey taxa (Fig. 7.7(a)).

We now seem close to a general conclusion, based both on circumstantial and direct experimental evidence, that predators in streams can exert definite top-down control of their prey. The limitations to this conclusion include the following:

1 Fluctuations in discharge or other disturbances can reset benthic densities and overwhelm the effects of predation (Lancaster *et al* 1990; Lancaster 1990).

2 Predator impact may be limited to those hydraulic conditions which favour the predators. Some recent experiments by Peckarsky *et al* (1990) have illustrated this nicely (Fig. 7.7(b)). Hansen *et al* (1991) showed also that predation by a triclad on blackfly larvae was mediated by flow in quite subtle ways. Not only did flow preferences differ between predator and prey (thus imposing partially separate microdistributions), but the ability of triclads successfully to complete prey capture declined at higher velocities. They explicitly linked this to the idea that there are refugia for prey in streams and rivers, provided by spatial variations in the forces of flow. Flow refugia are only one example of a whole range of circumstances, including substrate heterogeneity and temporal variations, which may act similarly (e.g. Hildrew & Townsend 1977). Refugia strongly stabilize predator/prey models (Begon *et al* 1990) and could thus be of dynamic significance in food webs.

3 Predators vary in their effectiveness. For instance, among the fish, salmonids seem to affect only rather large or otherwise vulnerable prey (e.g. Schofield *et al* 1988), while benthic feeders (Cottidae, Cyprinidae) have more widespread effects (e.g. Gilliam *et al* 1989; Schlosser & Ebel 1989). Salmonidae may also be relatively effective in slow rather than in fast-flowing streams (Cooper *et al* 1990).

4 Prey differ in their vulnerability through a miscellany of devices such as shells, cases and retreats. Selection pressures imposed by predators on how, where and when species feed, and upon the evolution of antipredator devices, probably play an important role in structuring freshwater communities (Jeffries & Lawton 1984).

Finally, it is clear that many predators are markedly polyphagous in running waters and take prey roughly in proportion to their abundance. In these circumstances, while predators influence prey, it is possible that the population dynamics of individual prey taxa have little effect on predators. Such a circumstance will influence stability of food-web models in ways which have yet to be explored (Lawton 1989).

Interference, competition for space and 'self-damping' in river food webs

Interference competition, including pseudointerference (Free *et al* 1977), is strongly stabilizing in population and food-web models (Lawton 1989). How widespread is it in river food webs? Some well-studied fractions of communities seem

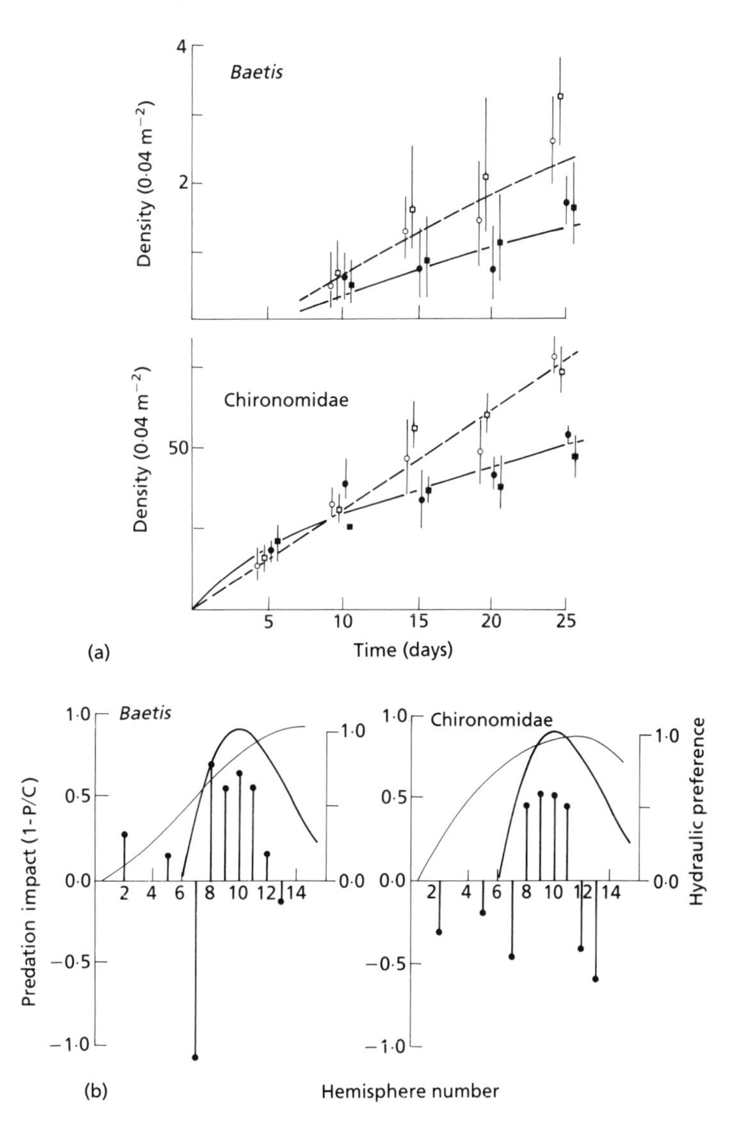

Fig. 7.7 (a) Lancaster (1990) shows the colonization by two prey taxa of four streamside channels with (solid symbols) or without (open symbols) predatory stoneflies (*Doroneuria baumanni*). Colonization curves are for with-predator (solid lines) or no-predator (broken lines) treatments. (b) In enclosure experiments, Peckarsky *et al* (1990) found that the impact of the predatory stonefly *Dinocras cephalotes* on two prey taxa varied along a hydraulic gradient (measured using the standard hemispheres of Statzner and Müller (1989), higher hemisphere numbers indicating greater forces of flow). Curves show the hydraulic preferences of predator (solid) and prey (broken). Predator impact (solid points) is measured as a relationship between mean prey density in cages with (P) or without (C) predators.

completely free of signs of structuring by competition for time, space or food (Tokeshi 1986; Tokeshi & Townsend 1987; Hildrew & Townsend 1987). Yet interference competition seems widespread among 'sessile' species of the stony benthos and there are examples of both interspecific and intraspecific competition for space among filter feeders, algal grazers and predators.

Various kinds of caddis larvae build semipermanent retreats and defend territories around them (Edington & Hildrew 1982). These include the filter-feeding Hydropsychidae, which are aggressive and stridulate (Matczak & Mackay 1990), the predatory Polycentropodidae (Hildrew & Townsend 1980), and the herbivorous Psychomyiidae and Hydroptilidae (Hart 1985). These families all contain dominant, competitive species that are important 'players' in many stream communities. The predatory larva of the polycentropodid *Plectrocnemia conspersa* is numerous in some fishless, acidic English streams, where it is an important predator on the remaining benthos. There is some evidence that its population density is limited by competition for space, since supplementation of net-spinning sites led to clear local increases in density (Lancaster *et al* 1988; Fig. 7.8(a)). Territoriality and interference competition is not limited to caddis larvae, and has also been described in grazing, tube-dwelling chironomids and in odonate larvae (Wiley & Kohler 1984; Johnson 1991).

Such species may pre-empt space and compete strongly with other sessile or mobile species (McAuliffe 1984b; Dudley *et al* 1990). Co-existence and relative abundance may then be determined by disturbance that opens up gaps which can be exploited by the less competitive species. This seems to be the case with *Hydropsyche*, and the more opportunistic *Simulium*, in some Californian streams (Hemphill & Cooper 1983; Fig. 7.8(b)). Hildrew & Townsend (1987) speculated that the sessile, territorial lifestyle is sustainable only in streams in which such disturbance is limited in its severity; for instance, where there is stable substrate and moderate or infrequent fluctuations in discharge and water level. Sessile species are absent from many streams and are abundant where stable substrate is combined with abundant renewable food (e.g. Valett & Stanford 1987).

7.3 INTERACTION WEBS IN RUNNING WATERS

The experiments and processes described so far are all limited to some extent to small spatial or temporal scales and to relatively simple, pairwise interactions. However, there is great interest in community ecology generally in multilevel, complex or indirect interactions (Kerfoot & Sih 1987; Huang & Sih 1990), and their importance in the regulation of trophic structure. Some of these complex interactions are plainly demonstrated in lakes (Carpenter 1988) but have attracted less attention in streams and rivers. The 'interaction' webs (partial networks of strong and weak biotic interactions between species or groups of similar species in communities) which often result from such studies (e.g. Menge & Sutherland 1987) form a 'half-way house' between the study of pairwise interactions of neighbouring compartments in

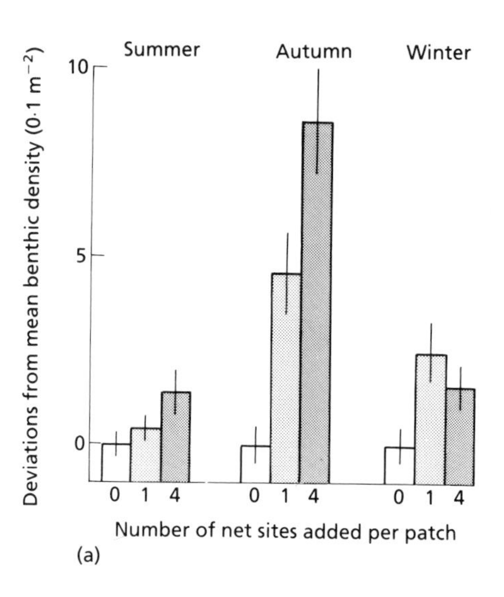

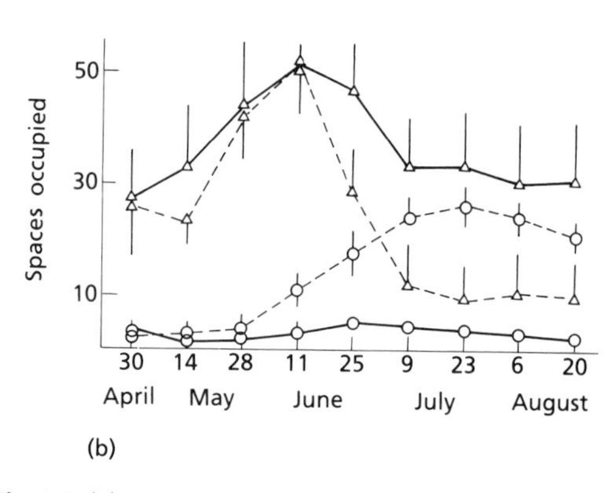

Fig. 7.8 (a) Lancaster *et al* (1988) supplemented net-spinning sites in small patches of streambed for the predatory caddis *Plectrocnemia conspersa*. Local density (measured as the deviation from that in the benthos) responded to the supplementation in each of three seasonal experiments. (b) Hemphill and Cooper (1983) regularly disturbed, by scrubbing, quadrats of bedrock in a Californian stream occupied by both *Simulium* and *Hydropsyche* larvae. Control quadrats were undisturbed. Counts are the mean (±SE) number of small spaces occupied by either insect between April and August. Notice that *Simulium* (△) showed a decline in the controls (broken line) whereas the larger, aggressive *Hydropsyche* (○) took up more spaces in these undisturbed areas.

the web, and the practically intractable problem of identifying every species and assessing every interaction.

Some authors have attempted large-scale, whole-river manipulations (despite the virtual impossibility of replication). Fertilization with phosphorus of the Kuparek River, a tundra stream in northern Alaska, was associated with an increase in algal biomass and productivity, an increase in bacterial activity in biofilms and an apparent increase in growth of blackfly larvae (Peterson *et al* 1985). Hershey and Hiltner (1988) subsequently found that the density of a filter feeding caddis, *Brachycentrus*, increased in the fertilized reach while that of blackflies declined, perhaps because of interference competition with the caddis. These results point to the possibility of impacts across several levels in the food web and of indirect interactions among animals.

Others have performed smaller scale, replicated experiments with several trophic levels using separate pools (Power *et al* 1985; McCormick 1990), enclosures (Hill & Harvey 1991) or replicated in-stream channels (Hart & Robinson 1990). Strongly positive results, in terms both of 'bottom-up' and 'top-down' control, have been obtained most often where the prevailing conditions are markedly pond-like. For instance, Power (1990) experimented with the food web around boulders in the Eel River, California at summer base flow. The web (Fig. 7.9(a)) consists essentially of four compartments: large fish, small predators (juvenile fish and invertebrates), tuft-weaving chironomids and the fast-growing macroalga Cladophora and its epiphytes.

Power (1990) found that the large fish depress the densities of the smaller predators; this releases the herbivorous chironomids from predation and, in turn, these reduce the Cladophora to a low prostrate form with few epiphytes. Removal of large fish results in high densities of smaller predators, fewer chironomids and tall, upright tufts of Cladophora with many epiphytes. These manipulations provide the first support in rivers for the Hairston *et al* (1960) and Fretwell (1977) model of community organization, in which alternate trophic levels are controlled by predation and competition (Fig. 7.9(b) and (c)). The main biological reason for this is the fact that the herbivore is invulnerable to the top carnivore, in this case the large fish which rarely eat the chironomids. Again, Power (1990) is careful to acknowledge that such interactions are likely to be detectable only in rivers where the frequency of disturbance does not exceed the capacity of the system to re-establish itself.

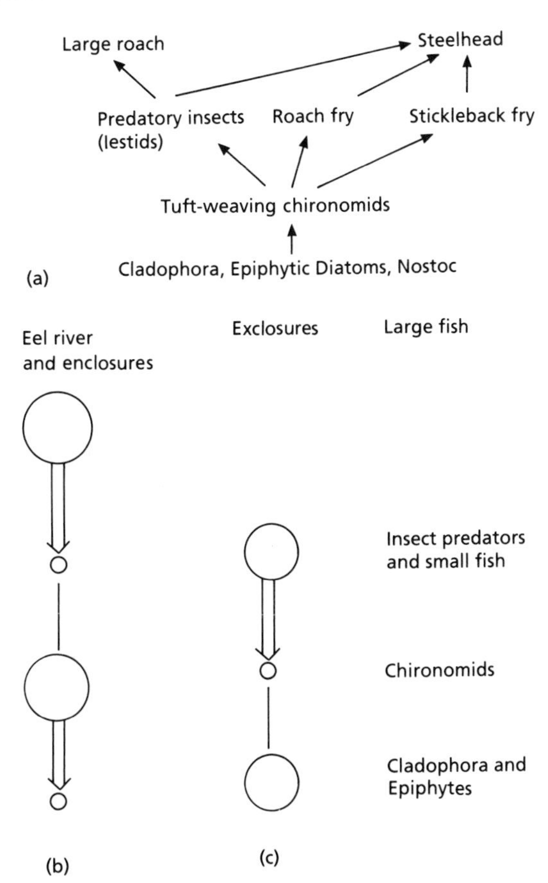

Fig. 7.9 (a) The 'summer boulder' food web in Eel River, California (Power 1990). (b) Power's exclusion of large fish changed the interaction web from a four-level to (c) a three-level system and altered the relative abundance of the elements (as indicated by the size of the circles).

In Fig. 7.10(a) a speculative interaction web for a hypothetical stony trout stream has been constructed. This draws on instances of strong interactions in the literature, many of them mentioned above, but in no single site have all these interactions been demonstrated. It is over-simplified in inumerable respects, not least because it represents only salmonid fish and ignores terrestrial linkages with birds and piscivorous mammals. The main features of this system are marked omnivory, self-damping by salmonid populations, strong interactions between algae and their grazers, and control of detritivores by their food and of large, vulnerable invertebrates (often predatory) by fish. Such a system will not be marked by obvious alternate level regulation by competition and predation, as is the Eel River. This is because it does not include a level of herbivorous/detritivorous consumers which are invulnerable to the dominant, but not to the subdominant, predator. It remains to be seen whether such an interaction web is realistic, but it obviously bears a close resemblance to the Menge and Sutherland (1987) model of community control. The main difficulty with this model comes with its assumption that physical disturbance affects sessile organisms least. These are rather less well represented in streams and, at least as far as animals go, seem rather vulnerable to physical disturbance.

Physiological stress, in the form of acidification, removes vulnerable species and whole levels in the food web of trout streams (Sutcliffe & Hildrew 1989). Fish are lost, as are many algal grazers. The web shown in Fig. 7.10(b) is derived mainly from the work of myself and collaborators in some southern English streams (e.g. Schofield *et al* 1988; Groom & Hildrew 1989; Lancaster *et al* 1991). The main features are marked omnivory, strong top-down control of prey by abundant, polyphagous invertebrate predators (which are themselves self-damped) and overall bottom-up control of productivity by detrital food quality.

Almost all stream ecologists are aware of the vulnerability of their systems to physical disturbance. Yet there is no very well-developed theoretical or empirical approach to the nature of disturbance and its effect on running waters: this is a very obvious growth area for research (Resh *et al* 1988). There is not much evidence for the wholesale loss of complete interaction elements in trout streams which are just fairly torrential and 'flashy'. It seems most likely that these kinds of hydraulic physical disturbances simply weaken species interactions, as shown in Fig. 7.10(c) and (d), while not altering the basic shape of the web overall. It should be kept in mind that all streams will be subject to reset by disturbance (Fisher *et al* 1982) and to periods when conditions are less extreme. Systems differ from each other in frequency of disturbance, predictability and mean conditions. The strong interactions illustrated will 'flicker' on and off in concert with the environment and, as Statzner (1987) has pointed out, different components of the community will have different recovery rates after reset. This makes the point that we need studies of interactions and food webs throughout the year, to take some account of seasonal and other temporal changes.

7.4 GAPS IN KNOWLEDGE AND FUTURE DEVELOPMENTS

Undoubtedly the most important lacuna in food-web research is in the area of the 'microbial food loop', in which stream research lags behind its marine equivalent by perhaps 10 years (Carlough & Meyer 1990). Research on the energetic base of streams has stressed three main sources and modes of transfer into food webs. These are: (i) terrestrial leaf litter and the detritus/microbe/shredder pathway; (ii) benthic algae grazed by macroinvertebrates; and (iii) dissolved organic matter, taken up in biofilms and grazed by invertebrates (see Winterbourn 1986 for a review). Groundwater, discharging up through bed sediments, is a possible source of the latter material (Hynes 1983). Recent research shows that biotic immobilization of dissolved free amino acids by sediments cores is very effective (Fiebig & Lock 1991).

A fourth pathway is for dissolved organics, either autochthonous or allochthonous in origin, to be taken up by suspended bacteria in suspension which could then be consumed by Protozoa. Carlough and Meyer (1990) estimated that suspended Protozoa cleared an average of 47% of the water column of the Ogeechee River per day. These Protozoa were then available to microfilterers such as *Simulium*. More generally there are some estimates of the densities both of suspended and sedimentary bacteria and Protozoa, the meiofauna, and of bacterivory by Protozoa

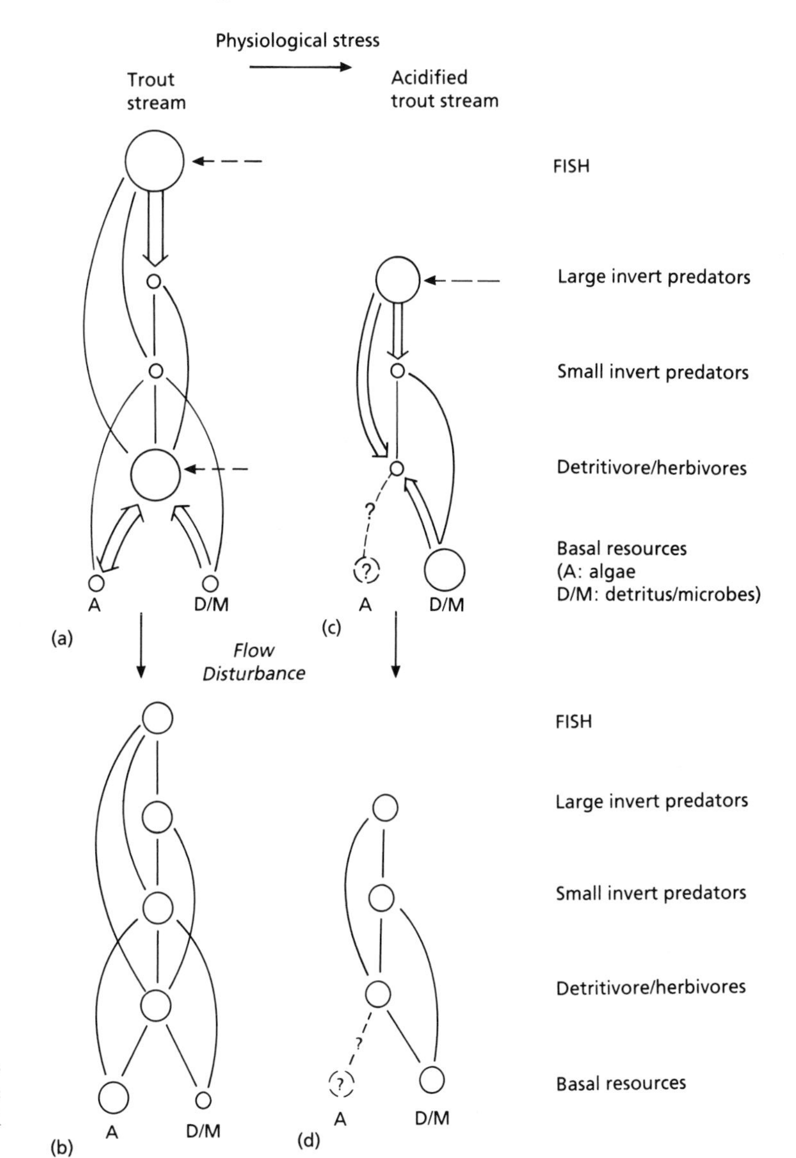

Fig. 7.10 Hypothetical interaction webs in (a) a stony trout stream, (b) a trout stream subject to frequent flow disturbance, (c) an acidified stony stream and (d) a similarly disturbed acid stream. Narrow solid lines represent trophic interactions; thick lines represent strong interactions with direction of main effect indicated by arrows; horizontal broken lines represent incidences of 'self-damping'. Size of circles crudely represents abundance of elements. The role of algae and their grazers in acid streams remains uncertain.

(e.g. Baldock & Sleigh 1988; Bott & Kaplan 1989, 1990; Carlough & Meyer 1989, 1990). The incorporation of dissolved organic matter into river food webs, probably through a microbial link, is a subject of the highest priority for research.

A food source that appears not often to be directly exploited in streams and rivers is that of living aquatic macrophytes (see Newman 1991). This is possibly due to chemical defences by macrophytes (Ostrofsky & Zettler 1986), although this subject is poorly explored in freshwaters. Aquatic angiosperms are a taxonomically scattered assemblage, consisting of a restricted number of representatives from a wide range of terrestrial groups (Sculthorpe 1967). They invaded freshwaters after the primarily aquatic insect orders did, and may be chemically well defended against the generalist detritivore/herbivores predominant amongst the really common aquatic insects. Newman (1991) points out that 63% of the freshwater phytophagous insect species are from primarily terrestrial insect orders (Lepidoptera, Coleoptera and Orthoptera). While aquatic macrophytes have clearly taken some of their terrestrial herbivores with them, species richness is restricted: about 45% of terrestrial insects are phytophagous compared with 11% of aquatic species. Perhaps this restriction can be ascribed to the patchy taxonomic representation of angiosperms in freshwaters. In effect there are large 'voids' of niche space separating 'islands' of plants.

Others argue that there is a coevolved mutualism between grazers, especially aquatic snails, and macrophytes (Bronmark 1985; Thomas 1987). Plants gain from the grazing of epiphytic algae from their surfaces, thus reducing shading, and snails gain food, shelter and access to dissolved organic molecules. A further possible role for mutualisms in freshwater food webs is now suggested between grazers and the epiphytic algae themselves (Underwood & Thomas 1990). Many algae seem to survive passage through grazer guts and actually benefit from grazing through access to mineral nutrients. The grazers would gain from the release of dissolved organics from algae in the gut. McCormick (1990) recently found that moderate herbivory enhanced algal growth in isolated stream pools under conditions of nitrogen limitation.

The role of parasites and diseases in stream and river food webs is often discounted (e.g. Statzner 1987). However, Cummins and Wilzbach (1988) have argued that pathogenic infections represent the primary regulator of stream invertebrate populations. They were able to muster rather little evidence but this is an area that surely merits further investigation. In the context of food webs, it would be particularly interesting if webs in running waters turned out generally to be composed of mixtures of conventional predators with microparasites and diseases. Such webs 'violate the assumptions of the (size) cascade model' (Lawton 1989).

Finally, one of the reasons river food webs may have appeared rather 'short' is that a potentially important link is almost always ignored—predation by streamside birds and mammals. For instance, one family of passerine birds, the Cinclidae or dippers, are particularly adapted to fast-flowing rivers and feed exclusively on small fish and stream invertebrates. The five species of dippers cover extremely large areas, including the Andes (two species), the Rocky Mountains (one), the Himalayas and western Pacific Islands (one), and the Palaearctic from western Europe to the Himalayas, north Africa and Persia (one) (Ormerod & Tyler 1991). Ormerod and Tyler estimated that the Eurasian dipper could account for almost the entire production of Cottidae in some small Welsh streams. These fish provide the birds with an essential source of calcium and are taken particularly in late summer. Similarly, hydropsychid caddis larvae are heavily preyed upon, making up between 12% and 26% of the diet of dippers on circumneutral streams, which could account for about 11–24% of hydropsychid production. We have already seen the dangers and discrepancies inherent in these kinds of calculations but this example surely demonstrates that land–water food web linkages should not be ignored.

More generally, the evident extension of river food webs into the riparian zone and beyond holds the possibility that, although there may be some compartmentalization of webs at habitat boundaries, isolation is by no means complete. River food webs, representing mainly interactions within the channel, are firmly 'nested' into the primarily terrestrial webs of river catchments. This is a community-level expression of the idea that running waters are part of catchment scale ecosystems.

7.5 FOOD WEBS AND RIVER MANAGEMENT

In areas where most river ecologists work, almost all running waters are profoundly affected by humans, and it seems that human impacts invariably alter the scope and influence of species interactions in food webs. The application of ideas in food-web ecology to river management and conservation, however, is in its infancy. Among the most obvious effects of humans are those on water quality (Calow *et al* 1990). The only substantial applications of community level phenomena to problems with water quality are in establishing methods of assessment, most obviously using various indices based on macroinvertebrates, algae or other groups. Community structure can also be used to establish 'targets' for environmental management and to judge conservation value. There has been little attempt, however, to establish the *processes* by which communities respond to pollution and thus underly those 'structures' (e.g. species composition and richness, relative abundance) which are taken to indicate various kinds of chemical stress. Emphasis has been strongly on the direct toxicity of pollutants to individual species while the possible ramifying effects through food webs have largely been ignored. I have tried to point out these effects, in the context of acidification, elsewhere (Sutcliffe & Hildrew 1989). Species introductions are another impact of humans which can be understood in the context of food webs. Many introductions appear to have little further ecological effect; a few have had profound impacts several links away in the web.

Perhaps the most interesting and pervasive of all are the ecological effects of a number of physical changes commonly wrought by engineering works on channel form and confinement (e.g. Brookes 1988). Contact and exchange between land and water is perhaps the most basic property of river systems, and its influence pervades almost every section of this book. The importance of land/water interactions, including those between the surface flow and groundwater, have been recognized particularly by Hynes (1975, 1983) and demonstrated most fruitfully at the ecosystem level of organization by Likens (1985). The extension of running-water food webs on to the land has been briefly mentioned above and finds its most spectacular examples in the great tropical floodplain rivers. In the Amazon, many adult fish, their eggs or fry migrate, or are swept by seasonal spates, out on to enormous inland floodplains where they may feed on prey, fruit, seeds or detritus of terrestrial origin (Welcomme 1979). Aquatic invertebrates may also share in this periodic use of floodplains (Gladden & Smock 1990). But most north temperate rivers, especially in Europe, have lost their floodplains to agriculture and urban development, and with them probably a large number of extra food web linkages. Many important land/water interactions remain, particularly where riparian vegetation is intact, and the importance of these edges or 'ecotones' is increasingly recognized (Naiman & Décamps 1990).

River engineering essentially confines rivers within their channels and tries to isolate them from the land. Ecological processes within the channels themselves are also altered by the engineers. Their task is essentially to prevent flooding and to remove water from catchments as soon as possible. Channels are straightened and deepened, and weedbeds and snags are removed to lessen hydraulic resistance (Brookes 1988). This reduces the surface area and physical complexity of the habitat and removes refugia from predation for vulnerable species and will alter the strength and outcome of species interactions.

Conventional river engineering, therefore, reduces the retention time of water and organic matter. By comparison, flow dead-zones in natural river channels, essentially unmixed parcels of non-flowing water, increase the average retention time of the river. Reynolds (1988) recognized that these features of flow, associated with channel features such as meanders, snags and holes in the bed, could at last provide an explanation for the persistence of phytoplankton in rivers. They could similarly account for the rich populations of water-column Protozoa in some rivers (Carlough & Meyer 1990). Flow dead-zones could also provide temporary refugia for invertebrates and fish during flow events, reducing catastrophic mortality and speeding up the return time of the system after disturbance (Hildrew *et al* 1991). In summary, most human impacts on water quality, channel form and flow must have simplified river food webs and increased the influence of disturbance.

7.6 CONCLUSIONS

Based on this brief survey of river food webs and the biotic interactions embedded in them, it would be difficult to conclude that dynamic constraints alone determine what trophic structures occur. While interactions are very important in certain circumstances, species dynamics are patchy in space and time and seem not to constitute 'closely coupled enemy–victim interactions'. Very speculatively, most interactions are probably weak most of the time. The static, trophic cascade model of Cohen and Newman (1985) therefore seems a much more profitable starting point. Recall that this proposes a trophic hierarchy such that species feed only on others below them. Warren and Lawton (1987) stress the role of body size in determining the order of vulnerability.

What seems to determine the trophic structure of Eel River (Power 1990) is the ecological disparity between top and bottom – the smallest prey are not vulnerable to the top predators. With strong interactions among elements, this produces a trophic structure close to the model of Hairston *et al* (1960) and Fretwell (1977). This does not seem obviously to be the case in many other stream and river food webs, where omnivory (species feeding on two or more levels in the web) is prominent. Perhaps river food webs are more generally short and broad because of a limited 'ecological disparity' between top and bottom, most obviously but not always based on size. Habitat structure, including refugia provided by complexity on vertical and horizontal spatial scales (Ward 1989), must also play a part in this 'disparity' and thus to the chain length and the degree of omnivory in food webs. Since so much of the anthropogenic change wrought on rivers removes physical complexity, food-web shapes and structures characteristic of 'pristine' systems could have been early casualties.

Finally, it is clear that this is an area where fundamental and applied ecology have related aims. We presently have a restricted ecological understanding of the basis of community structures in rivers. The need is to move from empiricism and rule of thumb to understanding and prediction within the constraints imposed by a very variable sytem. Both better food-web descriptions and studies on strong and weak interactions are badly needed.

ACKNOWLEDGEMENTS

I am very grateful to the colleagues who responded to my requests for help and for reprints and preprints of their work: Dave Allen, Paul L. Angermeier, Randi A. Hansen, David Hart, Walter Hill, Gary A. Lamberti, Jonathan T. Morales, Raymond M. Newman, Christopher G. Peterson, John Richardson, Amy D. Rosemond, Lane C. Smith, Dan Soluk and Phil Warren. I particularly thank Mary Power for the Eel River food web, and Jill Lancaster, John Lawton, Mike Winterbourn and an anonymous referee for their helpful reviews.

REFERENCES

Allan JD. (1983) Predator–prey relationships in streams. In: Barnes JR, Minshall GW (eds) *Stream Ecology: Application and Testing of General Ecological Theory*, pp 191–229. Plenum Press, New York. [7.2]

Arsuffi TL, Suberkropp K. (1988) Effects of fungal mycelia and enzymatically degraded leaves on feeding and performance of caddisfly (Trichoptera) larvae. *Journal of the North American Benthological Society* 7: 205–11. [7.2]

Baldock BM, Sleigh MA. (1988) The ecology of benthic Protozoa in rivers: seasonal variation in numerical abundance in fine sediments. *Archiv für Hydrobiologie* **111**: 409–21. [7.4]

Bärlocher F, Kendrick B. (1975) Leaf-conditioning by microorganisms. *Oecologia* **20**: 359–62. [7.2]

Begon M, Harper JL, Townsend CR. (1990) *Ecology: Individuals, Populations and Communities* 2nd edn. Blackwell Scientific Publications, Oxford. [7.2]

Biggs BJF, Close ME. (1989) Periphyton biomass dynamics in gravel bed rivers: the relative effects of flows and nutrients. *Freshwater Biology* **22**: 209–31. [7.2]

Bott TL, Kaplan LA. (1989) Densities of benthic Protozoa and nematodes in a Piedmont stream. *Journal of the North American Benthological Society* **8**: 187–96. [7.4]

Bott TL, Kaplan LA. (1990) Potential for protozoan grazing of bacteria in streambed sediments. *Journal of the North American Benthological Society* **9**: 336–45. [7.4]

Briand F. (1985) Structural singularities of freshwater food webs. *Verhandlungen der Internationalen Vereinigung für Theoretische und Angewandte Limnologie* **22**: 3356–64. [7.1]

Briand F, Cohen JE. (1987) Environmental correlates of food chain length. *Science* **238**: 956–60. [7.4]

Bronmark C. (1985) Interactions between macrophytes, epiphytes and herbivores: an experimental approach. *Oikos* **45**: 26–30. [7.4]

Brookes A. (1988) *Channelized Rivers: Perspectives for Environmental Management*. John Wiley & Sons, Chichester. [7.5]

Busch DE, Fisher SG. (1981) Metabolism of a desert stream. *Freshwater Biology* **11**: 301–7. [7.2]

Calow P, Armitage P, Boon P, Chave P, Cox E, Hildrew A *et al* (1990) *River Water Quality*. The Ecological Issues No. 1, British Ecological Society: Field Studies Council Publishers, Shrewsbury. [7.5]

Carlough LA, Meyer JL. (1989) Protozoans in two southeastern blackwater rivers and their importance to trophic transfer. *Limnology and Oceanography* **34**: 163–77. [7.4]

Carlough LA, Meyer JL. (1990) Rates of protozoan bacterivory in three habitats of a southeastern blackwater river. *Journal of the North American Benthological Society* **9**: 45–53. [7.4, 7.5]

Carpenter SR (ed.) (1988) *Complex Interactions in Lake Communities*. Springer-Verlag, New York. [7.3]

Cohen JE, Newman CM. (1985) A stochastic theory of community food webs. 1. Models and aggregated data. *Proceedings of the Royal Society of London, B* **224**: 421–48. [7.1, 7.6]

Cooper SD, Walde SJ, Peckarsky BL. (1990) Prey exchange rates and the impact of predators on prey populations in streams. *Ecology* **71**: 1503–14. [7.2]

Cummins KW, Wilzbach MA. (1988) Do pathogens regulate stream invertebrate populations? *Verhandlungen der Internationalen Vereinigung für Theoretische und Angewandte Limnologie* **23**, 637–41. [7.4]

DeNicola DM, McIntire CD, Lamberti GA, Gregory SV, Ashkenas LR. (1990) Temporal patterns of grazer–periphyton interactions in laboratory streams. *Freshwater Biology* **23**: 475–89. [7.2]

Dudley TL, D'Antonio CM, Cooper SC. (1990) Mechanisms and consequences of interspecific competition between two stream insects. *Journal of Animal Ecology* **59**: 849–66. [7.2]

Edington JM, Hildrew AG. (1982) *Caseless Caddis Larvae of the British Isles*. Scientific Publications of the Freshwater Biological Association, Ambleside. [7.2]

Feminella JW, Power ME, Resh VH. (1989) Periphyton responses to invertebrate grazing and riparian canopy in three northern California coastal streams. *Freshwater Biology* **22**: 445–57. [7.2]

Fiebig DM, Lock MA. (1991) Immobilization of dissolved organic matter from groundwater discharging through the stream bed. *Freshwater Biology* **26**: 45–55. [7.4]

Findlay S, Meyer JL, Smith PJ. (1986) Incorporation of microbial biomass by *Peltoperla* sp. (Plecoptera) and *Tipula* sp (Diptera). *Journal of the North American Benthological Society* **5**: 306–10. [7.2]

Fisher SG, Gray LJ, Grimm NB, Busch DE. (1982) Temporal succession in a desert stream ecosystem following flash flooding. *Ecological Monographs* **52**: 93–110. [7.3]

Free CA, Beddington JR, Lawton JH (1977) On the inadequacy of simple models of mutual interference for predation and parasitism. *Journal of Animal Ecology* **46**: 543–54. [7.2]

Fretwell SD. (1977) The regulation of plant communities by food chains exploiting them. *Perspectives in Biology and Medicine* **20**: 169–85. [7.3, 7.6]

Gee JRD. (1988) Population dynamics and morphometrics of *Gammarus pulex* L.: evidence of seasonal food limitation in a freshwater detritivore. *Freshwater Biology* **19**: 333–43. [7.2]

Gilliam JF, Fraser DF, Sabat AM. (1989) Strong effects of foraging minnows on a stream benthic invertebrate community. *Ecology* **70**: 445–52. [7.2]

Gladden JE, Smock LA. (1990) Macroinvertebrate distribution and production on the floodplains of two lowland headwater streams. *Freshwater Biology* **24**: 533–45. [7.5]

Gregory SV. (1983) Plant–herbivore interactions in stream systems. In: Barnes JR, Minshall GW (eds) *Stream Ecology: Application and Testing of General Ecological Theory*, pp 157–89. Plenum Press, New York. [7.2]

Groom AP, Hildrew AG. (1989) Food quality for detritivores in streams of contrasting pH. *Journal of Animal Ecology* **58**: 863–81. [7.2, 7.3]

Hairston NG, Smith FE, Slobodkin LB. (1960) Community structure, population control and competition. *American Naturalist* **94**: 421–5. [7.3, 7.6]

Hansen RA, Hart DD, Merz RA. (1991) Flow mediates predator–prey interactions between triclad flatworms and larval blackflies. *Oikos* **60**: 187–96. [7.2]

Hart DD. (1985) Causes and consequences of territoriality in a grazing stream insect. *Ecology* **60**: 404–14. [7.2]

Hart DD. (1987) Experimental studies of exploitative competition in a grazing stream insect. *Oecologia* **73**: 41–7. [7.2]

Hart DD, Robinson CT. (1990) Resource limitation in a stream community: phosphorus enrichment effects on periphyton and grazing. *Ecology* **71**: 1494–1502. [7.3]

Hemphill N, Cooper SD. (1983) The effect of physical disturbance on the relative abundance of two filter-feeding insects in a small stream. *Oecologia* **58**: 378–82. [7.2]

Hershey AE, Hiltner AL. (1988) Effect of a caddisfly on blackfly density: interspecific interactions limit blackflies in an arctic river. *Journal of the North American Benthological Society* **7**: 188–96. [7.3]

Hildrew AG, Townsend CR. (1977) The influence of substrate on the functional response of *Plectrocnemia conspersa* (Curtis) larvae (Trichoptera: Polycentropodidae). *Oecologia* **31**: 21–6. [7.2]

Hildrew AG, Townsend CR. (1980) Aggregation, interference and foraging by larvae of *Plectrocnemia conspersa* (Trichoptera: Polycentropodidae). *Animal Behaviour* **28**: 553–60.]7.2]

Hildrew AG, Townsend CR. (1987) Organization in freshwater benthic communities. In: Gee JHR, Giller PS (eds) *Organization of Communities: Past and Present*, pp 347–71. Symposia of the British Ecological Society, Blackwell Scientific Publications, Oxford. [7.1, 7.2]

Hildrew AG, Townsend CR, Hasham A. (1985) The predatory Chironomidae of an iron-rich stream: feeding ecology and food web structure. *Ecological Entomology* **10**: 403–13. [7.1]

Hildrew AG, Dobson MK, Groom A, Ibbotson A, Lancaster J, Rundle SD. (1991) Flow and retention in the ecology of stream invertebrates. *Verhandlungen der Internationalen Vereinigung für Theoretische und Angewandte Limnologie* **24**: 1742–7. [7.2, 7.5]

Hill WR, Harvey BC. (1991) Periphyton responses to higher trophic levels and light in a shaded stream. *Canadian Journal of Fisheries and Aquatic Sciences* **47**: 2307–14. [7.2, 7.3]

Hill WR, Knight AW. (1987) Experimental analysis of the grazing interaction between a mayfly and stream algae. *Ecology* **68**: 1955–65. [7.2]

Huang C, Sih A. (1990) Experimental studies on behaviorally mediated, indirect interactions through a shared predator. *Ecology* **71**: 1515–22. [7.3]

Hynes HBN. (1975) The stream and its valley. *Verhandlungen der Internationalen Vereinigung für Theoretische und Angewandte Limnologie* **19**: 1–15. [7.5]

Hynes HBN. (1983) Groundwater and stream ecology. *Hydrobiologia* **100**: 93–9. [7.4, 7.5]

Jeffries MJ, Lawton JH. (1984) Enemy free space and the structure of ecological communities. *Biological Journal of the Linnean Society* **23**: 269–86. [7.2]

Jeffries MJ, Lawton JH. (1985) Predator–prey ratios in communities of freshwater invertebrates: the role of enemy free space. *Freshwater Biology* **15**: 105–12. [7.1]

Johnson DM. (1991) Behavioral ecology of larval dragonflies and damselflies. *Trends in Ecology and Evolution* **6**(1): 8–13. [7.2]

Kerfoot WC, Sih A. (1987) *Predation: Direct and Indirect Impacts on Aquatic Communities*. University Press of New England, Hanover, New Hampshire. [7.3]

Lamberti GA, Moore JW. (1984) Aquatic insects as primary consumers. In: Resh VH, Rosenberg DM (eds) *The Ecology of Aquatic Insects*, pp 164–95. Praeger, New York. [7.2]

Lamberti GA, Feminella JW, Resh VH. (1987) Herbivory and intraspecific competition in a stream caddisfly population. *Oecologia* **73**: 75–81. [7.2]

Lamberti GA, Gregory SV, Ashkenas LR, Steinman AD, McIntire CD. (1989) Productive capacity of periphyton as a determinant of plant herbivore interactions in streams. *Ecology* **70**: 1840–56. [7.2]

Lancaster J. (1990) Predation and drift of lotic macroinvertebrates during colonization. *Oecologia* **85**: 457–63. [7.2]

Lancaster J, Hildrew AG, Townsend CR. (1988) Competition for space by predators in streams: field experiments in a net-spinning caddisfly. *Freshwater Biology* **20**: 185–93. [7.2]

Lancaster J, Hildrew AG, Townsend CR. (1990) Stream flow and predation effects on the spatial dynamics of benthic invertebrates. *Hydrobiologia* **203**: 177–90. [7.2]

Lancaster J, Hildrew AG, Townsend CR. (1991) Invertebrate predation on patchy and mobile prey in streams. *Journal of Animal Ecology* **60**: 625–41. [7.2, 7.3]

Lawton JH. (1989) Food webs. In: Cherrett JM (ed.) *Ecological Concepts*, pp 43–78. Symposia of the British Ecological Society, Blackwell Scientific Publications, Oxford. [7.1, 7.2, 7.4]

Likens GE (ed.). (1985) *An Ecosystem Approach to Aquatic Ecology: Mirror Lake and its Environment*. Springer-Verlag, New York. [7.5]

McAuliffe JR. (1984a) Resource depression by a stream herbivore: effects on distributions and abundances of other grazers. *Oikos* **42**: 327–33. [7.2]

McAuliffe JR. (1984b) Competition for space, disturbance, and the structure of a benthic stream community. *Ecology* **65**: 894–908. [7.2]

McCormick PV. (1990) Direct and indirect effects of consumers on benthic algae in isolated pools of an ephemeral stream. *Canadian Journal of Fisheries and Aquatic Sciences* **47**: 2057–67. [7.3, 7.4]

McCormick PV, Stevenson RJ. (1989) Effects of snail grazing of benthic algal community structure in different nutrient environments. *Journal of the North American Benthological Society* **8**: 162–72. [7.2]

Matczak TZ, Mackay RJ. (1990) Territoriality in filter-feeding caddisfly larvae: laboratory experiments. *Journal of the North American Benthological Society* **9**: 26–34. [7.2]

Menge BA, Sutherland JP. (1987) Community regulation: variation in disturbance, competition, and predation in relation to environmental stress and recruitment. *American Naturalist* **130**: 730–57. [7.1, 7.3]

Naiman RJ, Décamps H (eds). (1990) *The Ecology and Management of Aquatic–Terrestrial Ecotones*. Man and the Biosphere Series, Vol. 4, UNESCO, Parthenon Publishing Group, Paris. [7.5]

Newman RM. (1991) Herbivory and detritivory on freshwater vascular macrophytes by aquatic invertebrates: a review. *Journal of the North American Benthological Society* **10**: 89–114. [7.2, 7.4]

Ormerod SJ, Tyler SJ. (1991) Predatory exploitation by a river bird, the dipper *Cinclus cinclus* (L.), along acidic and circumneutral streams in upland Wales. *Freshwater Biology* **25**: 105–16. [7.4]

Ostrofsky ML, Zettler ER. (1986) Chemical defenses in aquatic plants. *Journal of Ecology* **74**: 279–87. [7.4]

Paine RT. (1988) On food webs: road maps of interactions or grist for theoretical development? *Ecology* **69**: 1648–54. [7.1]

Peckarsky BL. (1984) Predator–prey interactions among aquatic insects. In: Resh VH, Rosenberg DM (eds) *The Ecology of Aquatic Insects*, pp 196–254. Praeger, New York. [7.2]

Peckarsky BL, Penton MA. (1988) Why do *Ephemerella* nymphs scorpion posture: a 'ghost of predation past'? *Oikos* **53**: 185–93. [7.2]

Peckarsky BL, Horn SC, Statzner B. (1990) Stonefly predation along a hydraulic gradient: a field test of the harsh–benign hypothesis. *Freshwater Biology* **24**: 181–91. [7.2]

Peterson BJ, Hobbie JE, Hershey AE, Lock MA, Ford TE, Vestal JR *et al.* (1985) Transformation of a tundra river from heterotrophy to autotrophy by addition of phosphorus. *Science* **229**: 1383–6. [7.3]

Pimm SL. (1982) *Food Webs.* Chapman and Hall, London. [7.1, 7.2]

Pimm SL, Lawton JH, Cohen JE. (1991) Food web patterns, their causes and their consequences. *Nature* **350**: 669–74. [7.1]

Power ME. (1990) Effects of fish in river food webs. *Science* **250**: 811–14. [7.3, 7.6]

Power ME, Mathews WJ, Stewart AJ. (1985) Grazing minnows, piscivorous bass, and stream algae: dynamics of a strong interaction. *Ecology* **66**: 1448–56. [7.3]

Pringle CM, Naiman RJ, Bretschko G, Karr JR, Oswood MW, Webster JR *et al.* (1988) Patch dynamics in lotic systems: the stream as a mosaic. *Journal of the North American Benthological Society* **7**: 503–24. [7.1]

Resh VH, Brown AV, Covich AP, Gurtz ME, Li HW, Minshall GW *et al.* (1988) The role of disturbance in stream ecology. *Journal of the North American Benthological Society* **7**: 433–55. [7.1, 7.3]

Reynolds CS. (1988) Potamoplankton: paradigms, paradoxes and prognoses. In: Round FE (ed.) *Algae and the Aquatic Environment*, pp 285–311. Biopress, Bristol. [7.5]

Richardson JS. (1991) Seasonal food limitation of detritivores in a montane stream: an experimental test. *Ecology* **72**: 873–87. [7.2]

Rundle SD, Hildrew AG. (1990) The distribution of micro-arthropods in some southern English streams: the influence of physicochemistry. *Freshwater Biology* **23**: 411–31. [7.1]

Schlosser IJ, Ebel KK. (1989) Effects of flow regime and cyprinid predation on a headwater stream. *Ecological Monographs* **59**: 41–57. [7.2]

Schofield K, Townsend CR, Hildrew AG. (1988) Predation and the prey community of a headwater stream. *Freshwater Biology* **20**: 85–95. [7.2, 7.3]

Sculthorpe CD. (1967) *The Biology of Aquatic Vascular Plants.* St. Martin's Press, New York. [7.4]

Sinsabaugh RL, Linkins AE, Benfield EF. (1985) Cellulose digestion and assimilation by three leaf-shredding aquatic insects. *Ecology* **66**: 1464–71. [7.2]

Smock LA, Metzler GM, Gladden JE. (1989) Role of debris dams in the structure and functioning of low-gradient headwater streams. *Ecology* **70**: 764–75. [7.2]

Soluk DA. (1990) Postmolt susceptibility of *Ephemerella* larvae to predatory stoneflies: constraints on defensive armour. *Oikos* [7.2]

Statzner B. (1987) Characteristics of lotic ecosystems and consequences of future research directions. In: Schulze ED, Zwolfer H (eds) *Potentials and Limitations of Ecosystem Analysis*, pp 365–90. Ecological Studies 61, Springer, Berlin. [7.3, 7.4]

Statzner B, Müller R. (1989) Standard hemispheres as indicators of flow characteristics in lotic benthic research. *Freshwater Biology* **21**: 445–9. [7.2]

Sutcliffe DW, Hildrew AG. (1989) Invertebrate communities in acid streams. In: Morris R̈ Brown DJA, Brown JA (eds) *Acid Toxicity and Aquatic Animals*, pp 13–29. Seminar Series of The Society for Experimental Biology, Cambridge University Press, Cambridge. [7.3, 7.5]

Thomas JD. (1987) An evaluation of the interactions between freshwater pulmonate snail hosts of human schistosomes and macrophytes. *Philosophical Transactions of the Royal Society of London, B* **315**: 75–125. [7.4]

Tokeshi MW. (1986) Resource utilization, overlap and temporal community dynamics: a null model analysis of an epiphytic chironomid community. *Journal of Animal Ecology* **55**: 491–506. [7.2]

Tokeshi MW, Townsend CR. (1987) Random patch formation and weak competition: coexistence in an epiphytic chironomid community. *Journal of Animal Ecology* **56**: 833–45. [7.2]

Underwood GJC, Thomas JD. (1990) Grazing interactions between pulmonate snails and epiphytic algae and bacteria. *Freshwater Biology* **23**: 505–22. [7.4]

Valett HM, Stanford JA. (1987) Food quality and hydropsychid caddisfly density in a lake outlet stream in Glacier National Park, Montana, USA. *Canadian Journal of Fisheries and Aquatic Sciences* **44**: 77–82.

[7.2]
Ward JV. (1989) The four-dimensional nature of lotic ecosystems. *Journal of the North American Benthological Society* **8**: 2–8. [7.6]
Warren PH. (1990) Variation in food web structure: the determinants of connectance. *American Naturalist* **136**: 689–700. [7.1]
Warren PH, Lawton JH. (1987) Invertebrate predator–prey body size relationships: an explanation for upper triangular food webs and patterns in food web structure? *Oecologia* **74**: 231–5. [7.1, 7.6]
Waters TF. (1988) Fish production–benthos production relationships in trout streams. *Polskie Archiwum Hydrobiologii* **35**: 545–61. [7.2]
Welcomme RL. (1979) *Fisheries Ecology of Floodplain Rivers.* Longman, London. [7.5]
Wiley MJ, Kohler SL. (1984) Behavioral adaptations of aquatic insects. In: Resh VH, Rosenberg DM (eds) *The Ecology of Aquatic Insects*, pp 101–33. Praeger, New York. [7.2]
Winterbourn MJ. (1986) Recent advances in our understanding of stream ecosystems. In: Polunin N (ed.) *Ecosystem Theory and Application*, John Wiley & Sons, New York, pp 240–68. [7.2, 7.4]

8: Detritus Processing

L.MALTBY

8.1 INTRODUCTION

Detritus processing, the incorporation of non-living organic material into living biomass, plays a central role in the functioning of many flowing water systems. This account of detritus processing begins with a description of detritus (section 8.2), the organisms (detritivores and micro-organisms) involved in its processing (section 8.3) and the interactions between them (section 8.4). There is then a brief consideration of the enzymatic capabilities of invertebrate detritivores (section 8.5) before the relative importance of micro-organisms and detritivores in detritus processing is discussed (section 8.6). Finally, the implications of changes in water quality (section 8.7) and riparian vegetation (section 8.8) for detritus processing are considered.

8.2 FORMS AND SOURCES OF DETRITUS

Detritus has been described as an assemblage of living and dead material incorporating fungi, bacteria, microinvertebrates, algal cells and the substrate itself (Anderson & Cargill 1987). Although properties such as source, associated microbes, energy content, percentage ash and C:N ratios have been used to classify detritus (Cummins & Klug 1979), the most commonly used scheme is based on particle size. Detritus is divided into three major classes within which there are a number of subgroups: coarse particulate organic matter (CPOM, >1 mm), fine particulate organic matter (FPOM, >0.45 μm <1 mm) and dissolved organic matter (DOM, <0.45 μm) (Cummins 1974).

Detritus in streams may be either allochthonous, i.e. imported from the surrounding catchment (e.g. leaves, twigs, branches), or autochthonous, i.e. produced by the system itself (e.g. algae, dead aquatic animals and plants). The relative importance of allochthonous and autochthonous detritus will vary both spatially and temporally (Webster & Benfield 1986).

In wooded headwater streams, shading restricts primary production and CPOM is a major energy source. Inputs of allochthonous detritus into small streams flowing through mature forests may account for more than 95% of the total organic matter inputs (Triska *et al* 1982). In larger streams, where channels are wider and there is less shading, primary production is greater and allochthonous CPOM inputs are less important (Conners & Naiman 1984). Detrital material in middle- and high-order streams consists of FPOM transported from upstream regions as well as decomposing algae and macrophytes (Anderson & Cargill 1987). Consequently, whereas in small woodland streams the food of detritivores would be mostly allochthonous, in more open systems it would be primarily autochthonous. Evidence in support of this comes from studies of gut contents and stable isotope analysis (e.g. Rounick *et al* 1982; Ross & Wallace 1983; Winterbourn *et al* 1986).

The composition of the riparian vegetation influences both the quality and quantity of organic matter entering the aquatic environment. The quantity of organic detritus in a stream at any specific time is a balance between input and output processes (Swanson *et al* 1982). Small forested streams are extremely retentive and primarily export material in the form of FPOM or

DOM (Cummins *et al* 1983; Speaker *et al* 1984). Only in larger streams, where the formation of retention structures (e.g. log jams) is restricted or removed by the water flow, is CPOM such as leaf litter effectively transported (Triska *et al* 1982).

Coarse particulate organic matter

Particulate detrital inputs range from rapidly processed leaves, needles and twigs to large, slowly processed, woody debris. Although woody material has a lower food value than non-woody material, it may sustain organisms during periods when few leaves or needles are available. It is also important in that it increases the stability of the channel and retards the loss of other more palatable detritus (Bilby & Ward 1989; Winkler 1991).

The majority of studies on decomposition have concentrated on leaf material, in particular from deciduous trees. Breakdown rate of leaf litter varies between species such that leaves from non-woody plants break down significantly faster than leaves from woody plants. Within non-woody plants, leaves of submerged and floating macrophytes break down significantly faster than those from emergent macrophytes, terrestrial grasses or ferns (Webster & Benfield 1986).

Factors that affect breakdown rate include the level of nutrients, fibre and inhibitory chemicals present in the plant tissue. For example, leaves of woody plants with a high nitrogen and/or low lignin content break down faster than those with a low nitrogen and/or high lignin content (Kaushik & Hynes 1971; Sedell *et al* 1975; Suberkropp *et al* 1976). The breakdown rate of conifer needles is slower than that for deciduous leaves, due partly to their thick cuticle and the presence of inhibitory substances (Bärlocher & Oertli 1978a, 1978b).

Anderson and Cummins (1979) suggested that food items can be organized from low to high quality as follows: wood, terrestrial leaf litter, FPOM, decomposing vascular macrophytes, filamentous algae, diatoms and animal tissue. Such schemes define food quality in terms of the chemical properties of the organic material, i.e. nutrient content (nitrogen, protein) or abundance of refractory materials (lignin, cellulose). The observation that nitrogen and lignin content correlate well with breakdown rates has resulted in the suggested use of the lignin:nitrogen ratio as an index of leaf quality (Melillo *et al* 1982). However, although these may be convenient measures with which to categorize food items it must be remembered that the only true measure of food quality is the performance (i.e. growth, reproduction, survival) of the organism feeding upon it. As detritivores vary in their digestive capabilities (section 8.5), any particular type of organic material may be of high food quality for one organism but of low food quality for another.

The quality of the input of CPOM, defined in gross chemical terms, varies seasonally. In spring and summer it consists of high-nutrient pollen, flowers and insect excretory products, whereas in autumn it consists principally of leaves, and in winter and spring wood is the major input (Anderson & Sedell 1979). Most allochthonous input enters woodland streams during the autumn although leaves that fall on to the floodplain may not enter until major flooding occurs. Such flooding often occurs in spring and results in an input of particulate organic matter to the stream at a time when the material that entered in the autumn is greatly depleted. Under such circumstances floodplains may act as a store of organic material for the stream (Cummins *et al* 1983). For example, Cuffney (1988) calculated that 5.5 kg AFDW m^{-2} (ash-free dry weight) of particulate organic matter entered the Ogeechee River (USA) channel from the floodplain.

Fine particulate organic matter

FPOM can be produced by physical abrasion, microbial action, feeding by invertebrates, scouring of algae, bacteria and fungal spores from surfaces, flocculation of DOM or soil erosion. It is a mixture of shredded vascular plant material, faeces, algae and micro-organisms (Short & Maslin 1977; Anderson & Sedell 1979; Bowen 1984; Findlay & Arsuffi 1989). Models of detritus processing in streams (e.g. Boling *et al* 1975) suggest that small particles have undergone considerable processing and decomposition and should therefore contain the largest amount of refractory material such as lignin and cellulose

(Suberkropp *et al* 1976; Cummins & Klug 1979). However, contrary to these predictions, the percentage of lignin and cellulose and the C:N ratio decreases as particle size decreases (Ward 1986; Sinsabaugh & Linkins 1990). Possible explanations for this include: (i) naturally occurring FPOM comes from a variety of sources with different processing rates and therefore predicted patterns may be obscured (Ward 1986); (ii) as particle size decreases, surface area increases, thereby providing a larger surface for immobilizing bacteria and nutrients. Sinsabaugh and Linkins (1990) measured the microbial respiration of particles of differing sizes and found that below 2.5 mm, microbial oxygen consumption was inversely proportional to particle size.

Dissolved organic matter

DOM arises from a number of sources, both detrital and non-detrital (see Fig. 4.5). It may be 'leaked' by living algae and macrophytes, released during lysis of dead cells, or arise from microbial activities, groundwater seepages, throughfall from the canopy, and surface runoff. Miller (1987) emphasized the potential importance of non-detrital DOM as an energy source for microheterotrophs. He compared the biomass of heterotrophic micro-organisms on leaves incubated in control and darkened troughs in a stream. There were significantly more bacteria and fungi colonizing leaf litter incubated in the open troughs (Fig. 8.1) and this was interpreted as evidence in support of the hypothesis that non-detrital DOM, released by microalgae, was utilized by bacteria and fungi colonizing dead leaves. Compared with leaf leachate, which is released rapidly once the leaf enters the water (Kaushik & Hynes 1971), non-detrital DOM is less seasonal, being released throughout the lifespan of aquatic algae and macrophytes (Wetzel 1975; Bott & Ritter 1981), and consists mainly of low molecular weight labile compounds (Kaplan & Bott 1985). However, although there is little doubt that algae, when physiologically stressed, release DOM, figures of 40–50% of photosynthetically fixed carbon being released as DOC (e.g. Haack & McFeters 1982) may be overestimates. Berman (1990) suggested that a more realistic figure for actively growing, healthy algae, is less than 10%.

Some fungi (Bengtsson 1982) and invertebrates (Hipp *et al* 1986; Salonen & Hammer 1986; Thomas *et al* 1990) can assimilate DOM, although the main routes by which it is processed are: conversion to a particulate form by physicochemical processes (Bowen 1984; McDowell 1985); adsorption on to surfaces (Armstrong & Bärlocher 1989); or assimilation by bacteria (Mann 1988). Bacteria may be either suspended in the water column or associated with organic layers on stones. These organic layers, which consist of a matrix of polysaccharide slime containing fungi, bacteria, algae and FPOM, are an important site for the processing of organic material (Lock & Hynes 1976; Kaplan & Bott 1983, 1985; Lock *et al* 1984). They also provide a potential food source for detritivores (Rounick *et al* 1982; Rounick & Winterbourn 1983; Lock *et al* 1984; Bärlocher & Murdoch 1989). For example, the mayfly *Stenonema* acquires 47% of its daily carbon needs

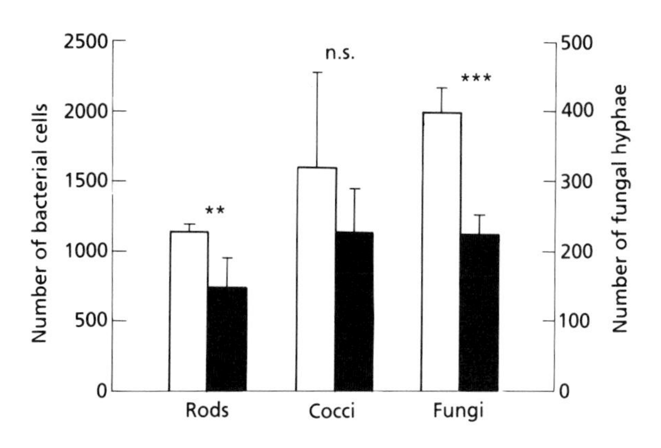

Fig. 8.1 Relative abundance of bacterial cells (median number/ten microscope fields) and fungal hyphae (median number/four 12-mm transects) accumulating *in situ* on leaf litter in troughs open to natural illumination (□) and from which natural illumination was excluded (■). Error bars represent 95% CL. ** significant difference at $P < 0.01$; *** significant difference at $P < 0.001$; n.s., no significant difference (after Miller 1987).

from bacteria associated with stone surface layers (Edwards & Meyer 1990) and the gut contents of *Tipula caloptera* and *Gammarus tigrinus* can release amino acids from organic layers (Armstrong & Bärlocher 1989; Bärlocher & Murdoch 1989).

DOM provides a continuous input of energy which can be utilized by invertebrates either after flocculation to FPOM or conversion to microbial biomass (Merritt *et al* 1984). In marine systems, considerable attention has been paid to the incorporation of DOM into bacterial biomass which is then grazed by protozoans (Porter *et al* 1985; Azam & Cho 1987). This 'microbial loop' has been less intensively studied in flowing freshwater systems even though recent studies have demonstrated its potential importance in the trophic dynamics and energy budgets of such systems (Bott & Kaplan 1990; Carlough & Meyer 1990). Whereas in marine systems free-living bacteria are mainly responsible for DOM uptake from the water column (Azam *et al* 1990), in rivers and streams attached bacteria are probably more important (Edwards & Meyer 1986). Bacteria suspended in the water column may be either consumed directly by macroinvertebrates or consumed by protozoans which in turn become prey for macroinvertebrate filter-feeders (Wallace *et al* 1987; Carlough & Meyer 1989). For example, 20–67% of the daily growth of simuliids in the Ogeechee River is due to incorporation of bacterial carbon (Edwards & Meyer 1987).

Interactions between DOM, bacteria, protozoans and macroinvertebrates are illustrated in Fig. 8.2.

8.3 ORGANISMS INVOLVED IN DETRITUS PROCESSING

Detritivores

Detritivores include representatives of a wide number of taxonomic groups including arthropods, molluscs, annelids and fish. However, the majority of detritivores are insects and, to a lesser degree, crustaceans (Table 8.1). Detritivores do not necessarily feed exclusively on one particular type of food but may switch diet either in response to changes in the abundance of a particular cat-

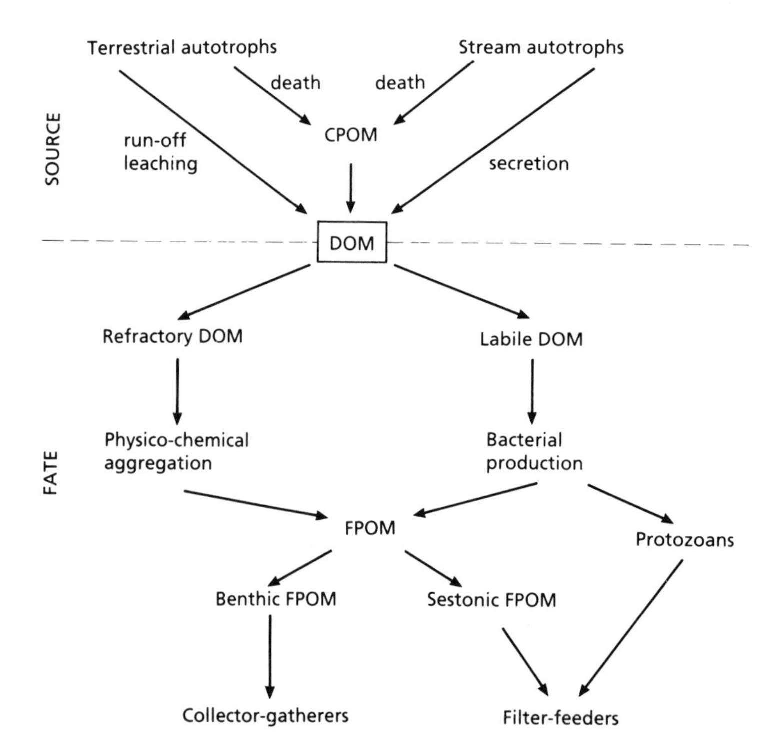

Fig. 8.2 Source and fate of dissolved organic matter (DOM) in streams. CPOM, coarse particulate organic matter; FPOM, fine particulate organic matter.

egory of detritus (Roeding & Smock 1989) or in response to changes in nutritional requirements during development (Fuller & Mackay 1981; Anderson & Cargill 1987). For example, in the few months before emergence, the diet of hydropsychids switches from being dominated by FPOM to being dominated by animal prey (Ross & Wallace 1983).

Freshwater animals have been categorized on the basis of the type of food they utilize and the way they acquire it (Chapter 5). The 'functional feeding groups' utilizing detritus are shredders, collector-filterers, collector-gatherers and gougers (Merritt & Cummins 1978; Anderson *et al* 1984). Shredders feed upon decomposing vascular plant material (CPOM). Their activity results in the production of FPOM either in the form of faeces or fragments. Invertebrates feeding on FPOM may either filter it out of the water column, in which case they are known as collector-filterers (e.g. simuliids, net-spinning caddisfly larvae), or feed on material that has been deposited on the streambed. These animals are known as collector-gatherers and include mayflies and chironomids. Gougers, which are sometimes incorporated within the shredder category (Cummins *et al* 1989), feed on wood and include elmid beetles (e.g. *Lara avara*) and caddisfly larvae (e.g. *Heteroplectron californicum*).

Micro-organisms

The types of heterotrophic microorganisms associated with detritus were described in Chapter 4. In general it is assumed that fungi are more important in the breakdown of CPOM than bacteria. This has been investigated by exposing leaf discs to bactericides and/or fungicides and either measuring their respiration rate (Triska 1970) or weight loss (Kaushik & Hynes 1971). For example, Triska (1970) measured the respiration rate of leaf discs recovered from streams and incubated with or without bactericides. The results from these studies are presented in Fig. 8.3. Throughout the year, fungal respiration was greater than bacterial respiration and this difference was greatest during the early stages of decomposition (i.e. winter). The relative abundance of fungi and bacteria on CPOM is dependent both upon the vegetation type and stage of decomposition (Fig. 8.4).

Table 8.1 Detritivorous freshwater arthropods

Class	Order	Family
Insecta	Trichoptera	Brachycentridae, Calamatoceridae, Glossosomatidae, Helicopsychidae, Hydropsychidae, Hydroptilidae, Lepidostomatidae, Leptoceridae, Limnephilidae, Molannidae, Oeconesidae, Odontoceridae, Philopotamyidae, Phryganeiidae Polycentropodidae, Psychomyiidae, Seriostomatidae
	Diptera	Ceratopogonidae, Chironomidae, Culicidae, Dixidae, Empididae, Ephydridae, Psychodidae, Ptychopteridae, Simuliidae, Stratiomyidae, Syrphidae, Tipulidae
	Ephemeroptera	Baetidae, Caenidae, Ephemerellidae, Ephemeridae, Heptageniidae, Leptophlebeidae, Potamanthidae, Siphlonuridae, Tricorythidae
	Plecoptera	Capniidae, Chloroperlidae, Leuctridae, Nemouridae, Peltoperlidae, Pteronarcidae, Taeniopterygidae
	Coleoptera	Elmidae, Helodidae, Hydrophilidae, Psephenidae, Ptilodactylidae, Scirtidae
	Hemiptera	Corixidae
Crustacea	Isopoda	Asellidae
	Amphipoda	Gammaridae

Based on information from Cummins (1974), Petersen and Cummins (1974), Merritt and Cummins (1978), Anderson and Sedell (1979), and Anderson and Cargill (1987).

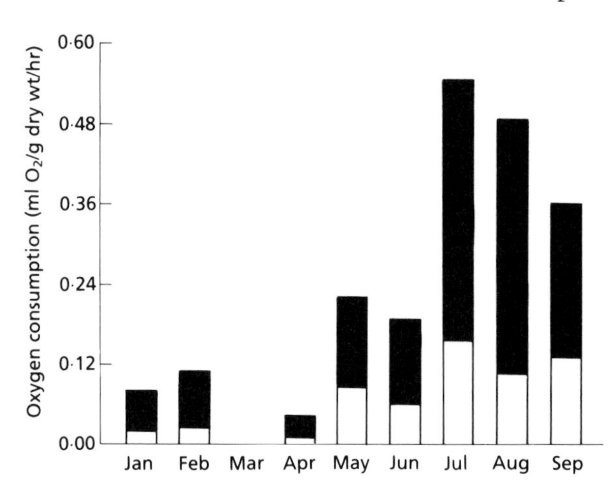

Fig. 8.3 Oxygen consumption (ml O_2/g dry weight/h) of alder leaf discs during decomposition. □, Bacterial respiration; ■, fungal respiration (after Triska 1970).

Bacteria may be more important in the latter stages of processing when particle size is reduced (Suberkropp & Klug 1976). Sinsabaugh and Linkins (1990) suggested that particles of less than 2–4 mm were too small to support the growth of fungal hyphae, and therefore microbial processing of these particles is dependent upon bacteria. Bacteria may also play an important role in the early decomposition of recalcitrant leaves, such as those of members of the Fagaceae (Iversen 1973; Davis & Winterbourn 1977), and are the major assimilators of DOM (section 8.2).

Within the fungi, the relative importance of 'terrestrial' over 'aquatic' groups is more difficult to assess owing to the different methodologies used by different workers (see Chapter 4). Many studies have identified fungal assemblages based on conidial production and have therefore immediately restricted their survey to aquatic hyphomycetes (e.g. Triska 1970; Padgett 1976). Other studies have used particle plating techniques but have incubated the leaf material on nutrient-rich agar at high temperatures, conditions well suited for 'terrestrial' hyphomycetes (e.g. Kaushik & Hynes 1971). Suberkropp and Klug (1976) performed a detailed study of the bacteria and fungi colonizing leaf material in Augusta Creek, Michigan. Several different techniques were used including direct observation, water incubation and particle plating using different agars and incubation temperatures. They concluded that fungi do dominate the microflora and that, although aquatic hyphomycetes are dominant, and therefore assumed most important, other groups such as Oomycetes and terrestrial hyphomycetes are also present. These results substantiated earlier conclusions drawn from a scanning electron microscopic study (Suberkropp & Klug 1974). The importance of different fungal species in decomposition depends on their enzymatic capability (Chapter 4, section 4.3) and their interaction with detritivores (section 8.4).

8.4 FUNGAL INVERTEBRATE INTERACTIONS

Importance of microbes as mediators of food quality

When CPOM (e.g. leaves) enters a water body, soluble compounds such as carbohydrates, amino acids and phenolic substances are leached out leaving the structural molecules behind. The DOM is either assimilated by bacteria or precipitated by physicochemical processes, whereas the CPOM is either consumed by animals or colonized by micro-organisms (Fig. 8.5). Early studies of leaf decomposition concluded that leaching may remove about 5–30% of the initial dry weight of leaves in 1–2 days depending on the species involved (e.g. Petersen & Cummins 1974). However, these studies used pre-dried leaves and more recent studies have found that, for fresh leaves, both leaching and rate of colonization by aquatic hyphomycetes are reduced (Gessner & Schwoerbel 1989; Bärlocher 1991). The rate at

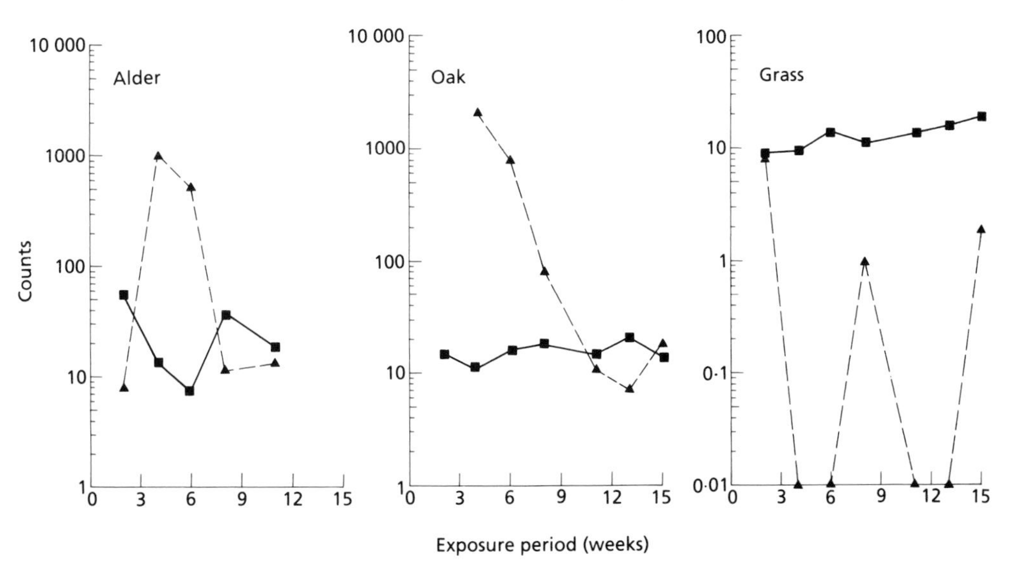

Fig. 8.4 Relative abundance of bacteria (■——■) and aquatic hyphomycete conidia (▲——▲) on different substrates. Data from Chamier (1987).

which energy flows through detritus-based communities is a function of the rate at which the refractory molecules remaining after leaching can be hydrolysed and assimilated by micro-organisms and detritivores. As mentioned in section 8.2, detritus consists not only of particulate organic matter but also contains micro-organisms. To understand energy flow in detritus-based systems it is important to know the relative importance of detrital structural molecules and microbial biomass to detritivore nutrition.

Many studies have confirmed that conditioned leaves (i.e. those colonized by micro-organisms) are more palatable to detritivores than unconditioned leaves (Kaushik & Hynes 1971; Kostalos & Seymour 1976; Rossi & Fano 1979; Golladay *et al* 1983; Bueler 1984) and that palatability is a function of: leaf type (Triska 1970; Kaushik & Hynes 1971, Bärlocher & Kendrick 1973a; Irons *et al* 1988), fungal species (Bärlocher & Kendrick 1973a; Marcus & Willoughby 1978; Fano *et al* 1982; Suberkropp *et al* 1983; Rossi *et al* 1983; Rossi 1985) and incubation time (Iversen 1973; Anderson & Grafius 1975; Arsuffi & Suberkropp 1984; Bueler 1984). When offered a choice, not only can detritivores discriminate between conditioned and sterile food, but they can also discriminate between leaf discs colonized by different fungal species, and can even discriminate between fungal patches on the same leaf (Arsuffi & Suberkropp 1985).

Detritivores exhibit preferences for different leaf types and tend to prefer those with the fastest rate of decay, although less preferred leaves can be made more palatable by inoculating them with a preferred fungus (Bärlocher & Kendrick 1973a). Therefore, not only are the leaf and micro-organisms important in themselves, but the interaction between them also influences food choice.

Several studies have attempted to correlate feeding preference with the physical (e.g. softness), chemical (e.g. nitrogen) or mycological (e.g. biomass, enzymatic capability) properties of inoculated leaf litter (Bärlocher & Kendrick 1973a; Suberkropp *et al* 1983; Bärlocher 1985). However, these studies have failed to provide a convincing explanation for the food choices observed, suggesting that other factors such as the accumulation of specific substances (e.g. lipids) and/or the breakdown of lignin or polyphenols may be important. Evidence that lipids may be important cues for detritivores comes from a study by Cargill and colleagues (1985) (see also Hanson *et al* 1983). Larvae of the caddisfly

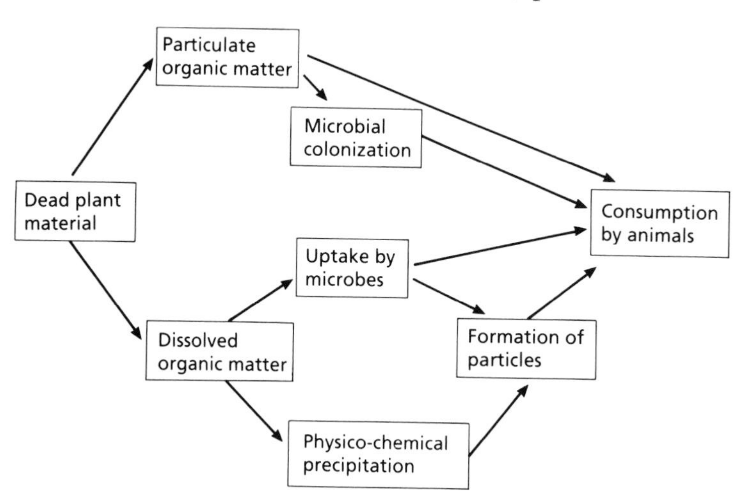

Fig. 8.5 Pathways of detritus processing in freshwater systems (after Mann 1988).

Clistoronia magnifica were reared on one of five diets: conditioned alder leaves, alder plus a mixture of fatty acids, alder and wheat grains, alder plus fungi, or fungi alone. Animals grew well and reproduced on a diet of alder plus wheat but did not complete their development on either alder alone or alder plus fatty acids. Larvae fed alder plus fungi or fungi alone developed successfully but failed to reproduce. The authors suggested that, in order to reproduce, *C. magnifica* needs to acquire large lipids stores during the last larval instar. These triglyceride stores could either be assimilated directly from the food or synthesized from carbohydrates. Wheat grains are a good quality food as they provide a rich source of soluble carbohydrates. This experiment highlights the fact that care must be taken when assessing food quality; just because a food promotes growth does not mean that the completion of the life cycle will be possible.

Preference rankings may also be the consequence of unpalatable species producing distasteful or toxic metabolites. There is some evidence that fungal preference changes through time, owing to the different processing rates of the fungi, and that leaves may become post-conditioned (*sensu* Boling *et al* 1975). For example, larvae of the caddisfly *Pschoglypha* were offered a choice of leaf discs inoculated with the fungus *Clavariopsis aquatica* and incubated for 7, 14 or 21 days. The consumption of food incubated for 7 days was three times greater than the consumption of leaf material incubated for 14 or 21 days, even though leaf material incubated for longer periods was softer and had a higher nitrogen content (Arsuffi & Suberkropp 1984). Whether such decreases in palatability with increasing incubation time are due to the accumulation of distasteful or toxic compounds or due to some other factor(s) is unknown.

Fungal preferences vary between taxa. For example, *Lemonniera aquatica* was not consumed by any of the three species of Trichoptera investigated by Arsuffi and Suberkropp (1984), although it was highly preferred by the mollusc *Helisoma trivolvis* (Arsuffi & Suberkropp 1989). The degree of selectivity also varies, with some detritivores (e.g. *Gammarus*) being much more selective than others (e.g. *Helisoma trivolvis, Pteronarcella badia*) (Arsuffi & Suberkropp 1989). Differences in food choice and degree of selectivity may be due to differences in digestive ability and mobility, more mobile species being more selective than relatively immobile ones (Bärlocher 1982). Other factors may also be important as food preferences vary both within (Christensen 1977) and between (Rossi *et al* 1983) populations of the same species.

Microbes as a food source

What is the relative importance of the microbes themselves as a food source, compared with the organic matter with which they are associated?

In feeding trials, the performance of detritivores fed fungal mycelia has been reported to be better than (Bärlocher & Kendrick 1973b; Rossi & Fano 1979; Cargill *et al* 1985), similar to (Kostalos & Seymour 1976; Marcus & Willoughby 1978; Arsuffi & Suberkropp 1988) or worse than (Bärlocher & Kendrick 1973b; Willoughby & Sutcliffe 1976; Marcus & Willoughby 1978; Sutcliffe *et al* 1981) that of animals fed decaying leaf material. However, it is difficult to compare the results from different studies as they have used different fungi, conditioning times and animals. For example, Sutcliffe *et al* (1981) found that the growth rate of the amphipod *Gammarus pulex* was lower when fed on a mycelial diet of *Lemonniera aquatica* or *Clavariopsis aquatica* compared with animals fed a diet of conditioned or unconditioned elm leaves. In contrast, Bärlocher and Kendrick (1973b) found that while some fungal species resulted in poor growth of *Gammarus pseudolimnaeus*, others supported growth rates two to five times greater than those for animals fed unconditioned leaves. When assessing the importance of micro-organisms in the diet of detritivores, differences in food quality must be taken into account. *Lemonniera aquatica* was one of the least preferred fungi by *Gammarus* sp. and *Gammarus pulex* (Arsuffi & Suberkropp 1989; Graça 1990) which may explain the poor performance of *G. pulex* fed *L. aquatica* mycelia (Sutcliffe *et al* 1981).

Not all fungal species increase the palatability of leaf material to the same extent and there is some evidence that animals show preferences for those diets affording the best survivorship and growth (Kostalos & Seymour 1976; Arsuffi & Suberkropp 1986), although this is not necessarily true in every case (Bärlocher & Kendrick 1973a, 1973b).

Although most of the work on fungal–invertebrate interactions has concentrated on aquatic hyphomycetes, other fungi are commonly found in freshwaters (Kaushik & Hynes 1971; Bärlocher & Kendrick 1974; Park 1974; Suberkropp & Klug 1976) and there is evidence that some of these (e.g. *Pythium*, *Cladosporium*, *Epicoccum*) can degrade plant material (Park & McKee 1978; Godfrey 1983). It has been argued that these so-called 'terrestrial' hyphomycetes are outcompeted by aquatic ones at the low temperatures prevalent in northern temperate streams during winter (see Chapter 8). However, 'terrestrial' hyphomycetes may be important in Mediterranean and tropical regions where higher water temperatures would favour them. In fact, Rossi and his co-workers, studying the trophic biology of detritivores in Italian freshwaters, have concentrated on 'terrestrial' hyphomycetes (Fano *et al* 1982; Rossi 1985; Basset & Rossi 1987).

Can microbial biomass present on particulate organic matter support detritivore populations?

Many detritivores have a rapid gut throughput time (e.g. <2 h for *Gammarus pulex* (Willoughby & Earnshaw 1982; Welton *et al* 1983)) and this has been taken as evidence that the digestive mechanism of these organisms is mainly one of stripping microbial nutrients rather than digestion of the more refractory detritus itself. Although the quantitative importance of micro-organisms has been demonstrated for some detritivores (Calow 1974; Bärlocher & Kendrick 1975; Arsuffi & Suberkropp 1986), other evidence is less clearcut. For example, although it is possible to rear simuliids on pure cultures of bacteria in the laboratory (Fredeen 1964), under field conditions the concentrations of bacteria may be too low to meet their nutritional requirements (Baker & Bradnam 1976; Edwards & Meyer 1987).

Radioactive isotopes have been used to estimate the contribution of microbial biomass to invertebrate energy budgets (Findlay & Tenore 1982; Lawson *et al* 1984; Findlay *et al* 1984, 1986a, 1986b). Lawson *et al* (1984), studying the dipteran larva *Tipula abdominalis*, concluded that the biomass of microbes on detritus did not influence larval growth but did influence consumption rate. They concluded that *Tipula* did require bacteria and fungi in their diet but not as a carbon or nitrogen source because 73–89% of their growth was derived from the leaf matrix itself. Results obtained for *Tipula* cannot be viewed as being representative of detritivores in general as they are unusual in having an alkaline midgut ($pH > 11$) which is thought to be an adaptation for protein digestion (Martin *et al* 1980; Bärlocher 1982, 1985). Findlay, Meyer and Smith (1986a) fed the isopod *Lircus* either oak or willow leaves inoculated with either the zygomycete *Mucor* or the hyphomycete. *Triscelophorus*. They found that the contribution of fungal carbon to the total carbon respired varied between treatments from about 1% for animals fed willow plus *Triscelophorus* to 57% for animals fed oak plus *Mucor*. Therefore, although the authors concluded from their study that fungi alone could not meet the carbon requirements of *Lircus*, this is obviously dependent upon the fungal species present. If other, more preferred, fungi were used, the animals' carbon requirements may have been met. In contrast to the studies of Findlay *et al* (1986a, 1986b) and Lawson *et al* (1984), Arsuffi and Suberkropp (1986) concluded that fungal biomass could account for a significant proportion of the growth of trichopteran larvae.

The inability of fungal biomass to support detritivore populations has also been suggested by others workers who believe that, because microbial biomass of detritus is low, the main role of microbes is as modifiers of leaf material (Iversen 1973; Cummins & Klug 1979; Bärlocher & Kendrick 1981). Estimates of fungal biomass may vary depending on the technique used (Newell *et al* 1986). Moreover, Findlay and Arsuffi (1989) observed that the contribution of microbial biomass to the amount of organic carbon in the leaf–microbe complex varied both within and between leaf types (Fig. 8.6). However, it was

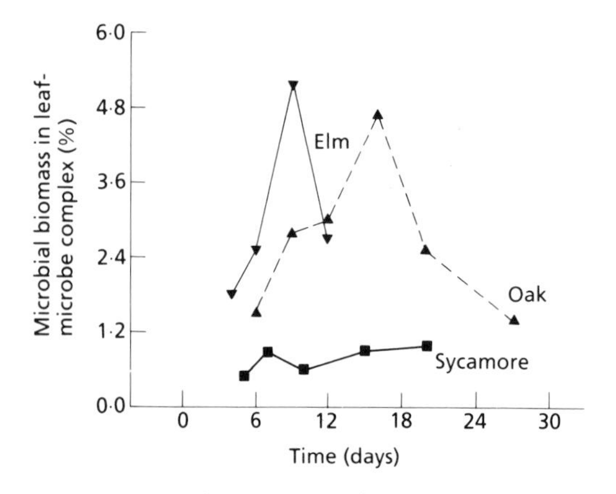

Fig. 8.6 Temporal variation in the contribution of microbial biomass to the amount of organic carbon in leaf–microbe complexes (after Findlay & Arsuffi 1989).

always less than 6% and, of this, fungi constituted between 82% and 96% of the microbial biomass on both oak and elm leaves. Although there is less microbial than leaf biomass in leaf–microbe complexes, micro-organisms are assimilated with a greater efficiency than the plant tissue itself. For example, although *Gammarus pseudolimnaeus* assimilates only 18% of elm and maple leaves, 67–93% of the energy in the fungi colonizing the leaf material is assimilated (Bärlocher & Kendrick 1975). Similarly, *Hyalella azteca* utilizes 60–90% of the bacterial biomass it ingests but assimilates only 5% of elm leaves (Hargrave 1970).

There is evidence, albeit equivocal, in support of the importance of microbes, especially fungi, both as modifiers of leaf material and as a food source. In an attempt to tease out whether food preferences exhibited by trichopteran larvae were due to the fungal biomass *per se* or due to the enzymatic action of the fungi on the leaves, Arsuffi and Suberkropp (1988) altered these two parameters by conditioning leaves in the presence or absence of glucose. The addition of glucose suppresses the production of extracellular polysaccharidases. Leaves that had been conditioned in the presence of glucose had a thick mycelial mat over their surface but they were not softened. In feeding trials, leaves conditioned in the presence of glucose were preferred to leaves conditioned without added glucose. This, and the observation that larvae feeding on the glucose-treated leaves fed almost exclusively on the fungal mat, led Arsuffi and Suberkropp (1988) to conclude that *Psychoglypha* larvae preferred fungal mycelium over the combination of fungal biomass and enzyme-modified leaf tissue.

There is no reason to assume that the relative importance of microbes as food will be the same for all detritivores. Evidence for the importance of microbes as modifiers of leaf material includes the observation that treating leaves with an extract of fungal enzymes increases their palatability to *Gammarus pseudolimnaeus* (Bärlocher & Kendrick 1973a, 1975) and, when offered a choice of conditioned leaves or pure cultures of *Tetrachaetum elegans*, *Gammarus minus* preferred conditioned leaves (Kostalos & Seymour 1976).

8.5 ENZYMATIC CAPABILITIES OF DETRITIVORES

As leaf material can be made more palatable by treatment with either fungal enzymes or hydrolysing agents (Bärlocher & Kendrick 1973a), it has been suggested that the importance of microorganisms to detritivores lies in their ability to break down leaf material into subunits that the detritivores can utilize. If this is true, the degree to which any specific detritivore relies on microbial conditioning will be dependent upon its own enzymatic capabilities.

The major component of autumn-shed leaves is cell wall material consisting of cellulose and pectin. There are three possible sources of cellulose-degrading enzymes associated with detritivores: (i) tissue-level synthesis; (ii) microbial symbionts; and (iii) 'acquired' enzymes.

All aquatic detritivores examined to date can degrade cellobiose, the basic subunit of cellulose; some are able to digest soluble cellulose (carboxymethyl cellulose, CMC); but only a few show enzymatic activity towards native crystalline cellulose (Bjarnov 1972; Monk 1976, 1977). Tissue-level synthesis of cellulases has been recorded for molluscs, annelids, crustaceans and insects (Calow & Calow 1975; Monk 1976; Schoenberg *et al* 1984; Chamier & Willoughby 1986; Chamier 1991), although some of the evidence is confusing, especially for members of the genus *Gammarus*. Whereas, Chamier and Willoughby (1986) stated that *Gammarus pulex* produces cellulases, chitinases and β-1,3-glucanase, Bärlocher and Porter (1986) suggested that similar enzymes, found in the gut of *Gammarus tigrinus*, arise from ingested fungi. For a third species, *Gammarus fossarum*, Bärlocher (1983) states that the release of reducing substances from microbially conditioned leaves was due to a combination of microbial and gut enzymes. Moreover, if freshly shed leaves were pretreated with 70% alcohol to remove soluble phenols, reducing sugars could be released by gut extracts of this species (Bärlocher 1982). Although it is certainly not beyond the bounds of possibility that members of a single genus differ in their enzymatic capabilities, another possible reason for the different results could be related to the methods used in the different studies. Chamier and Willoughby (1986) found it was necessary to incubate material at 37°C for 44 hours in order to detect enzymatic activity; no activity could be accurately measured at lower temperatures or over shorter time intervals. As the normal gut passage time of *Gammarus* at field temperatures (<20°C) is less than 2 hours, the ecological relevance of these results is questionable. Although the other studies (Bärlocher 1982; Bärlocher & Porter 1986) also used incubation times in excess of 2 hours (5 h and 14 h respectively), they did use realistic incubation temperatures. Martin *et al* (1981a), working on caddisfly larvae, could also only demonstrate the presence of enzymes active against CMC at elevated temperatures (i.e. 37°C).

Microbial symbiosis is thought to be important in the digestive physiology of the dipteran *Tipula abdominalis* (Klug & Kotarski 1980; Sinsabaugh *et al* 1985; Lawson & Klug 1989) whereas a range of other detritivores (Plectoptera, Trichoptera, Amphipoda) may utilize 'acquired' microbial enzymes (Martin *et al* 1981b; Bärlocher 1982; Sinsabaugh *et al* 1985; Bärlocher & Porter 1986).

Bärlocher (1982) has suggested that the digestive physiology of detritivores is related to their feeding behaviour and mobility. Shredders that specialize in digesting optimally conditioned leaf material (rich in microbes, high in enzymatic activity) must either be highly mobile or have access to other sources of food (e.g. *Gammarus* sp.). Relatively immobile species that are obligate detritivores must possess the digestive ability to break down the refractory material itself (e.g. *Tipula abdominalis*).

8.6 RELATIVE IMPORTANCE OF MICROBES AND INVERTEBRATES IN DETRITUS PROCESSING

Several studies have attempted to assess the relative importance of microbial and invertebrate activities in the processing of particulate organic material. As with other aspects of detritus processing, the majority of studies have concentrated on the breakdown of CPOM, in particular leaves. The results from these investigations are equivocal; whereas some conclude that invertebrates are not important in leaf breakdown (e.g. Mathews & Kowalczewski 1969; Kaushik & Hynes 1971; Reice 1978; Paul *et al* 1983; Leff & McArthur 1989), others have concluded that they are important (e.g. Petersen & Cummins 1974; Hart & Howmiller 1975; Iversen 1975; Minshall *et al* 1982; Kirby *et al* 1983; Oberndorfer *et al* 1984; Imbert & Pozo 1989).

Such differences may in part be due to the techniques and study sites used. Judgements as to the importance of invertebrates in detritus processing have been based on one of two types of study: either (a) manipulation of the intensity of invertebrate feeding (Herbst 1982; Newman *et al* 1987; Newman 1990); or (b) correlation between rate of decomposition and invertebrate density (Reice 1978; Minshall *et al* 1982; Kirby *et al* 1983; Leff & McArthur 1989). Some of these correlative studies have related decomposition rates to total invertebrate density, rather than just shredder density (e.g. Leff & McArthur 1989). Such an approach may confound any effect shredders may be having on decomposition rates with the attractiveness of leaf packs as shelters or as traps for FPOM.

Invertebrate feeding has been manipulated by either manipulating predation pressure (e.g. Oberndorfer *et al* 1984) or by manipulating densities of shredders in enclosures (e.g. Newman 1990) or within artifical streams (e.g. Petersen & Cummins 1974). For example, Newman (1990) manipulated the density of shredders (*Gammarus pseudolimnaeus*) in mesh bags containing watercress (*Nasturtium officinale*). The rate of breakdown was positively correlated with initial shredder density although the relationship was less pronounced in the coarser mesh bags, probably due to loss of particles in the size range from 1 mm to 200 μm (Fig. 8.7).

The most frequently employed approach is to place leaves in enclosures of different mesh sizes, thereby restricting access to invertebrates (Mathews & Kowalczewski 1969; Kaushik & Hynes 1971; Hart & Howmiller 1975; Iversen 1975; Cummins *et al* 1980; Imbert & Pozo 1989). Placing leaves in mesh bags not only restricts invertebrate processing but it may also inhibit microbial processing by reducing the flow of water through the bag, resulting in the establishment of

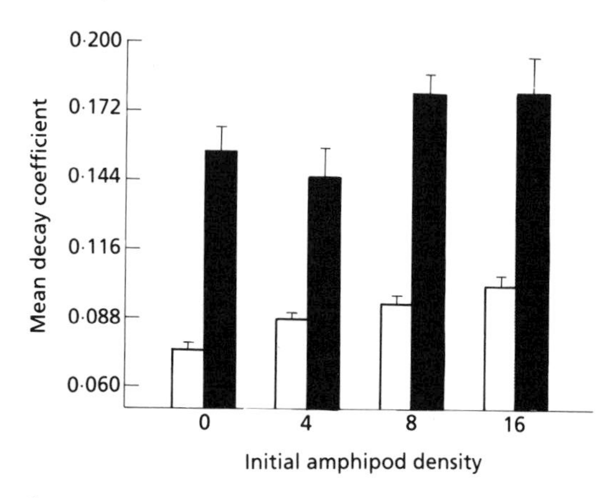

Fig. 8.7 Mean decay coefficients (+2 SE) of *Nasturtium officinale* in mesh bags stocked with *Gammarus pseudolimnaeus*. □, 200 μm-mesh; ■, 1 mm-mesh (after Newman 1990).

hypoxic or even anoxic conditions. Enclosing leaves in fine mesh bags will also influence loss rates due to fragmentation. In an attempt to distinguish between leaf weight loss due to invertebrate feeding or to fragmentation, Stewart and Davies (1989) devised a mesh bag, the body of which consisted of fine mesh (180 μm) and the top of coarse mesh (5 mm). This 'composite' mesh bag would therefore allow macroinvertebrates access but would retain fragments larger than 180 μm. The use of 'composite' mesh bags in combination with fine and coarse bags allows a distinction to be made between fragmentation and invertebrate feeding. In studies using *Cunonia capensis* and *Ilex mitis*, most of the breakdown of *Cunonia* was due to invertebrate feeding (no difference between composite and coarse mesh bags), whereas both invertebrate feeding and fragmentation was important for *Ilex* (Fig. 8.8).

Cummins and colleagues (1980) concluded that, whereas decomposition of leaf packs (groups of leaves loosely attached to an anchor, e.g. house brick) was representative of the breakdown of leaves in riffle areas, decomposition of leaves in mesh bags was more similar to the slower processing rates of leaves in more stagnant pool areas. In contrast, Mutch *et al* (1983) concluded that the processing rate of willow leaves was always greater for leaves in mesh bags than in leaf

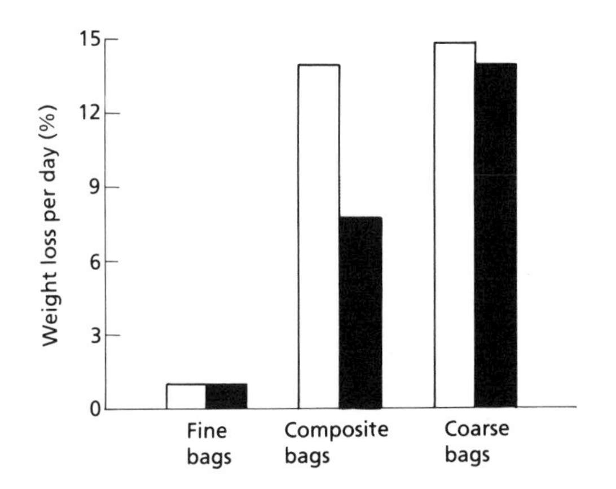

Fig. 8.8 Weight loss of *Cunonia capensis* (■) and *Ilex mitis* (□) leaf material from fine (180 μm), coarse (5 mm) or composite (180 μm and 5 mm) mesh bags (after Stewart & Davies 1989).

packs, irrespective of whether they were in riffles or pools.

The relative importance of macroinvertebrates in the breakdown of leaf litter varies between sites. Mathews and Kowalczewski (1969) concluded that macroinvertebrates were not important in litter breakdown in the River Thames (UK) and Reice (1978) could find no direct relationship between the number of individuals or species of macroinvertebrates in leaf packs and their rate of decomposition. However, both these studies were performed at sites where shredders were poorly represented in the natural community.

Therefore, although detritivores are known to feed selectively on conditioned leaf material, and some studies have calculated that 20–30% of litter processing is due to the activity of shredders (Petersen & Cummins 1974), results from field studies are far from conclusive.

8.7 EFFECT OF WATER QUALITY ON DECOMPOSITION

Many physicochemical factors have been shown to influence decomposition rates, including temperature, calcium, nitrogen, phosphorus and dissolved oxygen (Egglishaw 1968, 1972; Iversen 1975; Howarth & Fisher 1976; Reice 1977; Federle & Vestal 1980; Cummins *et al* 1983; Burton *et al* 1985). Recently, several studies have been designed to investigate the effect of anthropogenic changes in water quality on processing rates. These include studies into the effects of pH (Traaen 1980; Hildrew *et al* 1984; Burton *et al* 1985; Mackay & Kersey 1985; Chamier 1987; Mulholland *et al* 1987), chlorine (Newman *et al* 1987), acid mine drainage (Carpenter *et al* 1983; Gray & Ward 1983), heavy metals (Giesy 1978; Leland & Carter 1985), coal ash effluent (Forbes & Magnuson 1980) and pesticides (Wallace *et al* 1982; Fairchild *et al* 1983; Cuffney *et al* 1984, 1990). All these pollutants cause a reduction in the rate of processing of CPOM. Such a reduction may be due to either low microbial activity or changes in detritivore abundance and/or feeding behaviour.

Insecticides

Insecticides have been used as a means of quantifying the role of benthic invertebrates in decomposition (Wallace *et al* 1982; Cuffney *et al* 1984, 1990). The insecticide methoxychlor was applied to a headwater stream seasonally for 2 years. It caused a large increase in invertebrate drift and a reduction in leaf litter processing rates. Although shredder density was reduced after application of the insecticide, microbial respiration was not affected (Cuffney *et al* 1990). This study provided strong evidence for the importance of invertebrates over microbes in the processing of leaf litter in such streams.

Acidity

The inhibition of decomposition by low pH appears to occur at around pH 5 (Schindler 1980; Traaen 1980; but see McGeorge *et al* 1991). For example, Burton *et al* (1985) observed a reduction in leaf decomposition rate in acidified artificial streams (pH 4–4.3) compared with control streams (pH 7.2–7.5). However, this reduction was significant only after 6–7 months' exposure (Fig. 8.9). The effect of low pH on fungal communities is equivocal. Whereas, McKinley and Vestal (1982) recorded a decrease in fungi with increasing acidity such that they were virtually absent at pH 4, experimental acidification to pH

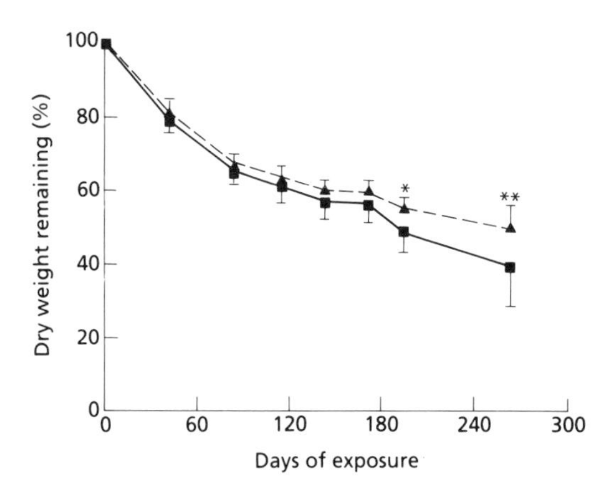

Fig. 8.9 Effect of acidification on the decomposition of *Acer sacchanum* leaves in artificial streams. Error bars represent 95% CL. ▲– – –▲, pH 4.0–4.3; ■——■, pH 7.2–7.5 (after Burton *et al* 1985).

4.3 resulted in an increase in fungal colonization of leaf litter (van Frankenhuyzen *et al* 1985). Studies on aquatic hyphomycetes have found that their decompositional abilities are strongly reduced at pH 4 (Thompson & Bärlocher 1989), although they grow best on solid media at pH 4–5 and in liquid medium at pH 5–6 (Rosset & Bärlocher 1985).

Field studies into the effect of acidity on decomposition have correlated reduction in decomposition with reduction in microbial activity (Chamier 1987; Mulholland *et al* 1987; Palumbo *et al* 1987). As increasing acidity reduces microbial activity, and given that microbes are important to the nutrition of detritivores, either as modifiers of organic material or as a food source (Section 8.4), it may be predicted that the density of shredders would also be reduced in acid streams. However, several workers have observed an increased abundance of shredders in acid streams (Hildrew *et al* 1984; Mackay & Kersey 1985; Mulholland *et al* 1987) even though acidity does reduce food quality (Groom & Hildrew 1989). Particulate organic matter is a finite resource for which microbes and detritivores compete (Bärlocher 1980). The conditioning of organic matter by microbes has costs as well as benefits for detritivores. Although microbes increase the nutritional quality of the substrate, if they are too abundant or active they may considerably reduce the period over which it is available. A stress-induced reduction in microbial processing rate may therefore be advantageous to shredders by extending or increasing the availability of CPOM.

Heavy metals

Studies of the effect of metals on decomposition rates have been concerned either with individual metals (Giesy 1978; Leland & Carter 1985) or with acid mine drainage which contains heavy metals such as iron and zinc (Carpenter *et al* 1983; Gray & Ward 1983; Maltby & Booth 1991). Leland and Carter (1985) observed an inhibition in the rate of leaf litter decomposition at 2.5 μg copper l^{-1} and Giesy (1978) recorded a reduction in microbial colonization and leaf decomposition at 5 μg cadmium l^{-1}. Acid mine drainage also reduces decomposition rates and microbial activity (Carpenter *et al* 1983; Gray & Ward 1983); the fungal group most affected being the aquatic hyphomycetes (Maltby & Booth 1991). Downstream of the input of acid mine effluents, iron precipitates out as ferric hydroxide. The reduction in fungal abundance could therefore be due either to a direct toxic effect of the metals or to a physical change in the leaf surface which either smothers the fungi or prevents their spores from germinating.

Heavy metals reduce both the rate of growth and spore production of aquatic hyphomycetes, sporulation being the most sensitive parameter (Abel & Bärlocher 1984; S. Bermingham, unpublished data). The sensitivity of fungi to heavy metals varies between species; species of aquatic hyphomycetes found downstream of a mine effluent appear to be less sensitive to metals than those only found upstream (S. Bermingham, unpublished data).

Micro-organisms colonizing detritus accumulate toxicants, including metals, which are then transferred along the food chain causing a reduction in the viability of detritivores and the organisms predating upon them (Patrick & Loutit 1976, 1978; Duddridge & Wainwright 1980; Pinkney *et al* 1985; Abel & Bärlocher 1988).

Chlorine

The effect of chlorine on the decomposition of *Potamogeton crispus* was investigated by Newman *et al* (1987) in the artificial stream system at Monticello (US EPA). Sites within streams were dosed with 10, 75 or 250 μg chlorine l^{-1} and weight loss of material from fine (210 μm) and coarse (1 mm) mesh bags was recorded after 4 and 11 days of exposure. Upstream sections of streams acted as controls. Microbial colonization and processing were reduced in both the 75-μg l^{-1} and 250-μg l^{-1} treatments during the first 4 days of exposure (Fig. 8.10). By 11 days there was no significant difference in either the rate of decomposition, bacterial density or microbial respiration rate between fine bags in either the control or dosed sites. There was, however, still a difference in the decomposition rates of leaves in coarse mesh bags, suggesting that feeding by macroinvertebrates was still inhibited. Chlorine appears to influence microbial decomposition by reducing colonization rates but not equilibrium densities. Reduced bacterial densities have been recorded below some chlorination sites (Maki *et al* 1986).

8.8 EFFECT OF CHANGES IN RIPARIAN VEGETATION ON DETRITUS PROCESSING

As stated in section 8.2, the riparian vegetation influences both the quantity and quality of allochthonous detritus entering streams and rivers. Given the high retention capacity of wooded streams and the distinct food preferences of detritivores, there is likely to be a strong link between the diversity and abundance of detritivores in streams and the riparian vegetation. In addition, the life cycles of detritivores, especially shredders, would be expected to be synchronized with the seasonal inputs of material. Therefore, changing the riparian vegetation may have considerable effects on both the structure and functioning of the adjacent aquatic community.

The structure or abundance of riparian vegetation may be altered as a result of many human activities. Stout and Coburn (1989) investigated the impact of road construction on leaf processing in a North American stream system. Two years after the construction work had been completed, leaf processing rates were still reduced downstream of the road compared with upstream reference sites. The authors concluded that the reduction in processing rates was due to a decrease in shredder abundance caused by the removal of riparian vegetation during road construction.

The principal human activity influencing riparian vegetation is forestry. Forestry practices not only alter the abundance and composition of the riparian vegetation but also affect water quality (Benfield *et al* 1991; Holopainan *et al* 1991). For example, streams are subjected to increased loadings of suspended solids, resulting in increased turbidity and sedimentation, as well as inputs of chemicals such as fertilizers and pesticides (Campbell & Doeg 1989).

Cummins *et al* (1989) have developed a model which predicts seasonal changes in shredder biomass based on information on the composition and percentage cover of riparian vegetation. These authors have suggested that the degree to which observations of shredder populations deviate from those predicted by the model could be used as a means of evaluating the impact of changes in riparian vegetation on stream invertebrates.

8.9 SUMMARY

When detritus enters a water body, soluble compounds are leached out, leaving the more recalcitrant structural molecules behind. The rate at which detritus is processed is therefore dependent upon the rate at which these structural polymers are assimilated into living biomass. The two groups of organisms involved in the processing of detritus are micro-organisms (bacteria and fungi) and detritivores (principally macroinvertebrates).

The relative importance of detritivores and micro-organisms in detritus processing is a matter for continuing debate. There is evidence both for the importance of detritivores over micro-organisms and *vice versa*. One possible reason for the differing results could be related to the sites used. Those studies that have concluded that detritivores are relatively unimportant in detritus

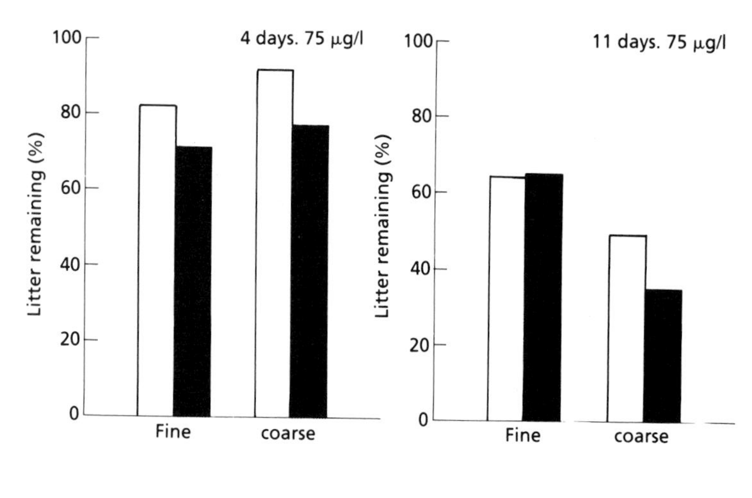

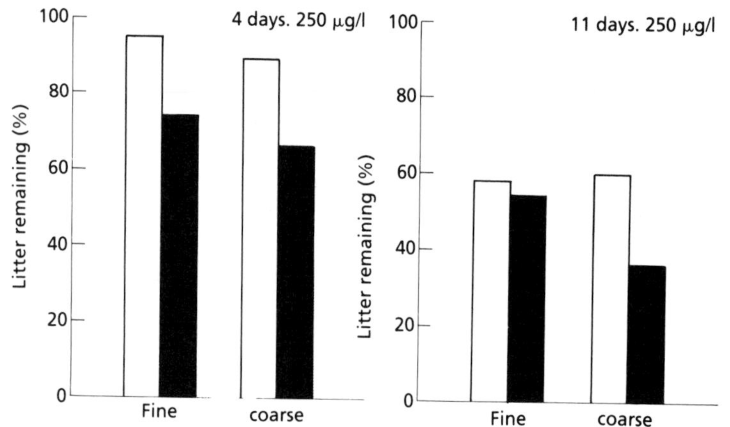

Fig. 8.10 Effect of chlorine on the decomposition of *Pomatogeton crispus* in coarse (1 mm) and fine (210 μm) mesh bags. □, Control sites; ■, dosed sites (after Newman *et al* 1987).

processing have tended to be conducted at sites where accumulations of CPOM and/or shredders are rare and therefore the invertebrate community would be ill adapted to cope with artificial inputs of CPOM.

Another controversial area concerns the interaction between micro-organisms and detritivores. Many studies have shown that detritivores exhibit clear preferences for microbially conditioned detritus; however, what is not clear, is whether microbes are important to detritivores as modifiers of detritus or as a food source. Evidence cited in favour of micro-organisms as modifiers of detritus include: low biomass of micro-organisms relative to detritus; fast gut passage times; and results from radiolabelling studies. Evidence cited in favour of micro-organisms as a food source comes primarily from laboratory studies of the performance (growth, reproduction) of detritivores on different diets. It is likely that the relative importance of micro-organisms to the nutrition of detritivores is dependent upon both the species of detritivore studied and the types of micro-organisms used in the experiments.

Anthropogenic activities that affect detritus processing include those resulting in changes in water quality and/or riparian vegetation. The impact of pesticides, acidity and heavy metals on detritus processing has been investigated. In all cases processing rates were decreased, although whether such decreases were due to low microbial activity or changes in detritivore abundance and/or feeding behaviour is open to debate.

The majority of investigations into detritus

processing in freshwaters have concentrated on northern temperate, upland wooded streams where the main source of detritus is allochthonous CPOM, primarily autumn shed leaves and, to a lesser extent, woody debris. Most information is therefore available on the processing of leaf litter. Fungi are the major micro-organisms colonizing leaf material during the early stages of decomposition. Bacteria are more important during the latter stages of decomposition and may be important in the processing of highly recalcitrant leaf material. Detritivores feeding on leaf litter are referred to as shredders. Their activity results in fragmentation of the leaf litter, which, together with their faecal material, produces FPOM which can be utilized by other members of the detrital food web. FPOM is colonized by bacteria and the macroinvertebrates feeding upon it (i.e. collectors) either filter it from the water column or gather it from the river bed.

A third category of organic matter in freshwaters is DOM, which may be either detrital or non-detrital in origin (Miller 1987). Most studies of DOM utilization have concentrated on its incorporation into organic layers on stones or its flocculation into FPOM. Another possible pathway is the incorporation of DOM into suspended bacteria which are then grazed by protozoans. This so-called 'microbial-loop', important in the trophic dynamics of standing-water systems (marine and freshwater), has been virtually ignored in studies of running-water systems.

REFERENCES

Abel TH, Bärlocher F. (1984) Effects of cadmium on aquatic hyphomycetes. *Applied and Environmental Microbiology* **48**: 245–51. [8.7]

Abel T, Bärlocher F. (1988) Uptake of cadmium by *Gammarus fossarum* (Amphipoda) from food and water. *Journal of Applied Ecology* **24**: 223–31. [8.7]

Anderson NH, Cargill AS. (1987) Nutritional ecology of aquatic detritivorous insects. In: Slansky F, Rodriguez JR (eds), *Nutritional Ecology of Insects, Mites, Spiders and Related Invertebrates*, pp 903–25. John Wiley & Sons, New York. [8.2, 8.3]

Anderson NH, Cummins KW. (1979) Influences of diet on the life-histories of aquatic insects. *Journal of the Fisheries Research Board of Canada* **36**: 335–42. [8.2]

Anderson NH, Grafius E. (1975) Utilization and processing of allochthonous material by stream Trichoptera. *Verhandlungen Internationale Vereinigung für Theoretische und Angewandt Limnologie* **19**: 3083–8. [8.4]

Anderson NH, Sedell JR. (1979) Detritus processing by macroinvertebrates in stream ecosystems. *Annual Review of Entomology* **24**: 351–77. [8.2, 8.3]

Anderson NH, Steedman RJ, Dudley T. (1984) Patterns of exploitation by stream invertebrates of wood debris. *Verhandlungen Internationale Vereinigung für Theoretische und Angewandt Limnologie* **22**: 1847–52. [8.3]

Armstrong SM, Bärlocher F. (1989) Adsorption and release of amino acids from epilithic biofilms in streams. *Freshwater Biology* **22**: 153–9. [8.2]

Arsuffi TL, Suberkropp K. (1984) Leaf processing capabilities of aquatic hyphomycetes: interspecific differences and influence on shredder feeding preferences. *Oikos* **42**: 144–54. [8.4]

Arsuffi TL, Suberkropp K. (1985) Selective feeding by stream caddisfly (Trichoptera) detritivores on leaves with fungal-colonized patches. *Oikos* **45**: 50–8. [8.4]

Arsuffi TL, Suberkropp K. (1986) Growth of two stream caddisflies (Trichoptera) on leaves colonized by different fungal species. *Journal of the North American Benthological Society* **5**: 297–305. [8.4]

Arsuffi TL, Suberkropp K. (1988) Effects of fungal mycelia and enzymatically degraded leaves on feeding and performance of caddisfly (Trichoptera) larvae. *Journal of the North American Benthological Society* **7**: 205–11. [8.4]

Arsuffi TL, Suberkropp K. (1989) Selective feeding by shredders on leaf-colonizing stream fungi: comparison of macroinvertebrate taxa. *Oecologia* **79**: 30–7. [8.4]

Azam F, Cho BC. (1987) Bacterial utilization of organic matter in the sea. In: Fletcher M, Gray TRG, Jones JG (eds) *Ecology of Microbial Communities*, pp 261–81. Cambridge University Press, Cambridge. [8.2]

Azam F, Cho BC, Smith DC, Simon M. (1990) Bacterial cycling of matter in the pelagic zone of aquatic ecosystems. In: Tilzer MM, Serruya C (eds) *Large Lakes, Ecological Structure and Function*, pp 477–88. Springer-Verlag, Berlin. [8.2]

Baker JH, Bradnam LA. (1976) The role of bacteria in the nutrition of aquatic detritivores. *Oecologia* **24**: 95–104. [8.4]

Bärlocher F. (1980) Leaf-eating invertebrates as competitors of aquatic hyphomycetes. *Oecologia* **47**: 303–6. [8.7]

Bärlocher F. (1982) The contribution of fungal enzymes to the digestion of leaves by *Gammarus fossarum* Koch (Amphipoda). *Oecologia* **52**: 1–4. [8.4, 8.5]

Bärlocher F. (1983) Seasonal variation of standing crop and digestibility of CPOM in a Swiss Jura stream. *Ecology* **64**: 1266–1272. [8.5]

Bärlocher F. (1985) The role of fungi in the nutrition

of stream invertebrates. *Botanical Journal of the Linnean Society* **91**: 83–94. [8.4]

Bärlocher F. (1991) Fungal colonization of fresh and dried leaves in the River Teign (Devon, England). *Nova Hedwigia* **52**: 349–57. [8.4]

Bärlocher F, Kendrick B. (1973a) Fungi and food preferences of *Gammarus pseudolimnaeus. Archiv für Hydrobiologie* **72**: 501–16. [8.4, 8.5]

Bärlocher F, Kendrick B. (1973b) Fungi in the diet of *Gammarus pseudolimnaeus. Oikos* **24**: 295–300. [8.4]

Bärlocher F, Kendrick B. (1974) Dynamics of the fungal populations on leaves in a stream. *Journal of Ecology* **62**: 761–91. [8.4]

Bärlocher F, Kendrick B. (1975) Assimilation efficiency of *Gammarus pseudolimnaeus* (Amphipoda) feeding on fungal mycelium or autumn-shed leaves. *Oikos* **26**: 55–9. [8.4]

Bärlocher F, Kendrick B. (1981) Role of aquatic hyphomycetes in the trophic structure of streams. In: Wicklow DT, Carroll GC (eds) *The Fungal Community. Its Organization and Role in the Ecosystem,* pp 743–60. Marcel Dekker, New York. [8.4]

Bärlocher F, Murdoch JH. (1989) Hyporheic biofilms—a potential food source for interstitial animals. *Hydrobiologia* **184**: 61–7. [8.2]

Bärlocher F, Oertli JJ. (1978a) Colonization of conifer needles by aquatic hyphomycetes. *Canadian Journal of Botany* **56**: 57–62. [8.2]

Bärlocher F, Oertli JJ. (1978b) Inhibitors of aquatic hyphomycetes in dead conifer needles. *Mycologia* **70**: 964–74. [8.2]

Bärlocher F, Porter CW. (1986) Digestive enzymes and feeding strategies of three stream invertebrates. *Journal of the North American Benthological Society* **5**: 58–66. [8.5]

Bassett A, Rossi L. (1987) Relationship between trophic niche breadth and reproductive capabilities in a population of *Proasellus coxalis. Functional Ecology* **1**: 13–18. [8.4]

Benfield EF, Webster JR, Golladay SW, Peters GT, Stout BM. (1991) Effects of forest disturbance on leaf breakdown in southern Appalachian streams. *Verhandlungen Internationale Vereinigung für Theoretische und Angewandt Limnologie* **24**: 1687–90. [8.8]

Bengtsson G. (1982) Patterns of amino acid utilization by aquatic hyphomycetes. *Oecologia* **55**: 355–63. [8.2]

Berman T. (1990) Microbial food-webs and nutrient cycling in lakes: changing perspectives. In: Tilzer MM, Serruya C (eds) *Large Lakes, Ecological Structure and Function,* pp 511–25. Springer-Verlag, Berlin. [8.2]

Bilby RE, Ward JW. (1989) Changes in characteristics and function of woody debris with increasing size of streams in Western Washington. *Transactions of the American Fisheries Society* **118**: 368–78. [8.2]

Bjarnov N. (1972) Carbohydrases in *Chironomus, Gammarus* and some Trichoptera larvae. *Oikos* **23**: 261–3. [8.5]

Boling RH, Goodman ED, van Sickle JA, Zimmer JO, Cummins KW, Petersen RC, Reice SR. (1975) Toward a model of detritus processing in a woodland stream. *Ecology* **56**: 141–51. [8.2, 8.4]

Bott TL, Kaplan LA. (1990) Potential for protozoan grazing of bacteria in streambed sediments. *Journal of the North American Benthological Society* **9**: 336–45. [8.2]

Bott TL, Ritter FP. (1981) Benthic algal production in a piedmont stream measured by ^{14}C and dissolved oxygen procedures. *Journal of Freshwater Ecology* **1**: 267–78. [8.2]

Bowen SH. (1984) Evidence of a detritus food chain based on consumption of organic precipitates. *Bulletin of Marine Science* **35**: 440–8. [8.2]

Bueler CM. (1984) Feeding preferences of *Pteronarcys pictetii* (Plecoptera: Insecta) from a small, acidic woodland stream. *Florida Entomologist* **67**: 393–401. [8.4]

Burton TM, Stanford RM, Allan JW. (1985) Acidification effects on stream biota and organic matter processing. *Canadian Journal of Fisheries and Aquatic Sciences* **42**: 669–75. [8.7]

Calow P. (1974) Evidence for bacterial feeding in *Planorbis contortus* Linn. (Gastropoda: Pulmonata). *Proceedings of the Malacological Society of London* **41**: 145–56. [8.4]

Calow P, Calow LJ. (1975) Cellulase activity and niche separation in freshwater gastropods. *Nature* **255**: 478–80. [8.5]

Campbell IC, Doeg TJ. (1989) Impact of timber harvesting and production on streams: a review. *Australian Journal of Marine and Freshwater Research* **40**: 519–39. [8.8]

Cargill AS, Cummins KW, Hanson BJ, Lowry RR. (1985) The role of lipids, fungi and temperature in the nutrition of a shredder caddisfly, *Clistoronia magnifica. Freshwater Invertebrate Biology* **4**: 64–78. [8.4]

Carlough LA, Meyer JL. (1989) Protozoans in two southeastern blackwater rivers and their importance to trophic transfer. *Limnology and Oceanography* **34**: 163–77. [8.2]

Carlough LA, Meyer JL. (1990) Rates of protozoan bactivory in three habitats of a southeastern blackwater river. *Journal of the North American Benthological Society* **9**: 45–53. [8.2]

Carpenter J, Odum WE, Mills A. (1983) Leaf litter decomposition in a reservoir affected by acid mine drainage. *Oikos* **41**: 165–72. [8.7]

Chamier A-C. (1987) Effect of pH on microbial degradation of leaf litter in seven streams of the English Lake District. *Oecologia* **71**: 491–500. [8.3, 8.7]

Chamier A-C. (1991) Cellulose digestion and metabolism in the freshwater amphipod *Gammarus pseudolimnaeus* Bousfield. *Freshwater Biology* **25**: 33–40. [8.5]

Chamier A-C, Willoughby LG. (1986) The role of fungi in the diet of the amphipod *Gammarus pulex* (L): an enzymatic study. *Freshwater Biology* **16**: 197–208. [8.5]

Christensen B. (1977) Habitat preference among amylase genotypes in *Asellus aquaticus* (Isopoda, Crustacea). *Hereditas* **87**: 21–6. [8.4]

Conners ME, Naiman RJ (1984) Particulate allochthonous inputs: relationships with stream size in an undisturbed watershed. *Canadian Journal of Fisheries and Aquatic Sciences* **41**: 1473–84. [8.2]

Cuffney TF. (1988) Input, movement and exchange of organic matter within a subtropical coastal blackwater river-floodplain system. *Freswater Biology* **19**: 305–20. [8.2]

Cuffney TF, Wallace JB, Lugthart GJ. (1990) Experimental evidence quantifying the role of benthic invertebrates in organic matter dynamics of headwater streams. *Freshwater Biology* **23**: 281–99. [8.7]

Cuffney TF, Wallace JB, Webster JR. (1984) Pesticide manipulation of a headwater stream: invertebrate responses and their significance for ecosystem processes. *Freshwater Invertebrate Biology* **3**: 153–71. [8.7]

Cummins KW. (1974) Structure and function of stream ecosystems. *BioScience* **24**: 631–41. [8.2, 8.3]

Cummins KW, Klug MJ. (1979) Feeding ecology of stream invertebrates. *Annual Review of Ecology and Systematics* **10**: 147–72. [8.2, 8.4]

Cummins KW, Spengler GL, Ward GM, Speaker RM, Ovink RM, Mahan DC, Mattingly RL. (1980) Processing of confined and naturally entrained leaf litter in a woodland stream ecosystem. *Limnology and Oceanography* **25**: 952–7. [8.6]

Cummins KW, Sedell JR, Swanson FJ, Minshall GW, Fisher SG, Cushing CE *et al.* (1983) Organic matter budgets for stream ecosystems: problems in their evaluation. In: Barnes JR, Minshall GW (eds) *Stream Ecology: Application and Testing of General Ecological Theory*, pp 299–353. Plenum Press, New York. [8.2, 8.7]

Cummins KW, Wilzbach MA, Gates DM, Perry JB, Taliaferro WB. (1989) Shredders and riparian vegetation. *BioScience* **39**: 24–30. [8.3, 8.8]

Davis SF, Winterbourn MJ. (1977) Breakdown and colonization of *Nothofagus* leaves in a New Zealand stream. *Oikos* **28**: 250–5. [8.3]

Duddridge JE, Wainwright M. (1980) Heavy metal accumulation by aquatic fungi and reduction in viability of *Gammarus pulex* fed Cd^{2+} contaminated mycelium. *Water Research* **14**: 1605–11. [8.7]

Edwards RT, Meyer JL. (1986) Production and turnover of planktonic bacteria in two southeastern blackwater rivers. *Applied and Environmental Microbiology* **52**: 1317–23. [8.2]

Edwards RT, Meyer JL. (1987) Bacteria as a food source for black fly larvae in a blackwater river. *Journal of the North American Benthological Society* **6**: 241–50. [8.2, 8.4]

Edwards RT, Meyer JL. (1990) Bactivory by deposit-feeding mayfly larvae (*Stenonema* spp.). *Freshwater Biology* **24**: 453–62. [8.2]

Egglishaw HJ. (1968) The quantitative relationship between bottom fauna and plant detritus in streams of different calcium concentrations. *Journal of Applied Ecology* **5**: 731–40. [8.7]

Egglishaw HJ. (1972) An experimental study of the breakdown of cellulose in fast-flowing streams. *Memorie Istituto Italiano di Idrobiologia* **29** (Suppl.): 405–28. [8.7]

Fairchild JF, Boyle TP, Robinson-Wilson E, Jones JR. (1983) Microbial action in detritus leaf processing and the effects of chemical perturbation. In: Fontaine TD, Bartell SM (eds) *Dynamics of Lotic Ecosystems*, pp 437–56. Ann Arbor Science, Michigan. [8.7]

Fano EA, Rossi L, Basset A. (1982) Fungi in the diet of three benthic invertebrate species. *Bollettino di Zoologia* **49**: 99–105. [8.4]

Federle TW, Vestal JR. (1980) Lignocellulose mineralization by arctic lake sediments in response to nutrient manipulation. *Applied and Environmental Microbiology* **40**: 32–9. [8.7]

Findlay SEG, Arsuffi TL. (1989) Microbial growth and detritus transformations during decomposition of leaf litter in a stream. *Freshwater Biology* **21**: 261–9. [8.2, 8.4]

Findlay S, Tenore K. (1982) Nitrogen source for a detritivore: detritus substrate versus associated microbes. *Science* **218**: 371–3. [8.4]

Findlay S, Meyer JL, Smith PJ. (1984) Significance of bacterial biomass in the nutrition of a freshwater isopod (*Lirceus* sp.). *Oecologia* **63**: 38–42. [8.4]

Findlay S, Meyer JL, Smith PJ. (1986a) Contribution of fungal biomass to the diet of a freshwater isopod (*Lirceus* sp.). *Freshwater Biology* **16**: 377–85. [8.4]

Findlay S, Meyer JL, Smith PJ. (1986b) Incorporation of microbial biomass by *Peltoperla* sp. (Plecoptera) and *Tipula* sp. (Diptera). *Journal of the North American Benthological Society* **5**: 306–10. [8.4]

Forbes AM, Magnuson JJ. (1980) Decomposition and microbial colonization of leaves in a stream modified by coal ash effluent. *Hydrobiologia* **76**: 263–7. [8.7]

Fredeen FJH. (1964) Bacteria as food for blackfly larvae (Diptera: Simuliidae) in laboratory cultures and in natural streams. *Canadian Journal of Zoology* **42**: 527–48. [8.4]

Fuller RL, Mackay RJ. (1981) Effects of food quality on the growth of three *Hydropsyche* species (Trichoptera:

Hydropsychidae). *Canadian Journal of Zoology* **59**: 1133–40. [8.3]

Gessner MO, Schwoerbel J. (1989) Leaching kinetics of fresh leaf-litter with implications for the current concept of leaf-processing in streams. *Archiv für Hydrobiologie* **115**: 81–90. [8.4]

Giesy JP. (1978) Cadmium inhibition of leaf decomposition in an aquatic microcosm. *Chemosphere* **6**: 467–75. [8.7]

Godfrey BES. (1983) Growth of two terrestrial microfungi on submerged alder leaves. *Transactions of the British Mycological Society* **81**: 418–21. [8.4]

Golladay SW, Webster JR, Benfield EF. (1983) Factors affecting food utilization by a leaf shredding aquatic insect: leaf species and conditioning time. *Holarctic Ecology* **6**: 157–62. [8.4]

Graça MAS. (1990) *Observations on the feeding biology of two stream-dwelling detritivores*: Gammarus pulex *(L) and* Asellus aquaticus *(L)*. Unpublished PhD thesis, University of Sheffield, Sheffield. [8.4]

Gray LJ, Ward JV. (1983) Leaf litter in streams receiving treated and untreated metal mine drainage. *Environment International* **9**: 135–8. [8.7]

Groom AP, Hildrew AG. (1989) Food quality for detritivores in streams of contrasting pH. *Journal of Animal Ecology* **58**: 863–81. [8.7]

Haack TK, McFeters GA. (1982) Nutritional relationships among microorganisms in an epilithic biofilm community. *Microbial Ecology* **8**: 115–26. [8.2]

Hanson BJ, Cummins KW, Cargill AS, Lowry RR. (1983) Dietary effects on lipid and fatty acid composition of *Clistoronia magnifica* (Trichoptera: Limniphilidae). *Freshwater Invertebrate Biology* **2**: 2–15. [8.4]

Hargrave BT. (1970) The utilization of benthic microflora by *Hyallela azteca* (Amphipoda). *Journal of Animal Ecology* **39**: 427–37. [8.4]

Hart SD, Howmiller RP. (1975) Studies on the decomposition of allochthonous detritus in two southern California streams. *Verhandlungen Internationale Vereinigung für Theoretische und Angewandt Limnologie* **19**: 1665–74. [8.6]

Herbst GN. (1982) Effects of leaf type on the consumption rates of aquatic detritivores. *Hydrobiologia* **89**: 77–87. [8.6]

Hildrew AG, Townsend CR, Francis J. (1984) Cellulolytic decomposition in streams of contrasting pH and its relationship with invertebrate community structure. *Freshwater Biology* **14**: 323–8. [8.7]

Hipp E, Mustafa T, Bickel U, Hoffman KH. (1986) Integumentary uptake of acetate and propionate (VFA) by *Tubifex* sp. a freshwater oligochaete. I. Uptake rate and transport kinetics. *Journal of Experimental Zoology* **240**: 289–97. [8.2]

Holopainen A-L, Huttunen P, Ahtiainen M. (1991) Effects of forestry practices on water quality and primary productivity in small forest brooks. *Verhandlungen Internationale Vereinigung für Theoretische und Angewandt Limnologie* **24**: 1760–6. [8.8]

Howarth RW, Fisher SG. (1976) Carbon, nitrogen and phosphorus dynamics during leaf decay in nutrient-enriched stream microecosystems. *Freshwater Biology* **6**: 221–8. [8.7]

Imbert JB, Pozo J. (1989) Breakdown of four leaf litter species and associated fauna in a Basque Country forested stream. *Hydrobiologia* **182**: 1–14. [8.6]

Irons JG, Oswood MW, Bryant JP. (1988) Consumption of leaf detritus by a stream shredder: influence of tree species and nutrient status. *Hydrobiologia* **160**: 53–61. [8.4]

Iversen TM. (1973) Decomposition of autumn-shed beech leaves in a springbrook and its significance for the fauna. *Archiv für Hydrobiologie* **72**: 305–12. [8.3, 8.4]

Iversen TM. (1975) Disappearance of autumn shed beech leaves placed in bags in small streams. *Verhandlungen Internationale Vereinigung für Theoretische und Angewandt Limnologie* **19**: 1687–92. [8.6, 8.7]

Kaplan LA, Bott TL. (1983) Microbial heterotrophic utilization of dissolved organic matter in a piedmont stream. *Freshwater Biology* **13**: 363–77. [8.2]

Kaplan LA, Bott TL. (1985) Acclimation of stream-bed heterotrophic microflora: metabolic responses to dissolved organic matter. *Freshwater Biology* **15**: 479–92. [8.2]

Kaushik NK, Hynes HBN. (1971) The fate of dead leaves that fall into streams. *Archiv für Hydrobiologie* **68**: 465–515. [8.2, 8.3, 8.4, 8.6]

Kirby JM, Webster JR, Benfield EF. (1983) The role of shredders in detrital dynamics of permanent and temporary streams. In: Fontaine TD, Bartell SM (eds) *Dynamics of Lotic Ecosystems*, pp 425–35. Ann Arbor Science, Michigan. [8.6]

Klug MJ, Kotarski S. (1980) Bacteria associated with the gut tract of larval stages of the aquatic cranefly *Tipula abdominalis* (Diptera; Tipulidae). *Applied Environmental Microbiology* **40**: 408–16. [8.5]

Kostalos M, Seymour RL. (1976) Role of microbial enriched detritus in the nutrition of *Gammarus minus* (Amphipoda). *Oikos* **27**: 512–16. [8.4]

Lawson DL, Klug MJ. (1989) Microbial fermentation in the hindguts of two stream detritivores. *Journal of the North American Benthological Society* **8**: 85–91. [8.5]

Lawson DL, Klug MJ, Merritt RW. (1984) The influence of the physical, chemical and microbiological characteristics of decomposing leaves on the growth of the detritivore *Tipula abdominalis* (Diptera: Tipulidae). *Canadian Journal of Zoology* **62**: 2339–43. [8.4]

Leff LG, McArthur JV. (1989) The effect of leaf pack composition on processing: a comparison of mixed and single species packs. *Hydrobiologia* **182**: 219–24. [8.6]

Leland HV, Carter JL. (1985) Effects of copper on production of periphyton, nitrogen fixation and processing of leaf litter in a Sierra Nevada, California, stream. *Freshwater Biology* **15**: 155–73. [8.7]

Lock MA, Hynes HBN. (1976) The fate of dissolved organic carbon derived from autumn-shed maple leaves (*Acer saccharum*) in a temperate hard water stream. *Limnology and Oceanography* **21**: 436–43. [8.2]

Lock MA, Wallace RR, Costerton JW, Ventullo RM, Charlton SE. (1984) River epilithon: towards a structural–functional model. *Oikos* **42**: 10–22. [8.2]

McDowell WH. (1985) Kinetics and mechanisms of dissolved organic carbon retention in a headwater stream. *Biogeochemistry* **2**: 329–53. [8.2]

McGeorge JE, Jagoe CH, Risley LS, Morgan MD. (1991) Litter decomposition in low pH streams in the New Jersey Pinelands. *Verhandlungen Internationale Vereinigung für Theoretische und Angewandt Limnologie* **24**: 1711–14. [8.7]

McKinley VL, Vestal JR. (1982) Effects of acid on plant litter decomposition in an arctic lake. *Applied and Environmental Microbiology* **43**: 1188–95. [8.7]

Mackay RJ, Kersey KE. (1985) A preliminary study of aquatic insect communities and leaf decomposition in acid streams near Dorset, Ontario. *Hydrobiologia* **122**: 3–11. [8.7]

Maki JS, LaCroix SJ, Hopkins BS, Staley JT. (1986) Recovery and diversity of heterotrophic bacteria from chlorinated drinking waters. *Applied and Environmental Microbiology* **51**: 1047–55. [8.7]

Maltby L, Booth R. (1991) The effect of coal-mine effluent on fungal assemblages and leaf breakdown. *Water Research* **25**: 247–50. [8.7]

Mann KH. (1988) Production and use of detritus in various freshwater, marine and coastal marine ecosystems. *Limnology and Oceanography* **33**: 910–39. [8.2, 8.4]

Marcus JH, Willoughby LG. (1978) Fungi as food for the aquatic invertebrate *Asellus aquaticus*. *Transactions of the British Mycological Society* **70**: 143–6. [8.4]

Martin MM, Martin JS, Kukor JJ, Merritt RW. (1980) The digestion of protein and carbohydrate by the stream detritivore *Tipula abdominalis* (Diptera, Tipulidae). *Oecologia* **46**: 360–4. [8.4]

Martin MM, Kukor JJ, Martin JS, Lawson DL, Merritt RW. (1981a) Digestive enzymes of larvae of three species of caddisflies (Tricoptera). *Insect Biochemistry* **11**: 501–5. [8.5]

Martin MM, Martin JS, Kukor JJ, Merritt RW. (1981b) The digestive enyzmes of detritus-feeding stonefly nymphs (Plecoptera; Pteronarcyidae). *Canadian Journal of Zoology* **59**: 1947–51. [8.5]

Mathews CP, Kowalczewski A. (1969) The disappearance of leaf litter and its contribution to production in the River Thames. *Journal of Ecology* **57**: 543–52. [8.6]

Melillo JM, Aber JD, Mutatore JF. (1982) Nitrogen and lignin control of hardwood leaf litter decomposition dynamics. *Ecology* **63**: 621–6. [8.2]

Merritt RW, Cummins KW. (1978) *An Introduction to the Aquatic Insects of North America*. Kendall/Hunt Publishing, Iowa. [8.3]

Merritt RW, Cummins KW, Burton TM. (1984) The role of aquatic insects in the processing and cycling of nutrients. In: Resh VH, Rosenberg DM (eds) *The Ecology of Aquatic Insects*, pp 134–63. Praeger Publications, New York. [8.2]

Miller JC. (1987) Evidence for the use of non-detrital dissolved organic matter by microheterotrophs on plant detritus in a woodland stream. *Freshwater Biology* **18**: 483–94. [8.2, 8.9]

Minshall GW, Brock JT, LaPoint TW. (1982) Characterization and dynamics of benthic organic matter and invertebrate functional feeding group relationships in the Upper Salmon River, Idaho (USA). *Internationale Revue de Gesamten Hydrobiologie* **67**: 793–820. [8.6]

Monk DC. (1976) The distribution of cellulase in freshwater invertebrates of different feeding habits. *Freshwater Biology* **6**: 471–5. [8.5]

Monk DC. (1977) The digestion of cellulose and other dietary components and pH of the gut in the amphipod *Gammarus pulex* (L.) *Freshwater Biology* **7**: 431–40. [8.5]

Mulholland PJ, Palumbo AV, Elwood JW, Rosemond AD. (1987) Effects of acidification on leaf decomposition in streams. *Journal of the North American Benthological Society* **6**: 147–58. [8.7]

Mutch RA, Steedman RJ, Berté SB, Pritchard G. (1983) Leaf breakdown in a mountain stream: a comparison of methods. *Archiv für Hydrobiologie* **97**: 89–108. [8.6]

Newell SY, Fallon RD, Miller JD. (1986) Measuring fungal-biomass dynamics in standing-dead leaves of a salt-marsh vascular plant. In: Moss ST (ed.) *The Biology of Marine Fungi*, pp 19–25. Cambridge University Press, Cambridge. [8.4]

Newman RM. (1990) Effects of shredding amphipod density on watercress *Nasturtium officinale* breakdown. *Holarctic Ecology* **13**: 293–9. [8.6]

Newman RM, Perry JA, Tam E, Crawford RL. (1987) Effects of chronic chlorine exposure on litter processing in outdoor experimental streams. *Freshwater Biology* **18**: 415–28. [8.6, 8.7]

Oberndorfer RY, McArthur JV, Barnes JR, Dixon J. (1984) The effect of invertebrate predators on leaf litter processing in an alpine stream. *Ecology* **65**: 1325–31. [8.6]

Padgett DE. (1976) Leaf decomposition by fungi in a tropical rainforest stream. *Biotropica* **8**: 166–78. [8.3]

Palumbo AV, Bogle MA, Turner RR, Elwood JW, Mulholland PJ. (1987) Bacterial communities in acidic and circumneutral streams. *Applied and Environ-*

mental Microbiology **53**: 337–44. [8.7]
Park D. (1974) Accumulation of fungi by cellulose exposed in a river. *Transactions of the British Mycological Society* **63**: 437–47. [8.4]
Park D, McKee W. (1978) Cellulolytic *Pythium* as a component of the river mycoflora. *Transactions of the British Mycological Society* **71**: 251–9. [8.4]
Patrick FM, Loutit MW. (1976) Passage of metals in effluents, through bacteria to higher organisms. *Water Research* **10**: 333–5. [8.7]
Patrick FM, Loutit MW. (1978) Passage of metals to freshwater fish from their food. *Water Research* **12**: 395–8. [8.7]
Paul RW, Benfield EF, Cairns J. (1983) Dynamics of leaf processing in a medium-sized river. In: Fontaine TD, Bartell SM (eds) *Dynamics of Lotic Ecosystems*, pp 403–23. Ann Arbor Science, Michigan. [8.6]
Petersen RC, Cummins KW. (1974) Leaf processing in a woodland stream. *Freshwater Biology* **4**: 343–68. [8.3, 8.4, 8.6]
Pinkney AE, Poje GV, Sansur RM, Lee CC, O'Connor JM. (1985) Uptake and retention of ^{14}C-Aroclor 1254 in the amphipod *Gammarus tigrinus*, fed contaminated fungus, *Fusarium oxysporum*. *Archives of Environmental Contamination and Toxicology* **14**: 59–64. [8.7]
Porter KG, Sherr EB, Sherr BF, Pace M, Sanders RW. (1985) Protozoa in planktonic food webs. *Journal of Protozoology* **32**: 409–15. [8.2]
Reice SR. (1977) The role of animal associations and current velocity in sediment-specific leaf litter decomposition. *Oikos* **29**: 357–65. [8.7]
Reice SR. (1978) Role of detritivore selectivity in species-specific litter decomposition in a woodland stream. *Verhandlungen Internationale Vereinigung für Theoretische und Angewandt Limnologie* **20**: 1396–1400. [8.6]
Roeding CE, Smock LA. (1989) Ecology of macroinvertebrate shredders in a low-gradient sandy-bottomed stream. *Journal of the North American Benthological Society* **8**: 149–61. [8.3]
Ross DH, Wallace JB. (1983) Longitudinal patterns of production, food consumption and seston utilization by net-spinning caddisflies (Trichoptera) in a south Appalachian stream (USA). *Holarctic Ecology* **6**: 270–84. [8.2, 8.3]
Rosset J, Bärlocher F. (1985) Aquatic hyphomycetes: influences of pH, Ca^{2+} and HCO_3^- on growth *in vitro*. *Transactions of the British Mycological Society* **84**: 137–45. [8.7]
Rossi L. (1985) Interactions between invertebrates and microfungi in freshwater ecosystems. *Oikos* **44**: 175–84. [8.4]
Rossi L, Fano AE. (1979) Role of fungi in the trophic niche of the congeneric detritivores *Asellus aquaticus* and *Asellus coxalis* (Isopoda). *Oikos* **32**: 380–5. [8.4]
Rossi L, Basset A, Nobile L. (1983) A coadapted trophic niche in two species of crustacea (Isopoda): *Asellus aquaticus* and *Proasellus coxalis*. *Evolution* **37**: 810–20. [8.4]
Rounick JS, Winterbourn MJ. (1983) The formation, structure and utilization of stone surface organic layers in two New Zealand streams. *Freshwater Biology* **13**: 57–72. [8.2]
Rounick JS, Winterbourn MJ, Lyon GL. (1982) Differential utilization and autochthonous inputs by aquatic invertebrates in some New Zealand streams: a stable carbon isotope study. *Oikos* **39**: 191–8. [8.2]
Salonen K, Hammer T. (1986) On the importance of dissolved organic matter in the nutrition of zooplankton in some lake waters. *Oecologia* **68**: 246–53. [8.2]
Schindler DW. (1980) Experimental acidification of a whole lake: a test of the oligotrophication hypothesis. In: Drabløs D, Tollan A (eds). *Proceedings International Conference on the Ecological Impact of Acid Precipitation, Norway 1980*, pp 370–4. SNSF Project. [8.7]
Schoenberg SA, Maccubbin AE, Hodson RE. (1984) Cellulose digestion by freshwater microcrustacea. *Limnology and Oceanography* **29**: 1132–6. [8.5]
Sedell JR, Triska FJ, Triska NS. (1975) The processing of conifer and hardwood leaves in two coniferous forest streams: I. weight loss and associated invertebrates. *Verhandlungen Internationale Vereinigung für Theoretische und Angewandt Limnologie* **19**: 1617–27. [8.2]
Short RA, Maslin PE. (1977) Processing of leaf litter by a stream detritivore: effect on nutrient availability to collectors. *Ecology* **58**: 935–8. [8.2]
Sinsabaugh RL, Linkins AE. (1990) Enzymic and chemical analysis of particulate organic matter from a boreal river. *Freshwater Biology* **23**: 301–9. [8.2, 8.3]
Sinsabaugh RL, Linkins AE, Benfield EF. (1985) Cellulose digestion and assimilation by three leaf-shredding aquatic insects. *Ecology* **66**: 1464–71. [8.5]
Speaker R, Moore K, Gregory S. (1984) Analysis of the process of retention of organic matter in stream ecosystems. *Verhandlungen Internationale Vereinigung für Theoretische und Angewandt Limnologie* **22**: 1835–41. [8.2]
Stewart BA, Davies BR. (1989) The influence of different litter bag designs on the breakdown of leaf material in a small mountain stream. *Hydrobiologia* **183**: 173–7. [8.6]
Stout BM III, Coburn CB Jr. (1989) Impact of highway construction on leaf processing in aquatic habitats of eastern Tennesse. *Hydrobiologia* **178**: 233–42. [8.8]
Suberkropp K, Klug MJ. (1974) Decomposition of deciduous leaf litter in a woodland stream. I. A scanning electron microscopic study. *Microbial Ecology* **1**: 96–103. [8.3]
Suberkropp K, Klug MJ. (1976) Fungi and bacteria associ-

ated with leaves during processing in a woodland stream. *Ecology* **57**: 707–19. [8.3, 8.4]

Suberkropp K, Godshalk GL, Klug MJ. (1976) Changes in the chemical composition of leaves during processing in a woodland stream. *Ecology* **57**: 720–7. [8.2]

Suberkropp K, Arsuffi TL, Anderson JP. (1983) Comparison of degradative ability, enzymatic activity and palatability of aquatic hyphomycetes grown on leaf litter. *Applied and Environmental Microbiology* **46**: 237–44. [8.4]

Sutcliffe DW, Carrick TR, Willoughby LG. (1981) Effects of diet, body size, age and temperature on growth rates in the amphipod *Gammarus pulex*. *Freshwater Biology* **11**: 183–214. [8.4]

Swanson FJ, Gregory SV, Sedell JR, Campbell GA. (1982) Land–water interaction: the riparian zone. In: Edmonds RL (ed.) *Analysis of Coniferous Forest Ecosystems in the Western United States*, pp 267–91. Hutchinson Ross, Stroudsburg, Pennsylvania. [8.2]

Thomas JD, Kowalczyk C, Somasundaram B. (1990) The biochemical ecology of *Biomaphalaria glabrata*, a freshwater pulmonate mollusc: the uptake and assimilation of exogenous glucose and maltose. *Comparative Biochemistry and Physiology* **95A**: 511–28. [8.2]

Thompson PL, Bärlocher F. (1989) Effect on pH on leaf breakdown in streams and in the laboratory. *Journal of the North American Benthological Society* **8**: 203–10. [8.7]

Traaen TS. (1980) Effects of acidity on decomposition of organic matter in aquatic environments. In: Drabløs D, Tollan A (eds) *Proceedings International Conference on the Ecological Impact of Acid Precipitation, Norway 1980*, pp 340–1. SNSF Project. [8.7]

Triska FJ. (1970) *Seasonal distribution of aquatic hyphomycetes in relation to the disappearance of leaf litter from a woodland stream*. Unpublished PhD thesis, University of Pittsburg, Pittsburg. [8.3, 8.4]

Triska FJ, Sedell JR, Gregory SV. (1982) Coniferous forest streams. In: Edmonds RL (ed.) *Analysis of Coniferous Forest Ecosystems in the Western United States*, pp 292–332. Hutchinson Ross, Stroudsburg, Pennsylvania. [8.2]

van Frankenhuyzen K, Geen GH, Koivisto C. (1985) Direct and indirect effects of low pH on the transformation of detrital energy by the shredding caddisfly, *Clistoronia magnifica* (Banks) (Limnephilidae. *Canadian Journal of Zoology* **63**: 2298–304. [8.7]

Wallace JB, Benke AC, Lingle AH, Parsons K. (1987) Trophic pathways of macroinvertebrate primary consumers in subtropical blackwater streams. *Archiv für Hydrobiologie* **74** (Suppl.): 423–51. [8.2]

Wallace JB, Webster JR, Cuffney TF. (1982) Stream detritus dynamics: regulation by invertebrate consumers. *Oecologia* **53**: 197–200. [8.7]

Ward GM. (1986) Lignin and cellulose content of benthic fine particulate organic matter (FPOM) in Oregon cascade mountain streams. *Journal of the North American Benthological Society* **5**: 127–39. [8.2]

Webster JR, Benfield EF. (1986) Vascular plant breakdown in freshwater ecosystems. *Annual Review of Ecology and Systematics* **17**: 567–94. [8.2]

Welton JS, Ladle M, Bass JAB, John IR. (1983) Estimation of gut throughput time in *Gammarus pulex* under laboratory and field conditions with a note on the feeding of young in the brood pouch. *Oikos* **41**: 133–8. [8.4]

Wetzel RG. (1975) *Limnology*. W.B. Saunders, Philadelphia. [8.2]

Willoughby LG, Earnshaw R. (1982) Gut passage times in *Gammarus pulex* (Crustacea, Amphipoda) and aspects of summer feeding in a stony stream. *Hydrobiologia* **97**: 105–17. [8.4]

Willoughby LG, Sutcliffe DW. (1976) Experiments on feeding and growth of the amphipod *Gammarus pulex* (L.) related to its distribution in the River Duddon. *Freshwater Biology* **6**: 577–86. [8.4]

Winterbourn MJ, Rounick JS, Hildrew AG. (1986) Patterns of carbon resource utilization by benthic invertebrates in two British river systems: a stable isotope study. *Archiv für Hydrobiologie* **3**: 349–61. [8.2]

Winkler G. (1991) Debris dams and retention in a low order stream (a backwater of Oberer Seebach – Ritrodat-Lunz study area, Austria). *Verhandlungen Internationale Vereinigung für Theoretische und Angewandt Limnologie* **24**: 1917–20. [8.2]

9: Primary Production

R.G.WETZEL AND A.K.WARD

9.1 INTRODUCTION

The roles of photosynthetic primary producers within lotic ecosystems have been the subject of past controversy. Early evaluations of the magnitude of primary productivity in comparison with other sources of organic matter, particularly importations of organic matter from the drainage basin (e.g. Hynes 1970; Fisher & Likens 1973), suggested stream primary production was essentially irrelevant. Often the rationale for such evaluations focused on the amount of readily utilizable particulate organic matter available for consumption by herbivorous organisms, particularly benthic invertebrates. The reason for this evaluation was because many of the early ecosystem level studies focused on small streams that were heavily shaded by forest canopies during much of the active growing season and, therefore, were determined to have a negative net primary productivity. Autochthonous primary productivity within the stream by algae and macrophytes was often found to be less than respiratory losses of organic matter on an annual basis. As a result, streams in general were often referred to as heterotrophic; that is, greater amounts of organic matter were being decomposed in running waters than were being produced autochthonously within them. In reviews of the subject by Wetzel (1975) and subsequently by Minshall (1978), Bott (1983) and Bott *et al* (1985), among others, a more expansive view of the importance of primary production was adopted as a broader array of lotic ecosystems was examined and a more robust conceptual framework of the longitudinal aspects of river drainages and organic matter inputs was formulated (Vannote *et al* 1980). More recently, even the importance of primary producers within heavily shaded, small, woodland streams has been recognized (e.g. Mayer & Likens 1987; Stock & Ward 1989).

Concurrent with the development of these concepts was an enlarged view of the 'boundaries' of stream ecosystems (Hynes 1974) and the importance of interfaces between the stream channel and other features of the stream catchment, specifically riparian (e.g. Cummins *et al* 1984), floodplain/wetland (e.g. Sedell & Froggatt 1984; Meyer 1990) and subsurface (hyporheic) regions (Grimm & Fisher 1984; Triska *et al* 1989a, 1989b). The existence and characteristics of these interfaces can clearly have a major impact on instream processes, including primary productivity. Therefore, a recent evolution in the conceptualization of the operation of stream and river ecosystems has concerned the coupled integration of dissolved and particulate nutrient/energy sources originating from outside the traditionally viewed stream channel with those produced and/or transformed within it. Much effort is now directed toward the sites of entry of this material, particularly from hyporheic groundwater through sediments to the overlying water and from riparian vegetation of adjacent land–water interface and floodplain regions.

Finally, much attention is correctly being directed to the fates of primary production within running waters. An evaluation of the factors important to the rates of primary production is inseparable from a discussion of decomposition and herbivory. As elaborated below, decomposition is the primary process of nutrient recycling that is supporting the primary production. Furthermore, herbivory can reduce as well as stimulate primary productivity. Although light availability, current, grinding action, and other physical factors remain as important regulators of the magnitude of primary productivity, the importance of microbial and higher-organism interactions are being recognized as integrating forces influencing both the primary productivity and the functional roles of primary producers in river ecosystems.

9.2 METHODS FOR MEASURING LOTIC PRIMARY PRODUCTIVITY AND THE MATTER OF SCALE

The methods used to estimate within-channel primary productivity have developed over many years in conjunction with the evolution and maturation of the total field of lotic ecology. Early estimates of lotic ecosystem metabolism, including primary production and respiration, used measurements of change in dissolved oxygen or carbon from pH in bulk streamwater measured at two stations over a diurnal period (e.g. Odum 1956; Hall & Moll 1975; reviewed by Wetzel & Likens 1991). Although this technique provides estimates of gross primary production in streams/rivers, it has been generally replaced by other methods because of problems in calculation of gas diffusion constants (particularly in small, rapidly flowing streams) and an inability to evaluate relative rates associated with subcomponents of the stream reach between the two stations (e.g. among habitats).

During the past 15 years, Plexiglas chambers equipped with submersible pumps to circulate streamwater over benthic substrata have been widely used as an *in-situ* technique to estimate various aspects of lotic, benthic metabolism (Bott *et al* 1978; Gregory 1980; Minshall *et al* 1983; Naiman 1983). The specific design and size of the chambers can be varied to accommodate the needs of the individual investigator, and changes in oxygen, pH, and total inorganic carbon or incorporation of $^{14}CO_2$ and $H^{14}CO_3$ into plant organic matter over a measured incubation period are used to estimate lotic primary productivity. The use of radiolabelled inorganic carbon is an extremely sensitive technique that probably best estimates net photosynthesis, as is assumed in other aquatic ecosystems, but does not provide information on community respiration (Wetzel & Likens 1991). However, since benthic samples are radioassayed after the end of the incubation period, different plant morphological types and/or taxa from the same chamber can be radioassayed separately or examined by autoradiography and their primary production rates compared (Ward *et al* 1985; Stock *et al* 1987; Stock & Ward 1989). Overall, *in-situ* chamber methods provide a means of estimating primary productivity of within-channel benthic habitats under conditions which reasonably simulate stream conditions and at a convenient scale for the operator. One disadvantage is that benthic substrata and streamwater placed in the chamber are removed from the renewal effects of overlying streamwater and underlying sediments, a problem which can, however, be ameliorated by short incubation times (Stream Solute Workshop 1990). A frequent assumption of this method is that values based on metabolic rates of samples covering a square metre or less of stream bottom can be extrapolated to much larger areas.

Currently, all ecosystem researchers, including lotic ecologists, face similar technological problems in addressing the need to make estimates of primary productivity (and other metabolic processes) at larger scales than have previously been routinely measured (Risser *et al* 1988). Several new methods are emerging with strong potential in this area, although published research on streams/rivers is generally lacking. Combinations of satellite, low-altitude aerial photography, and infrared analyses have been used to estimate primary production over large areas of ocean and terrestrial landscapes and could be applicable to large rivers and associated lateral wetland areas as well (e.g. Brown *et al* 1985; Eppley *et al* 1985; Aumen *et al* 1991). Estimates of net ecosystem production based on measurements of diel atmospheric gas flux over large areas have been used in forest ecosystems (e.g. Ryan 1990). Also, recent application of long-path Fourier-Transform Infrared Spectroscopy (FTIR) to ecosystem analysis has resulted in the ability to measure the flux of multiple gases over landscapes at spatial scales up to 1 km (Gosz *et al* 1988).

At the other end of the spatial spectrum, there is an increasing recognition of the importance of microbial biofilms in lotic ecosystems, including their composition, complexity and community interactions. More sophisticated technology with resolution in the 5–100-μm scale range is needed to analyse and interpret the roles of photosynthetic and other components of these communities. Relevant technologies include the use of oxygen and pH microelectrodes as well as combinations of scanning electron microscopy and autoradiography (e.g. Revsbech 1983; Carlton & Wetzel 1987; Burkholder *et al* 1990).

9.3 COMPARATIVE PRIMARY PRODUCTIVITY

Comparative analyses of primary productivity among diverse ecosystems are difficult because of large differences in the analytical methods, conversion factors and sampling frequencies employed (*cf.*, for example, Dickerman *et al* 1986). On a global basis, the primary productivity of inland waters is considered to be minor, because of the relatively small surface area covered by standing or running water (Woodwell *et al* 1978). In addition, the productivity of phytoplankton in any water is low; productivity cannot be increased because of light limitations, regardless of adequacy of nutrient and other growth factors, to levels approaching those of most terrestrial ecosystems (Wetzel 1983, 1990). The oft-cited comparative table of Woodwell *et al* (1978) is reiterated in Table 9.1; however, the values for inland waters are probably greatly underestimated. Among rivers, much of this underestimation has resulted from detailed estimations of the rates of primary productivity only by within-channel attached algae.

Productivity rates of attached algae In a detailed comparison of the rates of in-stream algal primary productivity from many geographical and geological regions of North America to those intensively studied sites in Table 9.2, rates were found to be similar (Bott *et al* 1985). Concentrations of chlorophyll *a* of benthic organisms spanned an order of magnitude (10–100 mg m^{-2})

Table 9.1 Net primary productivity of major plant communities of the earth based on biomass estimates

	Net primary productivity (g carbon m^{-2} year^{-1})
Aquatic ecosystems	
Swamp and marsh	1350.0
Marine, algal beds and reefs	1166.7
Marine, estuaries	714.3
Marine upwelling zones	250.0
Lakes and streams	200.0
Marine, continental shelf	161.7
Marine, open oceans	56.3
Terrestrial ecosystems	
Tropical rain forest	988.2
Tropical seasonal forest	720.0
Temperate evergreen forest	580.0
Temperate deciduous forest	542.9
Savanna grassland	406.7
Boreal forest	358.3
Woodland and shrubland	317.7
Cultivated land	292.9
Temperate grassland	266.7
Tundra and alpine meadow	62.5
Desert shrub	38.9
Rock, sand	1.3

After Woodwell *et al* (1978).

Table 9.2 Mean annual net primary productivity of in-stream community producers of North American streams

	Mean annual net community primary productivity (mg carbon m^{-2} day^{-1})
Eastern deciduous forest stream, coastal climate, Pennsylvania[a]	−27.0–246.8 ($\bar{x}$ 127.4)
Mesic hardwood forest stream, continental climate, Michigan[a]	−55.9–486.4 ($\bar{x}$ 207.3)
Cool, arid climate stream, much precipitation as snowfall, coniferous vegetation on north-facing slopes, sagebrush on south-facing slopes, Idaho[a]	48.2–524.7 ($\bar{x}$ 415.2)
Northern cool-desert stream open, no canopy, Idaho[b]	
Autochthonous	
Macrophytes	47.2–147.6 ($\bar{x}$ 93.4)
Periphyton	465.6–1748.0 ($\bar{x}$ 1004.0)
Allochthonous	0.7–16.9 ($\bar{x}$ 7.0)
Coniferous forest stream, coastal climate, Oregon[a]	−21.0–93.8 ($\bar{x}$ 45.1)

[a] Extracted from data of Bott (1983).
[b] Calculated from Minshall (1978) assuming 1 kcal = 4.6 g organic matter and 46.5% carbon in organic matter of aquatic plants (Westlake 1965).

with little geographical differentiation within a seasonal range of 1 mg m^{-2} to slightly in excess of 300 mg m^{-2}. Highest amounts were found in Great Plains or desert regions of high productivity.

Gross primary productivity (GPP) of within-stream benthic algae (Table 9.3), as well as community respiration (CR_{24}, consumption of dissolved oxygen in 24 h), commonly increases with downstream direction (e.g. Bott *et al* 1985). Photosynthetic efficiencies (percentage of irradiance (400–750 nm) utilized in existing rates of net or gross photosynthetic production) of benthic algae generally are below 4%, with averages near 1% and extremes of about 7% in regions of high nutrient enrichments. Photosynthetic efficiencies generally decline downstream with increasing light availability.

Accurate comparisons of the *annual* primary productivity of in-stream producers (periphyton, phytoplankton and macrophytes) simultaneously with the productivity of other producers of the saturated-soils of the floodplain and bank regions of river ecosystems do not exist. These data are important, however, to correct interpretations of composite ecosystem production and utilization of organic matter entering the ecosystem.

This underestimation has resulted, in part, from a too-narrow definition of the boundaries of standing and running-water ecosystems. That is, only the readily observed central or channel water was recognized as the water body 'proper', and the importance of adjacent interface zones and their interactions with the central water bodies were virtually ignored. In particular, recognition of the expansion of stream/river boundaries to include floodplain, riparian and other features in recent years has developed in parallel with a more integrated ecosystem paradigm of running waters, which gained great impetus from Hynes (1974). These concepts have roots in the earlier perceptions of Thienemann (1926) and Lindeman (1942) in which aquatic ecosystems were viewed as an integration of interdependent component parts and processes. Therefore, whether inland water bodies are labelled as lotic or lentic, all reside in and are closely coupled with their drainage basins through interface regions. As a result, addressing the primary productivity of stream/river *ecosystems* requires complete analyses of

Table 9.3 Example rates of in-stream primary productivity

Site	Mean primary productivity (g carbon m^{-2} day^{-1})	Technique	Reference
Morgans Creek, Kentucky	0.004–0.0007	Biomass change	Minshall (1967)
Walker Branch, Tennessee	0.008–0.011	Biomass change	Elwood & Nelson (1972)
Glade Branch, Virginia	0.010	^{14}C	Hornick *et al* (1981)
Piney Branch, Virginia	0.011	^{14}C	Hornick *et al* (1981)
Guys Run, Virginia	0.018	^{14}C	Hornick *et al* (1981)
Berry Creek, Oregon	0.099	Oxygen exchange in recirculating chamber	Reese (1967)
Yellow Creek, Alabama	0.0493	^{14}C	Lay & Ward (1987)
Little Schultz Creek, Alabama	0.0912	^{14}C	Lay & Ward (1987)
New Hope Creek, North Carolina	0.33–1.43	Diurnal oxygen curve	Hall (1972)
Deep Creek, Idaho	1.18	Oxygen exchange, upstream/downstream	Minshall (1978)
Catahoula Creek, Mississippi	1.85	Oxygen exchange, upstream/downstream	de la Cruz & Post (1977)
White Clay Creek, Pennsylvania	2.22	O_2/CO_2 exchange	Bott *et al* (1978)

the productivity both within and adjacent to the channel *per se*. Such comparative data from individual ecosystems are almost completely lacking. Further, although the expanded boundaries of lotic ecosystems can easily be recognized at the conceptual level, specific definitions of the exact physical dimensions in any given ecosystem can be more difficult. How should the boundaries of lotic ecosystems be defined, and how does that definition affect our perception of the importance and functions of primary production in streams and rivers?

9.4 FUNCTIONAL BOUNDARIES OF RUNNING-WATER ECOSYSTEMS

Most reported values of the primary productivity of streams/rivers have been restricted to those producers submersed within a traditionally viewed river channel, i.e. one that includes only plants associated with the visible, surface-flowing water and some of its benthic surfaces (Table 9.4). However, both small stream and large river ecosystems can have soils adjacent to the main channel that are water saturated some distance from the channel. In stream/river channels with gentle slopes this area can involve a region many times larger (>30 times) than the area of the open stream water. Therefore, contemporary concepts of the boundaries of stream ecosystems include lateral and subsurface regions around the main channel. Hence, stream ecosystems are 'bounded' by the subsurface, groundwater–stream-water interface rather than the main channel *per se* (Fig. 9.1). This region encompasses biologically active gradients between the two and is defined by the hydrological movement of water through the entire stream ecosystem (e.g. Triska *et al* 1989b). This perception of lotic boundaries has important implications for the functional aspects of metabolism of biota, particularly the primary producers, within the running waters as well as the dynamics

Table 9.4 Estimates of annual net primary productivity, estimated from seasonal maximum biomass, of aquatic macrophytes of river ecosystems

	g AFDW m^{-2} $year^{-1}$	g carbon m^{-2} $year^{-1}$ [d]
Temperate rivers[a]		
Emergent	320–3712	149–1726
Submersed	8–395	4–184
Polluted Wisconsin stream[b]		
Submersed	1162	540
Amazon River[c]		
'Floating meadows'	2430–4050	1130–1883

[a] Extracted from the review of Rodgers *et al* (1983) from numerous sources.
[b] Calculated from Madsen & Adams (1988).
[c] Calculated from Junk (1970).
[d] Using an average value of 46.5% of AFDW (ash-free dry weight = organic carbon (*cf.* Westlake 1965).

of biogeochemical cycling. In the lateral regions, flood-tolerant vegetation adapted to saturated soils contributes to the primary production of lotic ecosystems. Some plants, such as emergent and floating-leaved rooted macrophytes, may be semi-immersed in the channel water. Other more woody shrubs and riparian species, such as willow and alder, can be rooted in the subsurface, hyporheic zone adjacent to the channel, but are included in the region hydrologically defined above as part of the stream ecosystem. The organic matter produced by these plants is in part decomposed at or near the sites of production. Part of the organic matter produced is transported to the river water, both overland and within hyporheic groundwater.

Based on the above concept of the boundaries of lotic ecosystems, vegetation that should be included in comparative analyses of primary productivity of rivers includes all aquatic and amphibious higher and lower plants: (1) emergent wetland plants, riverine trees and shrubs, floating non-rooted macrophytes, floating-leaved rooted macrophytes, and submersed rooted macrophytes; (2) epiphytic algae and cyanobacteria associated with submersed aquatic macrophytes; (3) epipelic algae and cyanobacteria associated

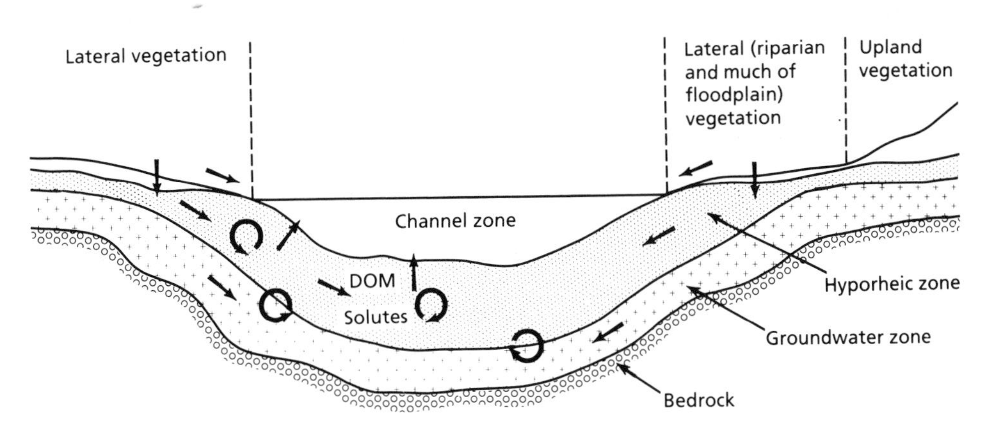

Fig. 9.1 Conceptualization of lateral and vertical boundaries of lotic ecosystems (modified extensively from Triska *et al* 1989b). The stream ecosystem boundary is defined as the hyporheic/groundwater interface and thereby includes a substantial volume beneath and lateral to the main channel. Vegetation rooted in the hyporheic zone is therefore part of stream ecosystem primary production. Arrows indicate flow pathways of dissolved organic matter and inorganic solutes derived from plant detritus within the stream ecosystem.

with particulate detritus and sediments in the understorey floodplain and bank wetlands; (4) epilithic algae and cyanobacteria associated with rock surfaces (from sand grains to boulders); and (5) phytoplankton and drift algae from epilithic and other attached sources.

The conceptual inclusion of all of these plants as part of lotic ecosystems is important because their productivity and metabolism have major effects on these systems. One of the most important effects of vegetation growing in lateral regions is the contribution of organic matter to the recipient stream, both in the particulate and dissolved form, a portion of which is metabolized microbially for potential utilization by higher trophic levels. Because of the importance of this organic matter to the metabolism, biogeochemical cycling of nutrients, and support of higher trophic levels within running-water systems, these organic matter sources drive subsequent metabolism and are functionally part of stream/river ecosystems. We contend that exclusion of this primary productivity by an arbitrary physical boundary and treatment as 'allochthonous organic matter loading' imposes an artificial demarcation that is incongruous with an integrated ecosystem perspective.

9.5 LONGITUDINAL AND LATERAL EFFECTS ON PRIMARY PRODUCTIVITY GRADIENTS

Longitudinal effects: light versus geomorphological characteristics

Much of the control of the observed primary productivity among running waters has been attributed to availability of solar insolation reaching the water and its attenuation within the water (e.g. Wetzel 1975; Minshall 1978; Bott 1983). As is the situation among standing waters, once nutrients are provided from both the water and the substrata in adequate supply to support maximum rates of photosynthesis, light availability is the dominant factor regulating photosynthesis. On the basis of several detailed comparative studies conducted in North America, a pattern of increased autotrophy with greater stream order was found (Minshall *et al* 1983; Naiman 1983). Variance from this general correlation is wide, however. Some of these differences are related to geomorphological characteristics of hydraulic energy as well as the influence of human activities, particularly from deforestation and agriculture, which have led to high inorganic loadings, sustained turbidity and light reductions.

Although the correlation of river primary productivity to stream order is direct, that relationship applies among relatively non-disturbed ecosystems (Fig. 9.2(a)). Much of the increase in autotrophy is associated with increasing contributions of periphytic algae. In non-disturbed river

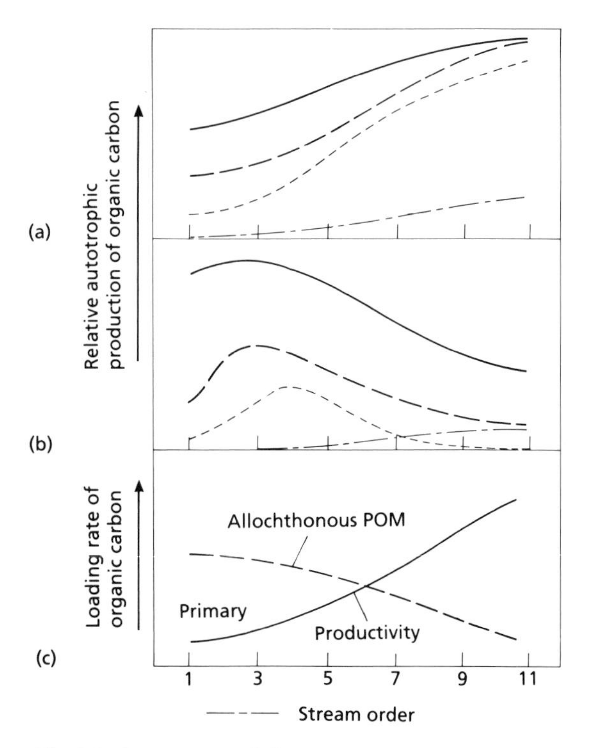

Fig. 9.2 Comparison of changes in patterns of autotrophic production of organic carbon with increasing stream order in (a) non-disturbed and (b) disturbed lotic ecosystems. (c) Longitudinal gradient of organic carbon loading from allochthonous (upland) versus lotic ecosystem primary productivity. ——, phytosynthetically derived *dissolved* organic carbon from wetlands and floodplain; — — — —, macrophytes (emergent and submersed); ----, periphyton; -—-—, phytoplankton.

ecosystems, as the slope decreases with increasing order, hydraulic energy dissipates to conditions where high water volumes are inadequate to transport much of the sediment loads. The result is widespread deposition with river braiding and meandering (e.g. Minshall *et al* 1985; Kellerhals & Church 1989; Sedell *et al* 1990), which leads to exponential increases of shallow substrata for colonization by algae. In addition, the sediment deposits within the streams commonly augment large bank regions of saturated soils and floodplains, and provide habitats for both submersed and emergent macrophytes and wetland vegetation. Similarly, particulate detrital deposits from high vegetative productivity within these land–water margins provide large surface areas which are rapidly colonized by microbial producers and degraders. Biogeochemical cycling of essential nutrients is very rapid. Furthermore, with the reduction of hydraulic energy, grinding action is reduced, light availability increases somewhat with lowered particulate loads, and the development and productivity of phytoplankton increase.

Because light availability is a major regulator of primary productivity of river ecosystems, this variable can be examined in relation to stream gradients. Unfortunately marked alterations in the light quality of most river ecosystems have occurred as land clearance, agriculture and other disturbances have increased (e.g. Sedell & Froggatt 1984; Triska 1984; Yount & Nieme 1990). Light available for photosynthesis *within* river channels *per se* is regulated by both canopy cover and turbidity (Fig. 9.3). Light availability is commonly low in small streams and rivers, at least seasonally during the active growing season, because of extensive vegetative canopy in both non-disturbed as well as disturbed river ecosystems. Inorganic and organic turbidity was considerably lower before extensive land clearance and agriculture than is now commonly the situation. Undisturbed rivers usually have low loads of turbidity and maintain high water transparency as size increases down-gradient (Fig. 9.3). A few examples of transparent waters in large river ecosystems still exist (e.g. portions of South America and Siberian USSR), but they are becoming increasingly rare. In general, inorganic turbidity increases with increasing drainage area (*cf.* Slaymaker 1988) and stream order.

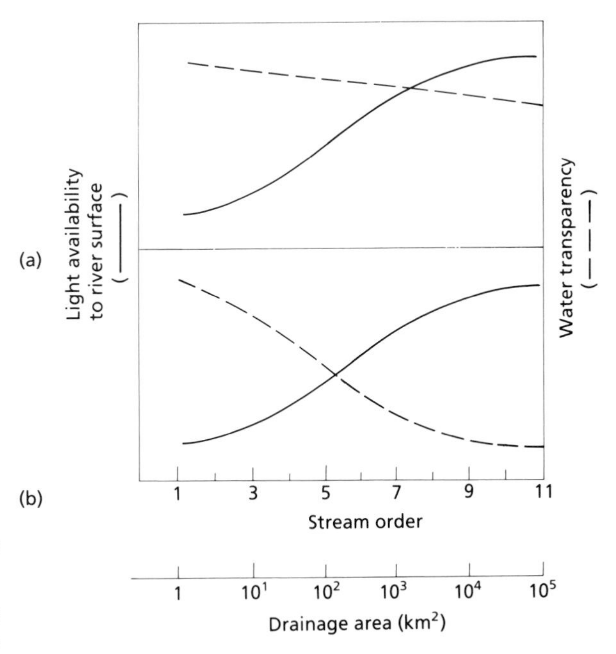

Fig. 9.3 Comparison of changes in patterns of light penetration to the stream water (water transparency) with increasing stream order in (a) preagricultural non-disturbed and (b) present/disturbed lotic ecosystems.

Longitudinal downstream gradients in resources result in a predictable structuring of the biota (Vannote *et al* 1980; Sedell *et al* 1989). Nutrient loadings usually increase as the sizes of the drainage basins and stream order increase, particularly among most river valleys that have been disturbed by agriculture. As a result, nutrient limitations to primary productivity of rivers generally decrease with increasing stream order. Primary productivity within the rivers, however, generally decreases with increasing stream order because of the light limitations (see Fig. 9.2(b)). In particular, the contributions by periphyton and submersed macrophytes decline with increasing size and turbidity.

Lateral contributions to autotrophic production of organic matter

The discussion above addresses contemporary syntheses that have emphasized the changes in photosynthetic metabolism observed along the length of stream gradients, i.e. with increasing stream order. Despite the reductions of *within*-riverwater primary productivity, we argue that, in general, the primary productivity of the river *ecosystem* increases with increasing stream order, particularly in non-disturbed but also in disturbed river systems (see Fig. 9.2(c)).

As the gradient slope decreases with increasing stream order and increased deposition of bedloads occurs, the percentage of the river valley containing saturated sediments and hydrosoils increases markedly. Colonization of these bank, wetland and floodplain areas by higher vegetation results in very high primary productivity (Table 9.5). Rates of primary productivity of these land–water interface areas are consistently high; in fact, they are among the highest of the biosphere. Decomposition of much of the particulate organic matter produced in these sites occurs at or near the sites of production. A very large portion of the organic matter synthesized by these high producers, however, is exported to the river water *per se* as dissolved organic matter.

Although there is a tendency to consider lateral development of stream ecosystems as increasing with increasing stream order, in fact lateral regions of both small streams *and* relatively large rivers can be restricted by geomorphological and/or human-induced features of the catchment. Reaches of lotic ecosystems are considered 'constrained' when the valley floor is narrower than two active stream channel widths (Sedell *et al* 1990). These river systems occur where natural geological or features constructed by humans constrict the valley floor and limit lateral mobility and adjacent plant communities. As a consequence, loading of particulate and dissolved organic matter from these lateral sources is relatively low, even though the total amount may constitute a significant portion of the total organic matter driving the ecosystem metabolism.

Unconstrained river reaches have valley floors that are wider than two active channel widths and lack major lateral geological or artificial constraints (Gregory *et al* 1991). These river ecosystems are characterized by active migrations of the channel that form extensive floodplains and often braided channels (Pinay *et al* 1990; Wissmar & Swanson 1990). The reduced erosive energy and depositional character results in widespread islands, peninsulas and inundated marginal environments with extensive development of fluvial and wetland vegetation (Fig. 9.4). The contemporary operational understanding of river ecosystems in three spatial dimensions and a fourth temporal dimension (Ward 1989) correctly incorporates the *dynamic scales* to account for the continual lateral expansion and contraction of river ecosystems. Clearly, the primary production through the floodplain to the valley walls loads the river channels with particulate and dissolved detrital organic carbon. Organic carbon produced in the floodplains is integrated with that produced within the channel *per se* (Table 9.6).

Table 9.5 Estimates of inputs of particulate organic matter to a 135-km reach of the New River (North Carolina, Virginia, West Virginia)

Source	Input (mT AFDW year^{-1})	Percentage of total input
Allochthonous		
Upstream and tributary	5893	53.8
Within study area	64	0.5
Autochthonous	3570	32.6
Periphyton	1435	13.1
Aquatic macrophytes		
Total particulate organic matter input	10962	

AFDW (ash-free dry weight). From Hill & Webster (1983).

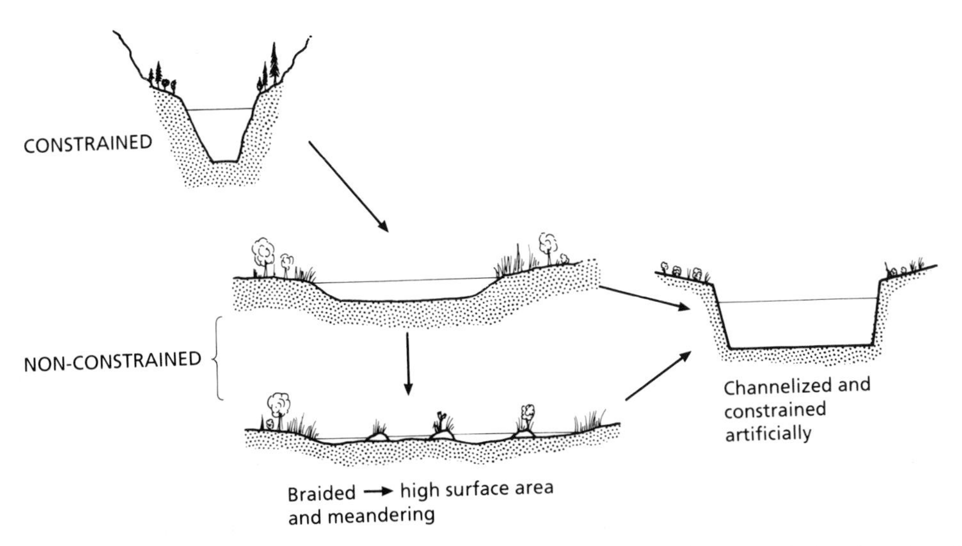

Fig. 9.4 Conceptualization of changes in extent of lateral vegetation and hyporheic zones (subsurface, dotted) in constrained, non-constrained and channelized lotic ecosystems.

9.6 FATES OF LOTIC PRIMARY PRODUCTION

Grazer pathways

Much of the research regarding fate of lotic primary production over the past decade has been directed toward consumption of within-channel, benthic primary production by macroinvertebrates and fish (e.g. Gregory 1983; Power *et al* 1985; Steinman *et al* 1987; Lamberti *et al* 1987; Power 1990; see also Chapters 5 & 7). Many of these studies have utilized elegant laboratory experiments in which the effects of various grazers on benthic periphyton communities grown on tiles have been examined. In general, these studies have shown that grazers can decrease periphyton biomass, decrease areal primary productivity, increase biomass specific primary productivity, and alter algal species and chemical

Table 9.6 Annual mean concentrations of organic matter in transport in the Ogeechee River

	Mean mg AFDW/l		Percentage of total organic matter	
Dissolved organic matter		25.4		96.32
Particulate organic matter				
Amorphous material (bacteria)	0.301		1.141	
Amorphous material (protozoans)	0.039		0.148	
Amorphous material (other)	0.521		1.976	
Vascular plant detritus	0.028		0.106	
Algae (mostly diatoms)	0.060		0.228	
Fungi	0.014		0.053	
Animals	0.002		0.008	
Total particulate organic matter		0.97		3.68
Total organic matter in transport		26.37		100.00

AFDW (ash-free dry weight). Modified from Benke & Meyer (1988).

composition. Since grazer consumption of periphyton production represents a first step in transfer of this energy resource to higher trophic levels, it is not surprising that so much emphasis has been in this direction and that results have frequently been used to infer that the majority of lotic primary production is transferred to higher trophic levels. However, definitive research addressing the quantity of benthic periphyton production that is transferred to grazers, as opposed to that which is unconsumed and enters detrital pathways (including dissolved and particulate components), has never been completed under natural conditions in streams or rivers. In cases where comparative, *in-situ* epilithic (algal plus bacterial) and grazer (snail) productivity data exist and have been extrapolated to estimate this quantity in streams with heavily grazed surfaces in forested catchments, results suggest that periphyton production was conservatively three to ten times that consumed by grazers (Stock & Ward 1989).

Compared with studies on lotic grazer consumption of benthic periphyton, substantially less research has been directed toward grazer impact on within-channel macrophytes, lateral vegetation and associated attached communities. As with lake littoral and wetland vegetation, most of this lotic plant material is probably not readily consumed directly because of the reduced palatability of the structural components. However, even in studies which have shown that substantial invertebrate production can be supported by river vascular plants (water lilies), the majority (80–90%) of plant production still entered detrital pathways (Wallace & O'Hop 1985).

Of the different types of plant communities that contribute to lotic ecosystem production (see above), we would predict that benthic periphyton communities would be most susceptible to consumption and direct transfer to higher trophic levels. However, although this material is important to the nutritional needs of the animals, most periphyton production in streams enters detrital pathways directly or in the form of incompletely digested animal egesta. Only minor portions of within-channel macrophytic or lateral vegetation would be directly consumed. Therefore, the majority of lotic *ecosystem* production enters detrital pathways and is modified primarily through microbial metabolism.

Nutrient regeneration and detrital pathways

Lotic primary production that is not consumed directly by higher trophic levels contributes both particulate and dissolved organic matter to stream/river ecosystems. Most lotic *ecosystem* primary production undergoes direct microbial degradation. However, the exact pathways and mechanisms of transformation, as well as the ultimate fate of much of this material, are still not well understood. This is particularly true of the dissolved organic fraction. In general the fate of particulate organic matter which falls directly into the stream channel in relatively constrained, upland streams has been studied to a much greater extent (e.g. Petersen *et al* 1989; review by Ward *et al* 1990) than the fate of unconsumed microbial primary production (e.g. epilithic, epipelic, epiphytic) in both constrained and unconstrained streams or rivers or macrophytes in laterally expansive regions.

In all stream ecosystems, a portion of the riparian vegetation will fall directly in the channel where it will be transformed by both biotic and abiotic processes or buried and stored within the sediment. However, most of the lateral vegetation will fall to the ground beside the main channel and be incorporated into subsurface soil regions, including hyporheic zones. Both organic matter stored and/or produced within the main channel, as well as that which feeds into the hyporheic zone from lateral regions, can fuel microbially active subsurface regions, where both aerobic and anaerobic processes further transform the detrital organic material. Some of this dissolved organic and solute material eventually re-emerges in the main channel, where it can serve as nutrient sources to within-channel communities, including primary producers (e.g. Crocker & Meyer 1987; Dahm *et al* 1987; Triska *et al* 1989b; Coleman & Dahm 1990; DeAngelis *et al* 1990).

As lateral regions of lotic ecosystems increase relative to the main channel, retention and recycling mechanisms associated with the increased complexity of the regions become more intense and circuitous. Although budgets incorporating transformation of total detrital material in laterally expansive regions of lotic ecosystems are lacking, we hypothesize that: (1) a substantial portion of the particulate organic matter will be retained and decomposed within the lateral region; (2) much of the detrital material will be converted to gaseous end-products within the lateral regions; and (3) dissolved organic matter will be the primary form of detrital material exported to the main channel. In general, we would expect community response to detrital inputs to be metabolically coordinated and oriented toward conservation, retention and recycling (*cf.* Wetzel 1990).

The basis for these hypotheses stems largely from research on lake littoral and wetland ecosystems, where retention and recycling processes occur at several levels (Wetzel 1990). Similar retention and recycling pathways probably exist for the laterally expansive regions of lotic ecosystems. Firstly, gases produced metabolically in photosynthesis and respiration as well as acquired nutrients are retained within individual plants. As the environment becomes more restrictive, such as extremely slow aqueous diffusion rates of nutrients and gases in nearly all sediment substrata, anatomical, morphological, and physiological plant adaptations for retention and re-utilization of essential resources increase in frequency and size. Examples include the retention, storage and re-utilization of inorganic carbon from respiration and photorespiration in internal gas spaces for subsequent recycling in photosynthesis. Limiting nutrients are largely translocated to and stored in perennating tissues during dormancy periods. Because much of the existing quantities of resources are retained and recycled, any of the external loadings of nutrients that are assimilated can be directed largely to net productivity.

Secondly, critical nutrients and organic carbon released by macrophytic primary producers are rapidly and efficiently assimilated and incorporated into the metabolism and productivity of microbiota associated with surfaces. Photosynthetic productivity of the attached algae can be extraordinarily high. In waters with submersed vegetation, the surface area for epiphytic microfloral colonization and development is increased and results in annual productivity values often far in excess of those of the supporting macrophytes. Again *recycling* of limiting essential gases, particularly oxygen and carbon dioxide, and dissolved nutrients within the attached community complex of algae, bacteria, fungi, protozoans, particulate detritus and inorganic precipitates, is the key to the commonly observed high sustained growth of attached microflora (Wetzel 1990). Intensive recycling of nutrients and gases in these microcommunities allows maintenance of photosynthetic biomass while minimizing losses and, as with the emergent macrophytes, allowing imported nutrients to be utilized primarily for net growth.

It should be noted that current velocities of water among these land–water gradients and over attached microbial communities can alter uptake rates of gases and nutrients. The diffusion rates across the boundary layer, however, can be enhanced only modestly by increased current velocities within the usual limnological range (e.g. Riber & Wetzel 1987). Even within dense macrophyte beds, flows are around and over the plants, and currents within the bed are largely slow and laminar (e.g. Losee & Wetzel 1992). As a result, most gaseous and nutrient fluxes within these attached communities are diffusion based and slow.

As a result of the above factors, much of the particulate organic matter in expansive lateral regions of lotic ecosystems is probably decomposed within the land–water interface regions to gaseous end-products. Nutrients released by the decomposition processes are usually rapidly assimilated by attached micro-organisms associated with particulate detritus, living plant tissues and inorganic substrata. Although irregular spates of variable magnitude can move particulate organic matter down-gradient, most exported organic matter is predominantly as dissolved organic substances. As the dissolved organic matter moves laterally towards the river channels, relatively energy-rich and chemically labile portions of the mixture of many organic compounds are utilized by attached microbiota selectively in conformance with kinetic models developed for polymer degradation (e.g. Cunningham & Wetzel 1989).

The more recalcitrant dissolved organic compounds can be partially degraded by ultraviolet photolysis as well as metabolism by the attached microflora (e.g. Mickle & Wetzel 1978, 1979; Wetzel 1991, 1992).

Dissolved organic compounds from peat drainage can affect nutrient availability and alter primary productivity. Humic complexes of aluminum and iron can bind phosphate ions, increase soil retention and reduce phosphorus availability for photosynthesis (e.g. Jackson & Schindler 1975). Phosphate sorbed to ferric iron–humic complexes may be released by ultraviolet-induced photoreduction of ferric iron to the ferrous state (Frankco & Heath 1982; Cotner & Heath 1990). In addition, colloidal humic–iron complexes can sequester iron and other micronutrients and possibly reduce primary productivity (Jackson & Hecky 1980). Dissolved humic compounds can also alter the reactivity of phosphate coprecipitation with calcium carbonate and other compounds (e.g. Otsuki & Wetzel 1972, 1973; Stewart & Wetzel 1981). These humic and fulvic acids can have variable effects on algal photosynthesis and productivity (Stewart & Wetzel 1982; Devol *et al* 1984). Interactive effects probably occur by complexation and restriction of extracellular enzyme activities and reduction of availability of essential nutrients from organic compounds (see discussion below). Where inorganic nutrient availability is relatively high, the effects of humic compounds would be minimal.

The dissolved organic compounds of wetland/littoral and periphytic origins can function in two important ways in river ecosystems. Decomposition of the large reservoir of dissolved organic matter, although slow (1–2% per day), is collectively a major component of energy dissipation of the ecosystem (Wetzel 1983, 1984). Second, release of large amounts of dissolved organic compounds by wetland/littoral macrophytes and attached microbiota is seasonal and coupled to complex interactions among photosynthetic production, nutrient availability, phased senescence, plant tissue characteristics, and conditions, including hydrology, of decomposition (Wetzel 1990). Many of these relatively recalcitrant humic compounds complex chemically with extracellular enzymes of algae and bacteria (e.g. Wetzel 1991, 1992). In this manner, enzymatic reactions and metabolic kinetics of river planktonic and periphytic organisms can be regulated in both positive and negative ways.

REFERENCES

Aumen NG, Miller DE, Crist CL. (1991) Contribution of mudflat primary production to a reservoir carbon budget. *Verh Internat Verein Limnol* **24**: 1304–08. [9.2]

Benke AC, Meyer JL. (1988) Structure and function of a blackwater river in the southeastern USA. *Verh Internat Verein Limnol* **23**: 1209–18. [9.5]

Bott TL. (1983) Primary productivity in streams. In: Barnes JR, Minshall GW (eds) *Stream Ecology. Application and Testing of General Ecological Theory*, pp 29–53. Plenum Press, New York. [9.1, 9.3, 9.5]

Bott TL, Brock JT, Cushing CE, Gregory SV, King D, Petersen RC. (1978) A comparison of methods for measuring primary productivity and community respiration in streams. *Hydrobiologia* **60**: 3–12. [9.2, 9.8]

Bott TL, Brock JT, Dunn CS, Naiman RJ, Ovink RW, Petersen RC. (1985) Benthic community metabolism in four temperate stream systems: an inter-biome comparison and evaluation of the river continuum concept. *Hydrobiologia* **123**: 3–45. [9.1, 9.3]

Brown O, Evans RH, Gordon HR, Smith RC, Baker KS. (1985) Blooming off the US coast: a satellite description. *Science* **229**: 163–7. [9.2]

Burkholder JM, Wetzel RG, Klomparens KL. (1990) Direct comparison of phosphate uptake by adnate and loosely attached microalgae within an intact biofilm matrix. *Applied and Environmental Microbiology* **56**: 2882–90. [9.2]

Carlton RG, Wetzel RG. (1987) Distribution and fates of oxygen in periphyton communities. *Canadian Journal of Botany* **65**: 1031–7. [9.2]

Coleman RL, Dahm CN. (1990) Stream geomorphology: effects on periphyton standing crop and primary production. *Journal of the North American Benthological Society* **9**: 293–302. [9.6]

Cotner JB Jr, Heath RT. (1990) Iron redox effects on photosensitive phosphorus release from dissolved humic materials. *Limnology and Oceanography* **35**: 1175–81. [9.6]

Crocker MT, Meyer JL. (1987) Interstitial dissolved organic carbon in sediments of a southern Appalachian headwater stream. *Journal of the North American Benthological Society* **6**: 159–67. [9.6]

de la Cruz A, Post AH. (1977) Production and transport of organic matter in a woodland stream. *Archives of Hydrobiology* **80**: 227–38. [9.3]

Cummins KW, Minshall GW, Sedell JR, Cushing CE,

Petersen RC. (1984) Stream ecosystem theory. *Verhandlungen Internationale Vereinigung für Theoretische und Angewandt Limnologie* **22**: 1818–27. [9.1]

Cunningham HW, Wetzel RG. (1989) Kinetic analysis of protein degradation by a freshwater wetland sediment community. *Applied and Environmental Microbiology* **55**: 1963–7. [9.6]

Dahm CN, Trotter EH, Sedell JR. (1987) Role of anaerobic zones and processes in stream ecosystem productivity. In: Averett RC, McKnight DM (eds) *Chemical Quality of Water and the Hydrologic Cycle*, pp 157–78. Lewis Publ., Chelsea, Michigan. [9.6]

DeAngelis DL, Mulholland PJ, Elwood JW, Palumbo AV, Steinman AD. (1990) Biogeochemical cycling constraints on stream ecosystem recovery. *Environmental Management* **14**: 685–97. [9.6]

Devol AH, Santos AD, Forsberg BR, Zaret TM. (1984) Nutrient addition experiments in Lago Jacaretinga, Central Amazon, Brazil: 2. The effect of humic and fulvic acids. *Hydrobiologia* **109**: 97–103. [9.6]

Dickerman JA, Stewart AJ, Wetzel RG. (1986) Estimates of net annual aboveground production: sensitivity to sampling frequency. *Ecology* **67**: 650–9. [9.3]

Elwood JW, Nelson DJ. (1972) Periphyton production and grazing rates in a stream measured with a P^{32} material balance method. *Oikos* **23**: 295–303. [9.3]

Eppley RW, Stewart E, Abbott MR, Heyman U. (1985) Estimating ocean primary production from satellite chlorophyll: introduction to regional differences and statistics from the southern California bight. *Journal of Plankton Research* **7**: 57–70. [9.2]

Fisher SG, Likens GE. (1973) Energy flow in Bear Brook, New Hampshire: an integrative approach to stream ecosystem metabolism. *Ecological Monographs* **43**: 421–39. [9.1]

Francko DA, Heath RT. (1982) UV-sensitive complex phosphorus: association with dissolved humic material and iron in a bog lake. *Limnology and Oceanography* **27**: 564–9. [9.6]

Gosz JR, Dahm CN, Risser PG. (1988) Long-path FTIR measurement of atmospheric trace gas concentrations. *Ecology* **69**: 1326–30. [9.2]

Gregory SV. (1980) *Effects of light, nutrients, and grazing on periphyton communities in streams.* Dissertation. Oregon State University, Corvallis, Oregon. [9.2]

Gregory SV. (1983) Plant–herbivore interactions in stream systems. In: Barnes JR, Minshall GW. (eds) *Stream Ecology: Application and Testing of General Ecological Theory*, pp 157–89. Plenum Press, New York. [9.6]

Gregory SV, Swanson FJ, McKee WA. (1991) An ecosystem perspective of riparian zones. *BioScience* **41**: 540–41.

Grimm NB, Fisher SG. (1984) Exchange between surface and interstitial water: Implications for stream metabolism and nutrient cycling. *Hydrobiologia* **111**: 219–28. [9.1]

Hall CAS. (1972) Migration and metabolism in a temperate stream ecosystem. *Ecology* **53**: 585–604. [9.3]

Hall CAS, Moll R. (1975) Methods of assessing aquatic primary productivity. In: Lieth H, Whittaker RH (eds) *Primary Productivity of the Biosphere*, pp 19–53. Springer-Verlag, New York. [9.2]

Hill BH, Webster JR. (1983) Aquatic macrophyte contribution to the New River organic matter budget. In: Fontaine TD, Bartell SM (eds) *Dynamics of Lotic Ecosystems*, pp 273–82. Ann Arbor Science, Ann Arbor, Michigan. [9.5]

Hornick LE, Webster JR, Benfield EF. (1981) Periphyton production in an Appalachian mountain trout stream. *American Midland Naturalist* **106**: 22–36. [9.3]

Hynes HBN. (1970) *The Ecology of Running Waters.* Liverpool University Press, Liverpool. [9.1]

Hynes HBN. (1974) The stream and its valley. *Verh Internat Verein Limnol* **19**: 1–15. [9.1, 9.3]

Jackson TA, Hecky RE. (1980) Depression of primary productivity by humic matter in lake and reservoir waters of the boreal forest zone. *Canadian Journal of Fisheries and Aquatic Sciences* **37**: 2300–17. [9.6]

Jackson TA, Schindler DW. (1975) The biogeochemistry of phosphorus in an experimental lake environment: evidence for the formation of humic–metal-phosphate complexes. *Verhandlungen Internationale Vereinigung für Theoretische und Angewandt Limnologie* **19**: 211–21. [9.6]

Junk W. (1970) Investigations on the ecology and production biology of the 'floating meadows' (Paspalo-Echinochloetum) on the middle Amazon. Part 1. The floating vegetation and its ecology. *Amazoniana, Kiel* **2(4)**: 449–95. [9.4]

Kellerhals R, Church M. (1989) The morphology of large rivers: characterization and management. In: Dodge DP (ed.) *Proceedings of the International Large River Symposium*, pp 31–48. Canadian Special Publication of Fisheries and Aquatic Sciences 106. Ottawa, Ontario. [9.5]

Lamberti GA, Ashkenas LR, Gregory SV, Steinman AD. (1987) Effects of three herbivores on periphyton communities in laboratory streams. *Journal of the North American Benthological Society* **6**: 92–104. [9.6]

Lay JA, Ward AK. (1987) Algal community dynamics in two streams associated with different geological regions in the southeastern United States. *Archives of Hydrobiology* **108**: 305–24. [9.3]

Lindeman RL. (1942) The trophic–dynamic aspect of ecology. *Ecology* **23**: 399–418. [9.3]

Losee RF, Wetzel RG. (1992) Littoral flow rates within and around submersed macrophyte communities.

Freshwater Biology (in press) [9.6]
Madsen JD, Adams MS. (1988) The seasonal biomass and productivity of the submerged macrophytes in a polluted Wisconsin stream. *Freshwater Biology* **20**: 41–50. [9.4]
Mayer MS, Likens GE. (1987) The importance of algae in a shaded headwater stream as food for an abundant caddisfly (Trichoptera) *Journal of the North American Benthological Society* **6**: 262–9. [9.1]
Meyer JL. (1990) A blackwater perspective on riverine ecosystems. *BioScience* **40**: 643–51. [9.1]
Mickle AM, Wetzel RG. (1978) Effectiveness of submersed angiosperm–epiphyte complexes on exchange of nutrients and organic carbon in littoral systems. II. Dissolved organic carbon. *Aquatic Botany* **4**: 317–29. [9.6]
Mickle AM, Wetzel RG. (1979) Effectiveness of submersed angiosperm–epiphyte complexes on exchange of nutrients and organic carbon in littoral systems. III. Refractory organic carbon. *Aquatic Botany* **6**: 339–55. [9.6]
Minshall GW. (1967) Role of allochthonous detritus in the trophic structure of a woodland springbrook community. *Ecology* **48**: 139–49. [9.3]
Minshall GW. (1978) Autotrophy in stream ecosystems. *BioScience* **28**: 767–71. [9.1, 9.3, 9.5]
Minshall GW, Petersen RC, Cummins KW, Bott TL, Sedell JR, Cushing CE, Vannote RL. (1983) Interbiome comparison of stream ecosystem dynamics. *Ecological Monographs* **53**: 1–25. [9.2, 9.5]
Minshall GW, Cummins KW, Petersen RC, Cushing CE, Bruns DA, Sedell JR, Vannote RL. (1985) Developments in stream ecosystem theory. *Canadian Journal of Fisheries and Aquatic Sciences* **42**: 1045–55. [9.5]
Naiman RJ. (1983) The annual pattern and spatial distribution of aquatic oxygen metabolism in boreal forest watersheds. *Ecological Monographs* **53**: 73–94. [9.2, 9.5]
Odum HT. (1956) Primary production of flowing waters. *Limnology and Oceanography* **2**: 85–97. [9.2]
Otsuki A, Wetzel RG. (1972) Coprecipitation of phosphate with carbonates in a marl lake. *Limnology and Oceanography* **17**: 763–7. [9.6]
Otsuki A, Wetzel RG. (1973) Interaction of yellow organic acids with calcium carbonate in fresh water. *Limnology and Oceanography* **18**: 490–4. [9.6]
Petersen RC Jr, Cummins KW, Ward GM. (1989) Microbial and animal processing of detritus in a woodland stream. *Ecological Monographs* **59**: 21–39. [9.6]
Pinay G, Décamps H, Chauvet E, Fustec E. (1990) Functions of ecotones in fluvial systems. In: Naiman RJ, Décamps H (eds) *The Ecology and Management of Aquatic–Terrestrial Ecotones*, pp 141–69. Parthenon Publishing Group, Carnforth, UK. [9.5]
Power ME. (1990) Resource enhancement by indirect effects of grazers: armored catfish, algae, and sediment. *Ecology* **71**: 897–904. [9.6]
Power ME, Matthews WJ, Stewart AJ. (1985) Grazing minnows, piscivorous bass, and stream algae: dynamics of a strong interaction. *Ecology* **66**: 1448–56. [9.6]
Reese WH. (1967) *Physiological ecology and structure of benthic communities in a woodland stream*. PhD dissertation, Oregon State University, Corvallis, Oregon. [9.3]
Revsbech NP. (1983) *In situ* measurement of oxygen profiles of sediments by use of oxygen microelectrodes. In: Gnaiger E, Forstner H (eds) *Polarographic Oxygen Sensors*, pp 265–73. Springer-Verlag, Berlin. [9.2]
Riber HH, Wetzel RG. (1987) Boundary layer and internal diffusion effects on phosphorus fluxes in lake periphyton. *Limnology and Oceanography* **32**: 1181–94. [9.6]
Risser PG, Rosswall T, Woodmansee RG. (1988) Spatial and temporal variability of biospheric and geospheric processes: a summary. In: Rosswall R, Woodmansee RG, Risser PG (eds) *Scales and Global Change*, pp 1–10. John Wiley & Sons, New York. [9.2]
Rodgers JH Jr, McKivitt ME, Hammerlund DO, Dickson KL. (1983) Primary production and decomposition of submergent and emergent aquatic plants of two Appalachian rivers. In: Fontaine TD, Bartell SM (eds) *Dynamics of Lotic Ecosystems*, pp 283–301. Ann Arbor Science, Ann Arbor, Michigan. [9.4]
Ryan S. (1990) Diurnal CO_2 exchange and photosynthesis of the Samoa tropical forest. *Global Biogeochemical Cycles* **4**: 69–84. [9.2]
Sedell JR, Froggatt JL. (1984) Importance of streamside forests to large rivers: the isolation of the Willamette River, Oregon, USA from its floodplain by snagging and streamside forest removal. *Verhandlungen Internationale Vereinigung für Theoretische und Angewandt Limnologie* **22**: 1828–34. [9.1, 9.5]
Sedell JR, Richey JE, Swanson FJ. (1989) The river continuum concept: a basis for the expected ecosystem behavior of very large rivers? In: Dodge DP (ed.) *Proceedings of the International Large River Symposium*, Canadian Special Publication of Fisheries and Aquatic Sciences 106. [9.5]
Sedell JR, Reeves GH, Hauer FR, Stanford JA, Haukins CP. (1990) Role of refugia in recovery from disturbances: modern fragmented and disconnected river systems. *Environmental Management* **14**: 711–24. [9.5]
Slaymaker O. (1988) Slope erosion and mass movement in relation to weathering in geochemical cycles. In: Lerman A, Meybeck M (eds) *Physical and Chemical Weathering in Geochemical Cycles*, pp 83–111. Kluwer Academic Publishing, Dordrecht. [9.5]
Steinman AD, McIntire CD, Gregory SV, Lamberti GA, Ashkenas LR. (1987) Effects of herbivore type and

density on taxonomic structure and physiognomy of algal assemblages in laboratory streams. *Journal of the North American Benthological Society* **6**: 175–88. [9.6]

Stewart AJ, Wetzel RG. (1981) Dissolved humic materials: photodegradation, sediment effects, and reactivity with phosphate and calcium carbonate precipitation. *Archives of Hydrobiology* **92**: 265–86. [9.6]

Stewart AJ, Wetzel RG. (1982) Influence of dissolved humic materials on carbon assimilation and alkaline phosphatase activity in natural algal–bacterial assemblages. *Freshwater Biology* **12**: 369–80. [9.6]

Stock MS, Ward AK. (1989) Establishment of a bedrock epilithic community in a small stream: microbial (algal and bacterial) metabolism and physical structure. *Canadian Journal of Fisheries and Aquatic Sciences* **46**: 1874–83. [9.1, 9.2, 9.6]

Stock MS, Richardson TD, Ward AK. (1987) Distribution and primary productivity of the epizoic macroalga *Boldia erythrosiphon* (Rhodophyta) in a small Alabama stream. *Journal of the North American Benthological Society* **6**: 168–74. [9.2]

Stream Solute Workshop (1990) Concepts and methods for assessing solute dynamics in stream ecosystems. *Journal of the North American Benthological Society* **9**: 95–119. [9.2]

Thienemann A. (1926) Der Nahrungskreislauf im Wasser. *Verhandlungen der Deutschen Zoologischen Gesellschaft* **31**: 29–79. [9.3]

Triska FJ. (1984) Role of wood debris in modifying channel geomorphology and riparian areas of a large lowland river under pristine conditions: a historical case study. *Verhandlungen Internationale Vereinigung für Theoretische und Angewandt Limnologie* **22**: 1876–92. [9.5]

Triska FJ, Kennedy VC, Avanzino RJ, Zellweger GW, Bencala KE. (1989a) Retention and transport of nutrients in a third-order stream: channel processes. *Ecology* **70**: 1877–92. [9.1]

Triska FJ, Kennedy VC, Avanzino RJ, Zellweger GW, Bencala KE. (1989b) Retention and transport of nutrients in a third-order stream in northwestern California: hyporheic processes. *Ecology* **70**: 1893–905. [9.1, 9.4, 9.6]

Vannote RL, Minshall GW, Cummins KW, Sedell JR, Cushing CE. (1980) The river continuum concept. *Canadian Journal of Fisheries and Aquatic Sciences* **37**: 130–7. [9.1, 9.5]

Wallace JB, O'Hop J. (1985) Life on a fast pad: waterlily leaf beetle impact on water lilies. *Ecology* **66**: 1534–44. [9.6]

Ward AK, Dahm CN, Cummins KW. (1985) *Nostoc* (Cyanophyta) productivity in Oregon stream ecosystems: invertebrate influences and differences between morphological types. *Journal of Phycology* **21**: 223–7. [9.2]

Ward GM, Ward AK, Dahm CN, Aumen NG. (1990) Origin and formation of organic and inorganic particles in aquatic systems. In: Wotton RS (ed.) *The Biology of Particles in Aquatic Systems*, pp 27–56. CRC Press, Boca Raton, Florida. [9.6]

Ward JV. (1989) The four-dimensional nature of river ecosystems. *Journal of the North American Benthological Society* **8**: 2–9. [9.5]

Westlake DF. (1965) Some basic data for investigations of the productivity of aquatic macrophytes. *Memorie dell'Istituto Italiano di Idrobiologia* **18**(Suppl): 229–48. [9.3, 9.4]

Wetzel RG. (1975) Primary production. In: Whitton BA (ed.) *River Ecology*, pp 230–47. Blackwell Scientific Publications, Oxford. [9.1, 9.5]

Wetzel RG. (1983) *Limnology* 2nd edn. Saunders College Publishing, Philadelphia. [9.3, 9.6]

Wetzel RG. (1984) Detrital dissolved and particulate organic carbon functions in aquatic ecosystems. *Bulletin of Marine Sciences* **35**: 503–9. [9.6]

Wetzel RG. (1990) Land–water interfaces: metabolic and limnological indicators. *Verhandlungen Internationale Vereinigung für Theoretische und Angewandt Limnologie* **24**: 6–24. [9.3, 9.6]

Wetzel RG. (1991a) Extracellular enzymatic interactions in aquatic ecosystems: storage, redistribution, and interspecific communication. In: Chróst RJ (ed.) *Microbial Enzymes in Aquatic Environments*, pp 6–28. Springer-Verlag, New York. [9.6]

Wetzel RG. (1992) Gradient-dominated ecosystems: sources and regulatory functions of dissolved organic matter in freshwater ecosystems. In: Kairesalo T, Jones RI (eds) *Dissolved Organic Matter in Lacustrine Ecosystems: Energy Source and System Regulator. Hydrobiologia* **229**: 181–98. [9.6]

Wetzel RG, Likens GE. (1991) *Limnological Analyses* 2nd edn. Springer-Verlag, New York. [9.2]

Wissmar RC, Swanson FJ. (1990) Landscape disturbances and lotic ecotones. In: Naiman RJ, Décamps H. (eds) *The Ecology and Management of Aquatic–Terrestrial Ecotones*, pp 65–89. Parthenon Publishing Group, Carnforth, UK. [9.5]

Woodwell GM, Whittaker RH, Reiners WA, Likens GE, Delwiche CC, Botkin DB. (1978) The biota and the world carbon budget. *Science* **199**: 141–6. [9.3]

Yount JD, Niemi GJ. (1990) Recovery of lotic communities and ecosystems from disturbance—a narrative review of case studies. *Environmental Management* **14**: 547–69. [9.5]

10: The Sampling Problem

R.H.NORRIS, E.P.McELRAVY
AND V.H.RESH

10.1 INTRODUCTION

Aims of this chapter

Decisions about sampling requirements are generally acknowledged as an important component of river and stream studies. Sampling decisions, however, are often based on tradition rather than on optimizing the data obtained per unit effort. To us, this is the essence of the sampling problem.

In this chapter we shall review general principles and considerations in sampling river and stream environments and emphasize some essential topics that have not been previously examined or about which confusion still exists. These will cover: sampling requirements for determining values of estimates (e.g. densities) compared with detecting differences among sites or times; the use of transformations; and case histories analysing factors influencing sampling design.

Much has been written on optimizing study designs and readers are referred to Cochran (1963), Eberhardt (1976, 1978), Green (1979), Resh (1979), Waters and Resh (1979), Bernstein and Zalinski (1983), Millard and Lettenmaier (1985), Millard *et al* (1985), Norris and Georges (1986), and Waters and Erman (1990). We have also provided a summary of selected references on methods to study algae, macrophytes, bacteria and fungi, macroinvertebrates and fish to direct the reader into these subject areas (Table 10.1).

Setting objectives and the role of sampling

Statements on the need for good study design with clearly stated objectives have been repeated so often as to be tedious (e.g. Eberhardt 1963, 1976, 1978; Green 1979; Waters & Resh 1979; Ellis & Lacey 1980; Rosenberg *et al* 1981; Cairns & Pratt 1986; Norris & Georges 1986; Cooper & Barmuta 1991). No doubt much of this repetition is seen as necessary because poorly stated objectives are still commonly encountered. Although there are always logistic or economic constraints on what or how much can be measured, the relationships among variables being measured and the problem of interest must be clear; little is gained from sampling irrelevant variables. A discussion of how different types of objectives may control sampling is provided by Ellis and Lacey (1980).

There are some questions that sampling cannot hope to cover: it cannot be used to provide judgements on 'good' or 'bad' scales, provide direct measures of aesthetic values or environmental ambience, or directly decide on what is, for example, 'wild and scenic'. Sampling can be used to: detect environmental change through space and time; make comparisons with pre-established standards; establish relationships between variables by statistical inference, experimentation or manipulation; develop predictions of effects or on relationships; and provide estimates of environmental variables such as population size or community structure. Commonly, the questions to be answered by sampling fall into two groups: (1) estimates of variables of interest such as the population density at one site; and (2) comparisons among sites or times. This important distinction has been ignored in discussions on the number of replicate sample units needed in a study. For example, estimates of fish production for calculation of harvestable yield may require estimates of totals or means, but often problems

Table 10.1 Sources of references for sampling structural and functional components of rivers that may be affected by disturbance for five groups of organisms

Measure	Taxonomic category: Algae	Macrophytes	Bacteria/fungi	Invertebrates	Fish
Standing stock, biomass	Aloi 1990 Trotter & Hendricks 1979 Blum 1956	Dennis & Isom 1984 Downing & Anderson 1986	Kogure *et al* 1984 Bott 1973 Karl 1985	Elliott 1977 Merritt *et al* 1984	Nielsen & Johnson 1983 Lagler 1978 Robson & Spangler 1978
Transport, drift	Power 1990 Lamberti & Resh 1987 Stevenson 1983	Dawson 1980 Dawson 1988	Webster *et al* 1987	Waters 1972 Brittain & Eikeland 1988	Harvey 1987 Gale & Mohr 1978
Production	Aloi 1990 Karl 1985	Dennis & Isom 1984 Hill *et al* 1984	Karl 1985 Moriarty 1986	Benke 1984 Rigler & Downing 1984	Chapman 1978
Taxonomic richness	Aloi 1990 Holder-Franklin 1986 Blum 1956	Dennis & Isom 1984	De Long *et al* 1989 Karl 1985 Bohlool & Schmidt 1980 Holder-Franklin 1986	Resh & McElravy 1992	Haedrich 1983 Lowe-McConnell 1978 Lundberg & McDade 1990 Strauss & Bond 1990.
Structural, functional, trophic diversity	Steinman & McIntire 1986	den Hartog 1982 den Hartog & van der Velde 1988 Best 1988	Karl 1985 Atlas 1985 Niels & Bell 1986	Cummins 1973 Merritt & Cummins 1984	Bowen 1983 Windell & Bowen 1978
Nutrient cycling	Mulholland *et al* 1990	Carpenter & Lodge 1986 Denny 1980	Webster & Benfield 1986 Atlas 1985 Boylen & Soracco 1986	Elwood *et al* 1983 Webster & Benfield 1986	Cederholm & ,Peterson 1985 Richey *et al* 1975
Life-history patterns	Bold & Wynne 1985	Dennis & Isom 1984 Kautsky 1988	Andrews 1986 Andrews 1991	Butler 1984	Baltz 1990
Size spectra	Steinman & McIntire 1986 Malone 1980	Wright *et al* 1981	Fry 1988 Andrews 1991	Mundie 1971 Clifford & Zelt 1972	Lagler 1978
Biotic interactions (within category)	Bothwell 1988 Peterson *et al* 1983	Carpenter & Lodge 1986	Atlas 1985 Boylen & Soracco 1986	Peckarsky 1984 Power *et al* 1988	Crowder 1990

of environmental concern will be comparisons of differences between (or among) sites or times. Parts of the sampling design may be similar for both types of questions (e.g. devices used), but the hypotheses tested, the number of sample unit replicates (i.e. the sample size), and the data and analysis needs may be quite different.

There will always be some differences between

the information obtained from measurements of the small part of the environment that has been 'sampled' compared with what actually exists in the total environment. This is the error associated with the data collected, and some information is needed about this error to estimate the risk of drawing wrong conclusions after examining the data obtained from sampling. 'The risk of being wrong' forms the basis for hypothesis testing, and it is needed whether data are being tested in the formal statistical sense or simply using scientific judgement. The validity of using scientific judgement, often based on years of experience and training, and possibly on extensive historical information, is often undervalued in favour of 'more rigorous' statistical approaches (see Berry 1987; Fryer 1987; Miller 1989). Statistical hypothesis testing usually is a more impartial approach (provided that the study has been properly designed) than using scientific judgement alone, particularly in situations requiring unequivocal results that are necessary for some management practices.

Sampling in rivers differs from many other environments because the unidirectional flow creates problems that are otherwise uncommon, e.g. difficulties in finding controls and replicating samples units. Commonly, upstream sites are used as the control against which changes at downstream sites are tested, and the problems with this design have received some attention (Eberhardt 1978; Hurlbert 1984; Cooper & Barmuta 1992; Norris & Georges 1992). Environmental conditions may be quite different at upstream and downstream sites, and upstream conditions may affect results downstream; in addition, the treatments themselves often are not properly replicated (Hurlbert 1984; Cooper & Barmuta 1992; Norris & Georges 1992). The following are also important: catchments usually will be more different from each other than sites within catchments; the same sites vary over time; and once a parcel of water has passed a site its effects will not be repeated. Because of these difficulties, interpretation based on scientific judgement alone may carry more weight with regard to rivers than other habitat types.

10.2 GENERAL PRINCIPLES

Study types

Studies in lotic systems are generally performed in one of two ways: (1) as mensurative studies, where space and time are the main experimental variables or treatments; and (2) manipulative studies, usually involving some intervention by the experimenter to control one or more external factors relative to the experimental units (Hurlbert 1984).

Interpretation of mensurative studies is mostly by inference, often (but not always) based on statistical analysis. Data from these studies will be from what Eberhardt (1963) terms 'unconfined populations'; if analysed by ANOVA, a random-effects model is usually used because of many uncontrolled factors. In these situations, homogeneity of variances is a critical factor (often achieved only after transformation), and sampling will need to be random or stratified random where only features of interest are sampled randomly. Any inferences made about the populations can be supported with probability statements, and thus be less affected by a poor choice of mathematical or ecological model (Eberhardt 1963). Functional relationships may not be apparent from this approach, and interpretation will depend on ingenuity and foresight in designing the study as well as expert knowledge of the field (Eberhardt 1963). This latter point has been criticized by some authors (e.g. Hurlbert 1984), but knowledge and expertise are usually prerequisites for adequate interpretation, regardless of sampling design. Indeed, where statistical testing is not done, it is the only basis for interpretation.

Analysis of data from manipulative studies is often based on a fixed model ANOVA where heterogeneity of variances may be of little concern (Eberhardt 1963). Manipulative studies of disturbance effects in streams are constrained by ethical and practical difficulties; nevertheless, many have been performed. The issues involved have been discussed in detail by Resh *et al* (1988), who concluded that studies involving severe degradation should meet high standards in statistical design and measurement procedures to minimize the likelihood of inconclusive results and the need for repetition. The same high standards are also needed for mensurative studies, and thus design requirements are similar in both cases. Data needs and methods of analysis have been covered in detail by Cooper and Barmuta (1992), who discussed manipulative experimental designs, and by Norris and Georges (1992), who considered analysis options more generally.

Design of many studies on rivers is difficult because of temporal problems that occur independent of disturbance, and because sites cannot be replicated easily. A commonly used design for assessing disturbance is to locate sites upstream and downstream of the disturbance, and then to replicate sampling units (or measurements) on each sampling occasion before and after the disturbance occurs. This would appear to be a simple two-way ANOVA with two factors, each with two levels: area (control and disturbed) by time (before and after) (Barmuta 1987). However, treatments within the design are not properly replicated because the replicate sampling units are only nested within each of the cells (Barmuta 1987). Eberhardt (1976) called this 'pseudodesign', and Hurlbert (1984) popularized it as 'pseudoreplication' because it is possible that some factor other than that being tested may affect the downstream impact area but not the control. Thus, the results of tests of significance need to be qualified in that they depend on the validity of using upstream sites as a control for downstream sites (Eberhardt 1978).

When sampling units at each site and time have not been replicated, it is possible to use samples through time as replicates (Eberhardt 1976; Stewart-Oaten *et al* 1986). However, this practice should be used with caution. Although it is valid in the sense that significant results can be believed, statistical testing in ANOVA that uses samples through time as replicates will be inefficient unless true trends in time are identical in magnitude and direction for all sites (Norris & Georges 1992). Normally, this would not be expected in most studies; to ensure the best opportunity for detecting true differences between sites (avoiding type II errors, see below), it is necessary to replicate sampling units on each visit to each site (Norris & Georges 1992).

Habitat types and stratification

Habitat stratification is implicit in the design of any stream or river study that describes sampling as being done in a riffle, pool, or any kind of 'uniform' habitat. Random sampling in rivers is inevitably stratified random sampling. Why do we stratify our sampling efforts? Ideally it is to make the sampling universe reflect the population universe of the organisms under question so that precision of estimates is maximized for minimal sampling effort. Sampling should include both a spatial and a temporal component; migratory species, such as diadromus fish, emphasize this point.

In practice, stratification is done to reduce variability or to facilitate intersite comparisons; both of these approaches improve the efficiency of sampling. Reductions in sampling variability from stratification by depth, current, substratum composition, or individual and combined hydraulic features have been repeatedly demonstrated for benthos (Resh 1979) and other freshwater fauna and flora including aquatic mammals (Statzner *et al* 1988). Stratification may result from choice of sampling devices (e.g. glass slides, allowing preferential colonization of some algal taxa), but usually it is done by spatial delineation. Although hundreds of publications have discussed stream zones and distribution of benthos, a hierarchical framework, such as suggested by Frissel *et al* (1986), is generally used in practice. In their approach, a stream system (say 10^3 m) is subdivided into successively lower levels: stream segment (10^2 m), reach (10^1 m), pool/riffle (10^0 m) and microhabitat (10^{-1} m) are the most widely used levels.

Restriction of the sampling universe to a level below that of the population universe may have a logistical advantage (and a corresponding statistical justification) of maximizing precision of estimates, but there are risks and uncertainties associated with stratification. For example, how representative is the restricted sample? Does it comprise a consistent proportion of the population (with different size classes or stages, or in successive years) in study sites under comparison? Moreover, the temporal and spatial variability inherent in unstratified designs may clearly provide an explanation of factors explaining observed patterns. This latter reason explains why pilot studies most often involve transect or other designs that maximize variability, allowing strata to be selected for follow-up analyses.

Measurement types

Physical and chemical measures

Many physical and chemical methods have been highly standardized and often are tested with interlaboratory comparisons. Possibly the most widely accepted reference for standardization of physical, chemical and many biological methods is that from the American Public Health Association (1989). Because of the high degree of standardization, especially for commonly used physical and chemical measures, no attempt has been made to summarize the literature further. Neither have we attempted to deal with hydrometric measurements, although hydraulic regimes, in terms of frequency, intensity and predictability of extreme flows, are important when interpreting data from rivers (Resh *et al* 1988).

Many of the standard methods for physical and chemical data (e.g. American Public Health Association 1989) are seen as being inexpensive, fast and easy to perform; consequently they have found their way into legislation protecting water quality, with regulatory standards usually being set based on toxicity testing. Caution is urged in selecting these physical and chemical measures, because most are surrogates for measuring biological changes of interest, and many, such as pesticide determination, are neither inexpensive nor quick. This is not an argument in favour of one form of sampling over another, but rather that the variables and methods chosen should be relevant to the study problem.

Biological measures

Structural and functional components of rivers that may be affected by disturbance can be summarized in ten categories (Resh *et al* 1988), and we have presented methods of measurement relevant to each of these for major categories of organisms (Table 10.1). Vertebrates other than fish have not been included because we felt that these came under the umbrella of wildlife, rather than river, management; references to these can be found in more specific texts (e.g. Seber 1982).

10.3 VARIABILITY AND TRANSFORMATIONS

Variability

Variability in measurements is probably the most significant factor to be considered when implementing a sampling programme, and it has received attention from several authors, for example Eberhardt (1978) for studies on animal populations in general; Downing (1979), Resh (1979), Allan (1984), Schwenneker and Hellenthal (1984), Morin (1985), Norris and Georges (1985, 1986), and Canton and Chadwick (1988) for invertebrates; Downing and Anderson (1986) for macrophytes; Ridley-Thomas *et al* (1989) for periphyton; Kosinski (1984) on oxygen productivity; Morin *et al* (1987) on primary and secondary production estimates; Mahon (1980), and Raleigh and Short (1981) on fish; Clarke (1990) for suspended sediments; Crisp (1990) for stream temperature; Jordan (1989) for pH; and Kerekes and Freedman (1989) on seasonal variations in water chemistry.

Variability of measurements has two components to be considered when selecting methods and determining required levels of replication: precision and accuracy. Precision is a measure of the similarity of repeated measurements, and is most important when comparisons are being made. Accuracy refers to the closeness of a measurement to its true value, and is most important when estimates are sought. Whether estimating means (for population numbers, or the legal limit of a chemical concentration) or determining whether differences between sites or times exist, confidence in conclusions will depend on a knowledge of data variability (measures of precision). Adequate replication (at least at some stage in the study) in both sample collection and analysis can provide a measure of precision. Failure to include adequate replication can lead to estimates that are so imprecise as to be little better than guesses, or in an inability to decide whether differences exist and thus be a costly waste of resources. Likewise, too much replication of sampling units will provide more data than are needed to demonstrate the points of interest, and again wastes resources. Much work on variability has considered only the variance of an estimator (e.g. total numbers), rather than consideration of the variance in maximizing the power of a study design to detect change (Millard & Lettenmaier 1985). Decisions on precision and the power of designs to detect change are both issues that we will discuss in detail (section 10.4).

Alternatively, variability may be viewed as potentially valuable information rather than just as noise. Resh (1979) has given a detailed presentation of the lifecycle and behavioural features of a net spinning caddisfly that would contribute to sampling variability (or worse, bias) in a randomized design, but provide information in a restricted design. Thus, variability in space (e.g. habitat selection) or time (e.g. reproductive phases) may sometimes need to be maximized as being of direct interest. Likewise, variability may be important in elucidating mechanisms (Resh & Rosenberg 1989). Prior knowledge of variability may be also needed, such as that just described, so that sampling can avoid or minimize its impacts (as discussed by Eberhardt 1963, 1976).

Organisms are rarely distributed evenly or randomly through the environment. Clumping usually occurs because many environmental features to which organisms respond are unevenly distributed. Behavioural, reproductive and dispersal features of some species may also produce aggregation without the influence of environmental factors. Clumping is also dependent on the scale at which sampling is performed. Clumps of high densities may be unevenly distributed through an environment of otherwise low density, but distribution within clumps may be regular because of competition for resources. Aquatic organisms most commonly have a negative binomial distribution where the variance is greater than the mean (Elliott 1977; Eberhardt 1978). Likewise physical and chemical variables may be unevenly distributed in the environment, but less attention seems to be paid to the problem of sampling for them because streams are generally considered to be well mixed. An example would be if downstream transport of phosphorus is of interest. Most of the phosphorus may move in particulate form during high flow. How, for example, should one cope with particles that are too large for the necks of commonly used sampling bottles. Dead fish, birds and mammals (and other large particles) represent clumps of high concentrations of phosphorus commonly transported under such conditions but are usually ignored. The statistical distributions of physical and chemical data have not been considered in nearly as much detail as biological ones because usually only single collections are made, or if multiple collections are made they are combined before analysis, thus depriving the investigator of any measure of variability.

Need for transformation

A further problem is encountered when statistical tests are applied to data representing environmental attributes that are contagiously distributed: the possible need for transformation to meet the tests' assumptions. These include: (1) linearity between the variance of the means of replicated samples and the average variance of the values that make up each sample; (2) homogeneity of variances; and (3) the normal distribution of data. Other assumptions of randomness in sampling and independence of observations are generally achieved by sound sampling design (Green 1979), although independence may be a greater problem in rivers because of unidirectional flow. The assumptions that replicated values of a measure are normally distributed and variances are equal, are often beyond the control of the investigator at the time of data collection and must be dealt with during preliminary analysis. Violations of these assumptions are generally expressed as a relationship between the variance and the mean, which should be independent for normally distributed data. Such a relationship is common for counts of benthic macroinvertebrates (Downing 1979; Morin 1985; Norris & Georges 1986) and is generally in the form $S^2 = a\overline{Y}^b$ (Taylor 1961). If sampling units have been replicated at each site or time, a plot of the sample variances against the sample means, followed by a test of the correlation between the two, will indicate whether a transformation is necessary (Table 10.2).

Traditionally, the square-root transformation ($Y' = \text{SQRT}(Y + ½)$) is applied if the variance and mean are approximately equal (corresponding to a random spatial or temporal distribution), and a log transformation ($Y' = \log_{10}(Y + 1)$) is applied when the variance is consistently larger than the mean (corresponding to a clumped distribution) (Elliott 1977). In extreme cases no transformation will normalize such data. After the data have been transformed, their means and variances can be calculated again and plotted against each other to see if the transformation has been successful (Norris & Georges 1991).

Transformation — case studies

To demonstrate the effects of collections with low numbers of organisms (or many zeros) on the effectiveness of transformation, two data-sets have been used: counts of the bug *Aphelocheirus aestivalis* from two sites, the upper and lower Schierenseebach, Germany (a full description of sites and methods can be found in Statzner (1978) and Statzner *et al* (1988)), and counts of total numbers of invertebrates and the mayfly *Baetis tricaudatus* from three sites in a northern Californian coast range stream, Hunting Creek (see Resh & Jackson 1991 for site description and methods). Several transformations ($\log_{10}$, square root, fourth root, and Taylor's power (Taylor 1961)) have been applied to the above data. The Shapiro–Wilk statistic (SAS Institute 1987) was used to test for normality on each transformation, as well as the raw data. *Apelocheirus aestivalis* data were then grouped into twos to double the sample unit size, removing most zero counts from the new data-set, and the transformations and tests for normality were reapplied (Table 10.2). For the purposes of this example, and because the collections of *A. aestivalis* were made randomly, these data were arbitrarily grouped into lots of five for calculation of means and variances. Five replicates from each site and time were used for the Hunting Creek data. Log_{10} of variances were regressed against their corresponding means (also transformed $\log_{10}$) for the raw data. Significant relationships between means and variances were used to determine power transformations using Taylor's relationship as discussed below (Table 10.2). If the data were to be used for comparisons between sites, Taylor's power relationship would need to be calculated on the combined sites and a single transformation used, because all data would need to be handled consistently.

The presence of many zeros in a data-set results in the transformation's inability to normalize the data (Table 10.2). This problem is overcome in large degree by doubling the sample unit size, thereby eliminating zeros and increasing the count per collection. Similar findings are reported by Allan (1984), who also cited Anderson (1965) as stating that variance-stabilizing transformations could not be found for collections with a mean ($\bar{Y}$) < 3. Square root transformations ($(Y + ½)^{0.5}$) were frequently not strong enough to normalize the data (Table 10.2). Fourth root transformations (recommended by Downing 1979) and the commonly used $\log_{10}$ were most effective, with the latter slightly more so (Table 10.2). The highly recommended transformations based on Taylor's power relationship performed poorly except where sample sizes and/or means were large (e.g. total numbers in Hunting Creek; Table 10.2).

Allan (1984) suggested that at higher mean densities the fourth root and power transformations were favoured over the log, but with small means the log transformation was most efficient, even if it did not always eliminate the dependency of the variance on the mean. However, the findings reported here indicate that the logarithmic transformation was as least as good or better than the other transformations used in all examples (Table 10.2). The degree to which transformation alters data distribution must also be considered; for example the effect of not properly normalizing data with a transformation is to reduce the sensitivity of ANOVA, and increase the likelihood of concluding no effect exists when in fact one does (type II error) (Cochran 1947). In cases where the log transformation produced distributions still significantly different from normal (Table 10.2), observation of frequency plots of the data indicated that departure from normality was not severe. Thus, in many cases where a transformation is applied but the data distribution is still

Table 10.2 Effectiveness of transformations on normalizing distributions of *Aphelocheirus aestivalis* in collections from the Upper (site 1) and Lower (site 2) Schierenseebach, Germany, in samples of single-unit size (1*X*) or double-unit size (2*X*), and for three sites in Hunting Creek, northern California, based on five replicates from each site collected using a Surber sampler in mid-April from 1984 to 1990

				Raw[a]		$\log_{10}(Y+1)$		$(Y+½)^{0.5}$		$Y^{0.25}$		$Y_{1-0.5b}$	
Data-set	n	$\overline{Y}$	S^2	R^2	W	R^2	W	R^2	W	R^2	W	R^2	W
A. aestivalis													
Both sites 1*X*	135	9.5	244	0.852	0.561	0.007	0.938	0.558	0.858	0.069	0.884		[b]
				0.000	0.000	0.684	0.570	0.000	0.000	0.019	0.000		
Both sites 2*X*	66	19.3	502	0.872	0.674	0.684	0.979	0.809	0.891	0.373	0.967		[b]
				0.000	0.000	0.511	0.629	0.000	0.000	0.035	0.187		
Site 1 1*X*	67	5.0	29	0.628	0.776	0.122	0.927	0.204	0.934	0.592	0.813	0.069	0.693
				0.001	0.000	0.241	0.000	0.121	0.002	0.002	0.000	0.365	0.000
Site 1 2*X*[c]	33	10.1	55		0.849		0.963		0.958		0.970		0.064
					0.000		0.367		0.272		0.575		0.409
Site 2 1*X*	68	13.8	420	0.896	0.626	0.038	0.926	0.508	0.882	0.154	0.889	0.670	0.596
				0.000	0.000	0.525	0.000	0.006	0.000	0.185	0.000	0.000	0.000
Site 2 2*X*[c]	33	28.5	789		0.745		0.965		0.923		0.971		0.962
					0.000		0.427		0.026		0.568		0.364
Hunting Creek													
All sites and all years													
Total numbers	99	497.6	277006	0.562	0.767	0.096	0.977	0.129	0.024	0.032	0.994	0.015	0.978
				0.000	0.000	0.185	0.383	0.120	0.000	0.305	0.246	0.605	0.430
B. tricaudatus	99	49.3	4810	0.928	0.711	0.045	0.954	0.459	0.905	0.032	0.961	0.879	0.530
				0.000	0.000	0.368	0.006	0.001	0.000	0.453	0.026	0.000	0.000

[a] R^2 on raw data has been calculated on variances and means in groups of five for *A. aestivalis* and on sets of five replicates for Hunting Creek, which were then logged and regressed.
[b] Power transformations were not performed for the total *A. aestivalis* data-set because the exponent was close to 2 indicating $\log_{10}$ to be appropriate (see text).
[c] *Aphelocheirus aestivalis* regressions were not done for individual sites because the number of samples was too small to form enough groups for a meaningful analysis.
R^2 for regression of means against variances, W is the Shapiro–Wilk test of normality. Upper lines are R^2 and W values, lower lines are levels of significance, n = number of observations, $\overline{Y}$ = the arithmetic mean and s^2 = variance. Transformations: Raw = raw data, $\log_{10} = \log_{10}(Y+1)$, $(Y+½)^{0.5}$ = square root, $Y^{0.25}$ = fourth root, $Y^{1-0.5b}$ = power transformation from the variance to mean relationship, *Aphelocheirus aestivalis* site 1 $Y^{0.15}$, site 2 $Y^{0.06}$, Hunting Creek, total numbers $Y^{0.21}$, *Baetis tricaudatus* $Y^{0.05}$.

significantly different from normal the likelihood of a type II error is reduced, and the transformation may be judged acceptable.

Sampler unit size

It has often been stated that larger numbers of smaller (in area or volume) sampling units will provide estimates with better precision than if fewer numbers of larger ones had been collected (Elliott 1977; Downing 1979). The advantages may not be as great as first thought (Riddle 1989) because of the additional problem of the intractability of zero counts to transformation. This point is demonstrated when the sample unit size is doubled for *A. aestivalis*; this resulted in better normalization of the data (Table 10.2). Where costs have been considered it has been recommended that small sample unit areas be used when mean densities are high, and larger ones where they are low (Morin 1985). This also coincides with data needs for transformation that have been demonstrated here; in cases of low densities, larger sample unit sizes will tend to eliminate zeros. One way of determining minimum sample unit area is by selecting the size required to eliminate most zero counts.

A more profound problem exits for physical and chemical data where values are reported as below the detection limits (usually called censored data) of the methods or apparatus used for measurement, but not zero. Several methods have been proposed for dealing with this problem after the data have been collected (Tiku 1967; Helsel 1986; El-Shaarawi 1989; El-Shaarawi & Dolan 1989; Norris & Georges 1991). If transformations are required on such data, their effectiveness will depend on their distribution and the way in which the below-detection-limit results are handled. In pollution studies where high concentrations are usually of concern, below-detection-limit readings are unlikely to be a problem. A study designed to investigate the relationships of low concentrations of essential chemicals might need to consider sampling alternatives such as evaporation, chelation or some other method of concentration to overcome this difficulty.

Selecting a transformation

Taylor (1961, 1980) found that for many types of aggregated data (e.g. abundance data), the following power relationship exists between the variance and the mean:

$$S^2 = a\overline{Y}^b$$

and that $Y' = Y^{1-0.5b}$ or $Y' = (Y + c)^{1-0.5b}$ (c is a constant such as 0.5, as used for the square root transformation) is the appropriate transformation for breaking the relationship between the mean and the variance. The parameter b is obtained by linearizing the above equation with a log–log transformation and applying linear regression. When $b = 0$, the distribution is considered normal and no transformation is necessary (for a discussion see Taylor 1961; Elliott 1977). When $b = 2$, a log transformation is appropriate (Elliott 1977; Green 1979).

In all cases of the data analysed here, variances were much greater than their corresponding means (Table 10.2), indicating aggregated distributions; a negative binomial distribution is commonly found in this type of data (Taylor 1961; Elliott 1977). In all cases there were significant relationships between means and variances (Table 10.2) and these were used to calculate the appropriate transformation based on Taylor's power relationship.

Poor performance of transformations based on Taylor's power relationship (Table 10.2) may have occurred because a log–log transformation may yield a grossly biased estimate of b (Zar 1968; Sprugel 1983), and the degree of bias will depend in part on the amount of scatter about the power relationship. Caution should also be exercised when calculating this transformation because it is optimized for the sample at hand; in fact, what is required is a transformation that will correct the population from which the sample was drawn (Norris & Georges 1991). These problems, its poor performance (Table 10.2), and the difficulty of calculating the power relationship to obtain the transformation, suggest that its use should not be recommended.

One of the most widely used statistical analyses is ANOVA, which is based on a linear relationship between the variance of means of replicated samples and the average variance of the values that make up each sample. The observed variance among the means for different sites or times can be compared to that expected to arise from this relationship; the ratio of the two variances is then compared using an *F*-test. ANOVA involving two or more factors assumes additivity, which Eberhardt (1978) considers more important than homogeneity of variances (normality). Many of the factors affecting environmental variables may be multiplicative so that a logarithmic transformation will make them additive, thus conforming to the model for ANOVA. Thus, the efficiency of the logarithmic transformation demonstrated above, its ease of computation, and its attributes of creating additivity for the ANOVA model from multiplicative environmental relationships lead us to suggest that it is the most favoured transformation. However, a single transformation will probably not be appropriate at all densities, and care should be taken to check the effects of the transformation used (Morin 1985).

10.4 THE REPLICATION PROBLEM

Estimates and comparisons

One of the most widely discussed topics in sampling is: 'how many replicates should be taken'? This can vary greatly depending on such simple considerations as whether collections are made to provide estimates of numbers of organisms or concentrations (Elliott 1977; Downing 1979; Green 1979; Norris & Georges 1986), or to detect a desired difference among means (Sokal & Rohlf 1981; Schwenneker & Hellenthal 1984; Zar 1984; Norris & Georges 1986). Estimates often need to provide a tightly defined measure of the real number or amount present, using data that are untransformed. However, because changes between sampling occasions may often be great and comparisons are usually tested using data that have been transformed to meet the assumptions of statistical tests, comparisons may require less sampling effort than estimates. This section examines the consequences of choices made rather than specifics of study design (which are determined by study objectives).

Studies of an organism's demography, biomass or abundance, or the assessment of chemical concentrations with regard to standards, often require calculations of sample size using formulae that provide the number of replicates necessary for a desired precision around the mean (formulae 1 and 2, Table 10.3). Studies to determine the sample size required to detect a difference or a change of a desired magnitude usually make use of a second type of formula (formulae 3, 4 and 5, Table 10.3). This distinction has often been neglected, but it is important because it is usually unacceptable to transform data when an estimate of a mean is needed, whereas transformation is often appropriate and required when data from sites or times are being compared with statistical procedures. Some authors have suggested that estimate data can be transformed and then back-transformed, e.g. Allan (1984) who also provides worked examples of resulting asymmetrical confidence limits. A geometric mean (from back-transformation of logged data) will always be lower than or equal to (but never more than) the arithmetic mean. It will give an indication of where the number of organisms or concentration was most of the time or in most of the replicates most of the time. This may be appropriate for some studies (e.g. long-term resources needed to support a population). The investigator needs to make this decision on whether transformed data can be used before initiating the study because less replication is generally needed for transformed data (Norris & Georges 1986). Statistical properties of data are altered by transformation, and investigators need to be sure of the meaning of back-transformed results. The following discussion proceeds on the premise that estimates will usually not be transformed.

Additionally, all the formulae require some prior knowledge of the variability expected in the data. This should present little difficulty if the well accepted view on the need for a pilot study before committing full resources to a programme is adhered to. All formulae also present the problem of deciding on the desired level of precision, or the magnitude of the difference that the programme is desired to detect.

Table 10.3 Formulae for calculating sample sizes (n) for a required precision around the mean, and for testing for differences between means

(1) $n = \dfrac{S^2}{d^2\,\overline{Y}^2}$ (Elliott 1977, p. 129)

n = number of replications
S^2 = sample variance
d = the ratio of the standard error to the mean, e.g. $d = 0.2$ for SE ± 20% of $\overline{Y}$
$\overline{Y}$ = arithmetic mean

(2) $n = \dfrac{t^2\,S^2}{d^2\,\overline{Y}^2}$ (Elliott 1977, p. 130)

d = relative error as percentage CL of $\overline{Y}$, e.g. $d = 0.4$ for CL ±40% $\overline{Y}$
t = value from Student's t distribution with n degrees of freedom

(3) $n \geq 2\left(\dfrac{\sigma}{\delta}\right)^2 \left\{t_{\alpha|\nu|} + t_2(1-p)[\nu]\right\}^2$ (Sokal & Rohlf 1981, p. 263)

σ = true standard deviation
δ = difference between means expressed as a percent of $\overline{Y}$, e.g. = 20 for a 20% difference between means
P = desired probability that a difference will be found to be significant
ν = degrees of freedom of the sample standard deviation with a groups and n replications per group
$t_{\alpha|\nu|}$ and $t_{2(1-P)\,|\nu|}$ = values from a t table with ν degrees of freedom and corresponding to probabilities of α and $2(1-P)$, respectively. Note if $P = ½$ then $t_{1|\nu|} = 0$

(4) $n = \dfrac{2S_P^2\,(t_{\alpha(2),2(n-1)})^2 F_{\beta(1),2(n-1),\nu}}{d^2}$ (Zar 1984, p. 133)

d = relative error in terms of CL of $\overline{Y}$ expressed as an absolute value
S_P^2 = pooled variance with ν degrees of freedom
$1-\alpha$ = the confidence level of the desired confidence interval
$1-\beta$ = is the probability that the half-width of the confidence interval will not exceed d
F = critical values from an F table for numerator degrees of freedom of $2(n-1)$ and denominator degrees of freedom of ν
t = critical values from a t table with degrees of freedom $2(n-1)$

(5) $n = \dfrac{S^2\,(q_{\alpha,DF,k})^2\,F_{\beta(1),DF,\nu}}{d^2}$ (Zar 1984, p. 193)

d = half-width of the $1-\alpha$ confidence interval
$1-\beta$ = the probability that the half-width of the confidence interval will not exceed d
S^2 = estimated error variance with ν degrees of freedom
k = total number of means
DF = error degrees of freedom that the experiment would have with the estimated n (i.e. $DF = k(n-1)$) for a single factor ANOVA
q = critical values from a q table

Sample size for estimates

Precision for estimates and sample size required

The level of precision around a mean value indicates the degree of confidence that can be placed in it. Fortunately, decisions can be made on the desired precision *before* embarking on a sampling programme, and the number of replicates needed to achieve it can be easily calculated (formulae 1 and 2, Table 10.3).

The level of precision accepted for a study is

important from the point of view of available resources because the number of replicates needed to achieve a desired precision increase with its inverse square (see Elliott 1977). Thus, four times as many replicates would be needed to achieve a precision of ±20% of the mean as for ±40% of the mean. At large sample sizes ($n > 30$), the value of t included in formula 2 is close to two and the results it yields will be almost identical to those from formula 1 (Table 10.3). At smaller sample sizes, formula 2 should be used and solved iteratively. The number of replicates needed for a given level of precision is also affected by two other features: the size of the mean and the degree of aggregation (Resh 1979). Larger means need fewer replicates for a given degree of aggregation and level of precision (Morin 1985; Norris & Georges 1986). Thus, if larger sample unit (area or volume) sizes are used, fewer replicates are needed for the same precision (Downing 1979). Higher levels of aggregation in variables of interest (e.g. indicated by b in Taylor's power relationship, section 10.2) need more replicates to attain the desired precision.

Replication to achieve a desired precision has been the subject of much discussion with respect to benthic macroinvertebrates. Chutter and Noble (1966) corrected analyses carried out earlier by Needham and Usinger (1956) for benthic invertebrate data from Prosser Creek, California. They concluded that hundreds of replicates would be needed to achieve estimates of biomass and abundance with 95% CL of the mean that would be acceptable. Allan (1984) suggested that 10 to 20 replicates would be needed for moderate precision (95% CL ±30–55% of the mean) for counts of benthic macroinvertebrates, and 50 replicates for high precision (95% CL ±10–25% of the mean). Elliott's (1977) popular book has been used extensively by stream ecologists, and his worked example for calculating required sample sizes using a precision of CL ±40% has resulted in this figure generally being accepted as a reasonable precision. The question to be answered is not how many replicates are needed for a given level of precision, but rather what level of precision is needed to meet the study objectives? Subsequent to this decision, calculation of the required number of replicates is straightforward.

Levels of precision in stream data have been discussed more generally by Norris and Georges (1986) who also included typical levels of accuracy (closeness to the real value) commonly accepted for some physical and chemical variables. These range from ±10% for biochemical oxygen demand and total Kjeldahl nitrogen to ±2% for electrical conductivity. Acceptable precisions on chemical determinations (for repeated field measurements) may be much tighter than they are for biological measurements. If this is so, several replicate collections also may be needed for estimation of physical and chemical variables, a practice usually not done. Note that the above precision is based on repeated collections from the field, *not* repeated measurements in the laboratory on the same water sample; the latter gives a precision only for the method and apparatus, and not for the environment.

Choosing levels of precision for estimates

Levels of precision needed should not be blindly accepted from past work (e.g. 95% CL ±40% of the mean), but must be determined with scientific expertise to constitute a reasonable estimate of abundance or concentration, relative to the study objectives. Several factors may be considered when making this decision. First, some understanding will be needed of the natural variability of environmental features of interest. This may be obtained from past work, a pilot study, or drawn from the expertise of the investigator. For example, electrical conductivity is usually conservative in terms of variability, perhaps slowly increasing downstream, but it usually varies little through time or within sites. Alternatively, total nitrogen may be highly variable, being affected by nitrification and de-nitrification processes, temperature, oxygen concentrations and biological activity. Organism numbers may vary by orders of magnitude depending on lifecycle status, behavioural responses and natural environmental influences such as drought or flood, which may be independent of the effect of pollutants. Experienced scientists will have knowledge of what ranges can be expected for the variables to be measured; levels of precision can be set relative to these ranges. For example, in an unpolluted river where phosphorus concentrations are known to limit algal growth, estimates of a mean of 30 μg l^{-1} total P and 95% CL ±10 μg (33%) would probably be acceptable because a range of 10–40 μg l^{-1} might be considered usual, and unlikely to effect substantial changes in algal growth. If the threshold for algal growth in the same river was known to be near 40 μg l^{-1} P and a phosphorus discharge was to begin, replicates to achieve estimation of a mean of 40 μg l^{-1} 95% CL ±5 μg (12%) might be needed to provide data for stream-protection decisions. Alternatively, it may be deemed that algal growth will be controlled if the phosphorus concentrations remain below 50 μg l^{-1} for 90% of the time. In this case, the number of replicates needed would be calculated to yield a mean with the upper boundary for the 90% confidence interval at 50 μg l^{-1}. In the last two examples, as the mean approaches the concentrations considered important, tighter confidence limits may be needed and consequently larger numbers of replicates must be collected. Experienced scientists usually will be able to decide on how much variation will be important.

Precision of estimates—case studies

The number of replicates needed for precisions based on confidence limits of 20, 40, 80 and 100% have been calculated for the Schierenseebach and Hunting Creek data-sets described above. Calculations have been done across all sites, and because site differences may increase variability, the number of replicates needed may be somewhat inflated. Most analyses lose power if sample sizes are unequal and if assumptions are violated to some degree; other difficulties such as number of species being related to total numbers of organisms collected will also make interpretation difficult. Thus, calculations for sample size to yield a minimum level of precision are done with data from the most variable site, or across all sites as presented here. Because of large changes in t values, the calculation is inefficient for sample sizes <3; thus a value of three replicates has been reported as the minimum.

Levels of precision selected must meet study objectives as well as be realistic. A sample size of 265 for CL ±20% for the rare and endangered *A. aestivalis* (Statzner 1978) may exceed available resources, whereas 68 for CL ±40% may be reasonable (in fact equal to the number collected in the original study) (Table 10.4). To understand what this means we must consider the level of precision relative to the real values of the mean. At CL ±20% there will be a 95% probability that the real mean will lie somewhere between 8 and 12 animals per sample, and at CL ±40% it will lie somewhere between 5 and 14 animals per sample. Familiarity with the nature of the animal's habitat and its survival characteristics relative to density,

Table 10.4 Sample sizes needed to estimate population numbers of *Aphelocheirus aestivalis* in the Upper and Lower (combined) Schierenseebach, Germany, for single (1*X*) and double (2*X*) sample unit sizes, and *Baetis tricaudatus* and total numbers of benthic invertebrates in Hunting Creek, California. Calculations using formula 2 (Elliott 1977) and iterated, for raw and $\log_{10}$-transformed data

				Replicates for CLs			
Variable	n	Mean	Variance	±20%	±40%	±80%	±100%
A. aestivalis							
1*X* raw	135	9.452	244.2	265	68	19	13
1*X* $\log_{10}$	135	0.748	0.238	43	13	5	5
2*X* raw	68	18.765	497.84	139	37	11	8
2*X* $\log_{10}$	68	1.099	0.185	18	6	3	3
Hunting Creek							
Total number raw	99	497.58	277006	110	30	9	7
Total number $\log_{10}$	99	2.48	0.21	6	2	3	3
B. tricaudatus raw	99	49.32	4810	193	50	15	10
B. tricaudatus $\log_{10}$	99	1.29	0.44	28	9	4	3

together with the study objectives, should enable a decision to be made easily. The effect of doubling the sample unit size is approximately to halve the required number of replicates (Table 10.4). For example, 40 replicates per site (a sample unit size double that used in the original study) would be sufficient to yield mean estimates of abundance with 95% CL ±40% and would also avoid the transformation problems found with frequent zero counts (section 10.2). In the original study performed on this rare animal (Statzner *et al* 1988), the replication proposed here would have been appropriate.

Required numbers of replicates are quite high even at the CL ±40% level for Hunting Creek data, although estimates of total numbers of macroinvertebrates may need fewer replicates than required for individual taxa (Table 10.4). Numbers of replicates required to yield CL ±80% would have been reasonable (nine for total numbers and 15 for *Baetis tricaudatus*; Table 10.4) and likely to be affordable for an ongoing sampling programme. Hunting Creek sites are unimpacted and variability between sites, as well as between years (especially for *B. tricaudatus*), often exceeds 80% of individual means, and can be considered as 'natural variability'. If density estimates were made relative to natural variability, CL ±80% of the means might be sufficient to provide reasonable confidence in the data for each sampling occasion. In this case, precision could be selected relative to expected ranges of natural variability obtained either from a pilot study or from the experience of the investigators. Levels of precision suggested here are generally lower than other authors have considered (see above) but these are probably realistic relative to the magnitude of expected differences. If population numbers vary by an order of magnitude, there is little sense in sampling for precisions of CL ±5% or even CL ±40%.

Precision of estimates and transformed data

In many cases, an index of abundance or concentration may be all that is needed, and data may be transformed to provide this information. In such cases where the number of replicates needed for a required level of precision is calculated on the transformed data, many fewer replicates are needed for the same precision (Norris & Georges 1986; Table 10.4). Eberhardt (1978) suggested that it matters little what the units of an index are, and he argued against the need to back-transform. If data can be used when transformed, high levels of precision might be obtained with little effort (Table 10.4), a consideration that has been neglected in discussions of determining sample size for estimates (Elliott 1977; Downing 1979; Allan 1984).

Improving precision for estimates

There are several ways of improving precision without substantially changing allocated resources. Major improvements may be made by decreasing the sample unit size and collecting more replicates (Downing 1979, 1989). This will be so, provided that counts in each collection are not too small (Riddle 1989) and that frequent zeros are avoided (section 10.2). Sampling stratification also may reduce variability (section 10.2), and so improve precision. Such improvements may be more difficult for chemical sampling where the most straightforward way of improving precision is to increase the number of replicates collected.

There will be cases where acceptable levels of precision are difficult to achieve, for example obtaining estimates of a rare species; this is also true for highly toxic substances when their critical levels are below the detection limits of current methods or technology. These are problem areas that are necessarily expensive.

Determination of levels of precision for a sampling programme for estimates will include the following considerations.

1 *Determination of the ranges for the variables of interest from a pilot study, previous work, or the experience of the investigator.* Wider ranges may need less precise estimates.

2 *Determination of the biological or chemical consequences of decisions made based on the estimates provided.* If consequences are not expected even within wide ranges, or are unlikely to be serious, low precisions can be selected. Conversely, if consequences may occur in a narrow range and/or they are serious, corresponding high levels of precision are required.

3 *Determine how close estimates are likely to be to levels where an effect will occur.* Small differences require higher levels of precision than would otherwise be needed.

4 *Determination of the characteristics of the organisms or chemicals.* High reproductive potentials or high rates of gain/loss generally need lower levels of precision.

5 *Determination of whether the data can be used transformed.* High levels of precision may be obtained with lower numbers of replicates when using transformed data, but care should be exercised using back-transformed data.

Sample size for comparisons

Sample size comparison decisions

The objectives of many sampling programmes in rivers involve the determination of change, such as the differences between two or more sites or times. The problem of determining the number of replicates required to detect a given difference between means with a known chance of error is related to precision, but it has received less attention than the question of estimating mean concentrations or densities. Often means are compared using *t*-tests or ANOVA with data meeting the assumptions of these tests (section 10.3). Usually the only concern with comparisons are whether or not the means being compared are different, i.e. in which direction they differ (see below), and not with the absolute values. This is where confusion arises when calculating sample sizes required for comparisons, as compared with those done for estimates. Most discussions are concerned with calculating sample size to detect differences between means using a test such as ANOVA (e.g. Schwenneker & Hellenthal 1984) and most texts (e.g. Snedecor & Cochran 1967, Sokal & Rohlf 1981, Zar 1984) suggest doing this with untransformed data. However, with most environmental data, comparisons to test for significant differences between means are made with transformed data. Calculations for required sample size should therefore be done on the data in the form in which they will be analysed (Norris & Georges 1986).

Three decisions need to be made before calculating sample size for comparing means once the difference to be detected is known:

1 *The formula to be used.* There are two types of formulae, closely related, but with important differences in their calculation (formula 3, and formulae 4 and 5, Table 10.3): formula 3 (Sokal & Rohlf 1981) uses the expected percentage difference between two means; formulae 4 and 5 (Zar 1984) use the actual difference expected between means (which could be converted to a percentage). Formula 4 is for the two-sample case, and formula 5 is for three or more samples.

2 *The level of significance.* Both types of formulae include the level of significance (α), or the probability at which means will be accepted as being different. This is also the probability of drawing a conclusion that the means being compared are different when in fact they are not (type I error). Conventionally, environmental studies accept α as 0.05, but some cases may call for different levels (see Schwenneker & Hellenthal 1984).

3 *The power of the test.* Both formulae also include β ($1 - P$ in formula 3), which is the power of the test, or the probability of drawing a conclusion that the means being compared are not different when in fact they are (type II error). Unfortunately, little discussion is available to guide selection of the level for β.

The following discussion will consider points (1) and (3) using analyses on the Schierenseebach and Hunting Creek data-sets. All analyses on *A. aestivalis* have been done with the individual collections grouped into two to eliminate the problem of zeros with transformation. Logarithmic transformations were chosen based on previous arguments (section 10.2).

Choosing a formula to calculate sample size for comparisons

The initial choice of formula type (3, 4, or 5, Table 10.3) is determined by whether the investigator is able to express the difference between means that the test is to detect in absolute terms (formula 4 or 5, Table 10.3) or merely as a percentage (formula 3, Table 10.3). Generally, for specification of a difference in absolute terms, some estimate of at least two means is needed, using the methods we employed in the following study. Alternatively, if only one mean is available, the investigator needs to specify the departure to be detected. In our example of *A. aestivalis*, the means were 9.8 at site 1 upstream of the lake and 27.7 at site 2 downstream (see Table 10.2) giving a difference of 17.9. A number smaller than this could be included in the formula, with the degree of conservatism being determined by the investigator. If only the mean from site 2 was available, a decision would need to be made on how much difference would be important; perhaps this difference could be as low as 7 or 8. If only the mean from site 1 was available, the difference selected might have been lower (e.g. 2 or 3), unless the expected direction of change was an increase, in which case 7 or 8 might still be accepted. Clearly the size of the mean used, the expected direction of change, and the characteristics of the organism all affect decisions on the expected level of difference for use in the formula.

If a percentage difference is to be used (formula 3, Table 10.3), it could be estimated from both or either mean. Here we have a dilemma. The usual convention would be to calculate it as a percentage of the largest mean as follows:

$$\text{Percentage difference} = 100 - \left(\frac{\bar{Y}_{max}}{\bar{Y}_{max}} \times 100\right)$$

In our example the percentage difference, or decrease, would be about 65%. If an increase from the smallest mean was of interest, the calculation would be:

$$\text{Percentage increase} = \left(\frac{\bar{Y}_{max}}{\bar{Y}_{min}} \times 100\right) - 100$$

or about 182%. When considering difference and change as a percentage, the expected direction is critical and the two equate as follows:

Percentage increase on smallest mean	Percentage difference between means, or percentage decrease
10	9
20	17
50	33
100	50
200	67
300	75

Where small changes are considered important (20% or less), the difference between percentage increase and percentage decrease is trivial. However, the percentage difference (the decrease calculated using the formula above) is limited to 100% while the percentage increase is unlimited. The percentage to be included in formula 3 (Table 10.3) will depend on the expected direction of change and individual situations. For example, if changes in concentrations of toxic substances such as trace metals are of concern, an increase will usually be important. Alternatively, if low levels of dissolved oxygen are problematic, a decrease will be important. Most impacts on rivers cause a decrease in biota but some, such as the effects of nutrients on algal growth, effect increases. Calculations on the percentage difference should be done on control or reference sites for the expected direction of change. Of course, in some cases the direction of change cannot be specified, such as in baseline or general ecological studies where differences may occur in either direction. In these cases, a general estimate may be obtained across all sites and then the percentage changes or absolute differences determined relative to expected ranges. The smallest changes of interest should be included in the formula to determine sample size.

An important point following from this discussion is that tabulated values of *t* and *F* used in the formulae should be from a one-tailed test if the direction of change can be specified, and two-tailed if not. The test will be more powerful if the direction of change can be specified and consequently fewer replicates will be needed (e.g. *Aphelocheirus aestivalis* (see Table 10.6)).

Choosing the power of the test

Having accepted the significance level of α to be 0.05 (which is in general use), the next decision to be made is the level of β, which is the likelihood of making a type II error (Sokal & Rohlf 1981) and which is included in the formula. The level of β, or the power of the test, is related to α but is also dependent on the distributions of the data-sets being compared (Sokal & Rohlf 1981). The latter will affect it in three ways. First, where overlap of the tails of the distributions being compared is large, β will also be large; where it is small, β will also be small. Second, where the means of the data-sets being compared are far apart, β will be small; where they are closer it will be larger. Third, β is inversely proportional to sample size, being small at large sample sizes and large at small ones. Therefore levels of β may change depending on the nature of the data-sets being compared and the sample size, even though α remains constant at 0.05. Unfortunately, there is little discussion on how to determine what level of β should be included in the formulae to calculate sample size (Sokal & Rohlf 1981; Zar 1984), on what levels of β equate with levels of α for different types of data, or even, what levels are acceptable in biomonitoring work.

The most straightforward way to increase the power of a statistical test, such as ANOVA, when comparing means is to increase replication (sample size). This has the effect of reducing β relative to a fixed level of α and a fixed difference between means (Sokal & Rohlf 1981). Non-normally distributed data may also reduce the power of the test (increasing β) (Cochran 1947), but this interaction with the data distribution can be changed with transformation (section 10.2).

To assess the relationship of α and β with benthic data we drew 50 sets of collections at random from the *A. aestivalis* data-set for each of 10, 15, 17 and 20 replicates and performed ANOVA to test for significant differences (Table 10.5). The threefold difference between means in the full data-set (see Table 10.2) could be considered significant and this was verified by ANOVA for both raw and log-transformed data (sample size 33, Table 10.5). Therefore, where ANOVA failed to show significant differences for smaller sample sizes the results were considered to be type II errors. As expected, type II errors decreased as sample size increased for both raw and log-transformed data (Table 10.5). This analysis also demonstrates empirically that about 20 replicates are needed for raw data and about 10–15 for logged data to avoid a type II error and provide sufficient statistical power to detect the difference (Table 10.5). This is substantially lower than the estimates of required replicates calculated and presented above. Usually, fewer replicates are needed for transformed data (demonstrated empirically in Table 10.5) because homogeneity of variances, achieved through transformation, may increase the power of ANOVA (see below). The relationship between transformation and sample-size calculations are discussed below.

This type of analysis is not feasible when designing a sampling programme where the replication needed will be decided on a continuum from guesswork to use of sample-size formulae

Table 10.5 Tests for significant differences between means for a number of sample sizes of counts of *Aphelocheirus aestivalis*, Schierenseebach, Germany

Data type	n	F	P	$\bar{Y}_2$	$\bar{Y}_1$	Type II errors
Raw	5	3.33	0.103	30.6	9.06	5
	10	5.53	0.052	28.45	9.32	2
	15	7.34	0.025	29.86	10.10	2
	20	8.56	0.015	29.17	10.32	1
	33	13.26	0.0005	28.54	10.12	–
Log_{10}	5	3.18	0.095	1.341	0.947	5
	10	5.84	0.073	1.289	1.912	3
	15	11.24	0.015	1.355	0.946	0
	20	9.75	0.012	1.305	0.961	0
	33	16.21	0.0002	1.307	0.958	–

Values are means of 50 analyses, except for full sample of 33 which was one analysis. n = number of observations; F = F-value from ANOVA; P = probability level from ANOVA; $\bar{Y}_1$, $\bar{Y}_2$ = arithmetic means from sites 1 and 2, respectively.

based on some prior knowledge of data variability. Levels of β to include in the formulae were not available to us, and no convention has been established. To estimate levels of β we performed the *A. aestivalis* ANOVAs just mentioned (Table 10.5); the power (level of β) of each of these tests was determined using the approach of Zar (1984, p. 174), using an *a priori* level of α of 0.05. This was done separately for raw and log-transformed data because the relationships would be expected to be different for different data distributions (Sokal & Rohlf 1981). The values of β were then averaged (observations >0.7 were set to 0.7 for this calibration) yielding 0.30 for raw data and 0.22 for logged data. These were then used in formulae 3 and 4 for calculating sample size. We saw this as being a useful way of estimating levels of β at which the ANOVAs were operating. When these values were substituted into formulae 3 and 4 (Table 10.3) and the actual differences between means in the existing data also included (as percentages or absolute numbers), calculated sample sizes were equivalent to those that demonstrated differences at the 0.05 level of significance (i.e. 15–20, Table 10.6). This indicates that both formulae produce accurate estimates of sample size when levels of β are equivalent to those operating in the tests performed on the data used here. Average levels of β used here were larger than might have been suggested by worked examples in texts (Sokal & Rohlf 1981; Zar 1984), or those used in other work (e.g. Schwenneker & Hellenthal 1984). If those smaller levels had been used, calculated sample sizes would have been larger, possibly resulting in an inefficient sampling programme. In contrast, when sample sizes are small (e.g. $n = 5$ for Hunting Creek), β may be high. Comparisons across all three sites for total numbers of individuals and number of *B. tricaudatus* in the Hunting Creek data-set (Table 10.6) showed that about three-quarters of the tests had values of $\beta > 0.7$. The importance of determining, *a priori*, the levels of β as well as α during the design phase of a biomonitoring study is clearly demonstrated here.

Levels of β estimated here are likely to be different for data collected from different places and for different data types. Further testing would be needed to establish whether the mean values of β are general for benthic invertebrates. Similar analyses should also be performed for physical and chemical data to establish values of β to be used in formulae to calculate sample sizes for them.

Higher levels of β indicate greater chances of making a type II error, concluding that an effect did not exist when in fact it did. The levels we have used provide sample sizes with 30% (raw) and 22% (logged) chances of making a type II error. Risks as high as these of drawing an incorrect conclusion may be unacceptable in some cases, especially for assessment of environmental impact where it is usually important to detect an impact if one exists. In these cases smaller values of β may be included in the sample size formulae to reduce the risk to acceptable levels.

Variance, transformation and sample size for comparisons

Formulae 3, 4 and 5 (Table 10.3) all use an estimate of the variance in the numerator to calculate a required sample size. Thus, larger variances result in larger sample sizes if the differences to be detected remain equivalent. The best estimate of variance for ANOVA (and for use in the formulae) is the pooled variance (Snedecor & Cochran 1967 p. 101) across sites or times. Generally this will be smaller than a variance calculated for all sites because among-site differences will contribute in addition to within-site differences. There will be little difference between the pooled variance and total variance if the ANOVA assumption of homogeneity of variances is met. One of the reasons for transforming data is to meet this assumption (Section 10.3). Thus, data meeting the assumption of homogeneity of variances have a lower pooled variance than those not doing so. Consequently, when data are properly transformed, sample-size requirements are smaller than for the same data untransformed (Table 10.6), and calculations of required sample size should be done on the transformed data. Heterogeneity of variances will increase the likelihood of a type II error in ANOVA (Cochran 1947), but calculated sample-size requirements on such data will be larger, thus offsetting the problem (Table 10.6).

Table 10.6 Sample sizes needed to detect differences between means from three sites in Hunting Creek, northern California, and two sites in the lower Schierenseebach, Germany, using the formulae of Sokal and Rohlf (1981, p. 263) (S&R) and Zar (1984). The formula from Zar (1984, p. 133) was used for the two-sample case for *Aphelocheirus aestivalis* and the formula from Zar (1984, p. 193) was used for the Hunting Creek multisample case. Percentage differences are the actual differences between the smallest and largest means, α has been set at 0.05 and β at 0.30 for raw data and 0.22 for log-transformed data. Results of ANOVA (F) and corresponding significance levels (P) show the comparisons for which significant differences were found (underlined). $\overline{Y}$ = arithmetic mean, S_P^2 = pooled sample variance, CV = coefficient of variation.

Data type	Year	$\overline{Y}$	S_P^2	CV	Percentage difference	F	P	Sample size S&R	Sample size Zar
HUNTING CREEK, CALIFORNIA									
Total n (raw)	1985	289	39441	68.7	52.5	1.40	0.284	22	27
	1986	184	29240	92.9	68.6	1.15	0.351	24	29
	1987	251	14217	47.5	45.3	1.98	0.181	15	18
	1988	689	86126	42.6	69.7	8.71	0.005	6	7
	1989	963	561947	77.9	75.2	2.69	0.108	14	18
	1990	853	245831	58.2	36.7	0.96	0.409	32	39
Total n ($\log_{10}$)	1985	2.378	0.071	11.2	10.6	1.26	0.319	18	21
	1986	2.081	0.173	20.0	23.0	1.83	0.206	12	14
	1987	2.355	0.033	7.8	10.2	2.74	0.105	10	11
	1988	2.764	0.029	6.2	16.5	10.66	0.002	3	4
	1989	2.841	0.081	10.0	19.5	5.85	0.017	5	6
	1990	2.850	0.074	9.6	10.5	1.89	0.193	14	16
Baetis tricaudatus (raw)	1985	52	6073	151.0	93.6	1.35	0.295	33	40
	1986	32	1490	119.8	90.7	2.71	0.111	23	28
	1987	18	514	127.4	89.1	1.89	0.194	26	32
	1988	81	3691	75.1	69.6	2.31	0.141	16	19
	1989	90	6155	87.1	98.0	5.34	0.022	11	13
	1990	47	3871	132.8	95.2	3.45	0.066	25	31
Baetis tricaudatus ($\log_{10}$)	1985	1.327	0.195	33.3	60.9	8.76	0.004	6	7
	1986	1.228	0.136	30.0	56.8	8.34	0.006	5	7
	1987	0.995	0.158	40.0	67.6	7.56	0.008	6	8
	1988	1.625	0.305	34.0	52.1	5.46	0.021	8	9
	1989	1.520	0.155	25.9	72.6	22.58	0.000	3	4
	1990	1.347	0.092	22.5	61.6	18.60	0.000	3	4
LOWER SCHIERENSEEBACH, GERMANY									
Aphelocheirus aestivalis									
Total n (raw)		28.5	422.3	72.0	64.5	13.26	0.000		
Two-sided values of t								17	13
One-sided values of t								8	10
Total n ($\log_{10}$)		1.307	0.124	26.9	26.7	16.21	0.000		
Two-sided values of t								17	12
One-sided values of t								8	9

Smaller sample sizes and tighter confidence limits were obtained with transformed data (Table 10.6). Transformed data, plus the large differences between means in the test data, yielded the relatively small sample sizes needed to demonstrate differences between means using ANOVA (Table 10.6). Variability in the test data are well within the ranges of variability considered normal for benthic invertebrates (see Eberhardt (1978) for examples). The sample sizes suggested here are much smaller than those indicated by others (e.g. Schwenneker & Hellenthal 1984), but their use of raw data and stringent levels of β would have yielded much higher replicate numbers than might be necessary to demonstrate differences that are of interest.

Determination of required sample sizes for comparing means will include the following considerations:

1 *Determination of the difference or change to be detected.*

2 *Determination of the variance.* This should be a pooled variance across all sites, or times, calculated using data in the form in which they will be analysed. Where it is not possible to calculate a pooled variance and some other estimate of variance is used, conservative sample size estimates are likely to be generated.

3 *Determine if the direction of change can be specified.* This may affect calculation of the percentage difference to be accepted. Also, if the direction can be specified, values for α and β should be for one-tailed rather than two-tailed tests.

4 *Selection of the formula to be used.* This is based largely on whether differences between means are to be specified as a percentage or in absolute terms, and whether two or several means are to be compared. The percentage differences may not be affected as much as the absolute differences by subsequent changes in sampling methods.

5 *Selection of the significance or probability of a type I error at which the test will operate.* Conventionally, this is 0.05 for environmental data. Management and scientific needs may dictate other levels in special circumstances.

6 *Selection of the probability level that the test will find a difference when one exists (avoiding a type II error).* Different levels may be accepted for different data types: transformed, raw, chemical or biological. Results presented here suggest $\beta = 0.30$ for raw data on benthic invertebrates, and $\beta = 0.22$ for log-transformed data.

ACKNOWLEDGEMENTS

B. Statzner and J.A. Gore provided *Aphelocheirus aestivalis* data. Information for Table 10.1 was provided by T. Bott, D. Erman, G. Lamberti, F. Ligon, B. Orr, C. Dahm, T. Arsuffi and M. Power. A. Georges commented on the manuscript.

REFERENCES

Allan JD. (1984) Hypothesis testing in ecological studies of aquatic insects. In: Resh VH, Rosenberg DM (eds) *The Ecology of Aquatic Insects*, pp 485–507. Praeger Publishers, New York. [10.3, 10.4]

Aloi JE. (1990) A critical review of recent freshwater periphyton field methods. *Canadian Journal of Fisheries and Aquatic Sciences* **47**: 656–70. [10.1]

American Public Health Association, American Water Works Association, Water Pollution Control Federation (1989) *Standard Methods for the Examination of Water and Wastewater* 17th edn. American Public Health Association, Washington, DC. [10.2]

Anderson FS. (1965) The negative binomial distribution and sampling of insect populations. In: *Proceedings of the XIIth International Congress of Entomology*, p. 395. London. [10.3]

Andrews JH. (1986) *r*- and *K*-selection and microbial ecology. *Advances in Microbial Ecology* **9**: 99–147. [10.1]

Andrews JH. (1991) *Comparative Ecology of Microorganisms and Macroorganisms*. Brock/Springer Series in Contemporary Bioscience. Springer-Verlag, New York. [10.1]

Atlas RM. (1985) Applicability of general ecological principles to microbial ecology. In: Poindexter JS, Leadbetter ER (eds) *Bacteria in Nature, Vol 2. Methods and Special Applications in Bacterial Ecology*, pp 339–70. Plenum, New York. [10.1]

Baltz DM. (1990) Autecology. In: Schreck CB, Moyle PB (eds) *Methods for Fish Biology*, pp 585–607. American Fisheries Society, Bethesda, Maryland. [10.1]

Barmuta LA. (1987) Polemics, aquatic insects and biomonitoring: an appraisal. In: Majer JD (ed.) *The Role of Invertebrates in Conservation and Biological Survey*, pp 65–72. Western Australian Department of Land Management Report, Perth, Western Australia. [10.2]

Benke AC. (1984) Secondary production of aquatic insects. In: Resh VH, Rosenberg DM (eds), *The*

Ecology of Aquatic Insects, pp 289–322. Praeger Publishers, New York. [10.1]

Bernstein BB, Zalinski J. (1983) An optimum sampling design and power tests for environmental biologists. *Journal of Environmental Management* **16**: 35–43. [10.1]

Berry RJ. (1987) Scientific natural history: a key base to ecology. *Biological Journal of the Linnaean Society* **32**: 17–29. [10.1]

Best EPH. (1988) The phytosociological approach to the description and classification of aquatic macrophyte vegetation. In: Symoens JJ (ed.) *Vegetation of Inland Waters. Handbook of Vegetation Science 15/1*, pp 155–82. Kluwer Academic Publishers, Dordrecht. [10.1]

Blum JL. (1956) The ecology of river algae. *Botanical Reviews* **22**: 291–341. [10.1]

Bohlool BB, Schmidt EL. (1980) The immunofluorescence approach in microbial ecology. *Advances in Microbial Ecology* **4**: 203–41. [10.1]

Bold HC, Wynne MJ. (1985) *Introduction to the Algae*. Prentice-Hall, Englewood Cliffs, New Jersey. [10.1]

Bothwell ML. (1988) Growth rate responses of lotic periphyton diatoms to experimental phosphorus enrichment: the influence of temperature and light. *Canadian Journal of Fisheries and Aquatic Sciences* **45**: 261–70. [10.1]

Bott TL. (1973) Bacteria and the assessment of water quality. In: Cairns J Jr, Dickson KL (eds) *Biological Methods for the Assessment of Water Quality*, pp 61–75. A symposium presented at the 75th annual meeting, American Society for Testing and Materials. American Society for Testing and Materials Special Technical Publication 528. American Society for Testing and Materials, Philadelphia. [10.1]

Bowen SH. (1983) Quantitative description of the diet. In: Nielsen LA, Johnson DL (eds) *Fisheries Techniques*, pp 325–36. The American Fisheries Society, Bethesda, Maryland. [10.1]

Boylen CW, Soracco RJ. (1986) Autecological studies in microbial limnology. In: Tate RL III (ed.) *Microbial Autecology*, pp 183–231. John Wiley and Sons, New York. [10.1]

Brittain JE, Eikeland TJ. (1988) Invertebrate drift—a review. *Hydrobiologia* **166**: 77–93. [10.1]

Butler MC. (1984) Life histories of aquatic insects. In: Resh VH, Rosenberg DM (eds) *The Ecology of Aquatic Insects*, pp 24–55. Praeger Publishers, New York. [10.1]

Cairns J Jr, Pratt JR. (1986) Developing a sampling strategy. In: Isom BG (ed.) *Rationale for Sampling and Interpretation of Ecological Data in the Assessment of Freshwater Ecosystems*, pp 168–86. American Society for Testing and Materials, Philadelphia. [10.1]

Canton SP, Chadwick JW. (1988) Variability in benthic invertebrate density estimates from stream samples. *Journal of Freshwater Ecology* **4**: 291–7. [10.3]

Carpenter SR, Lodge DM. (1986) Effects of submersed macrophytes on ecosystems processes. *Aquatic Botany* **26**: 291–341. [10.1]

Cederholm CJ, Peterson NP. (1985) The retention of coho salmon (*Oncorynchus kisutch*) carcasses by organic debris in small streams. *Canadian Journal of Fisheries and Aquatic Sciences* **42**: 1222–5. [10.1]

Chapman DW. (1978) Production. In: Bagenal T (ed.) *Methods for Assessment of Fish Production in Fresh Waters* 3rd edn, pp 202–18. IBP Handbook 3, Blackwell Scientific Publications, Oxford. [10.1]

Chutter FM, Noble RG. (1966) The reliability of a method of sampling stream invertebrates. *Archiv für Hydrobiologie* **62**: 95–103. [10.4]

Clarke RT. (1990) Bias and variance of some estimators of suspended sediment load. *Journal des Sciences Hydrologiques* **35**: 253–61. [10.3]

Clifford HF, Zelt KA. (1972) Assessment of two mesh sizes for interpreting life cycles, standing crop, and percentage composition of stream insects. *Freshwater Biology* **2**: 259–69. [10.1]

Cochran WG. (1947) Some consequences when the assumptions for analysis of variance are not satisfied. *Biometrics* **3**: 22–38. [10.3, 10.4]

Cochran WG. (1963) *Sampling Techniques* 2nd edn, pp 413. John Wiley and Sons, New York. [10.1]

Cooper SD, Barmuta LA. (1992) Field experiments in biomonitoring. In: Rosenberg DM, Resh VH (eds) *Freshwater Biomonitoring Using Macroinvertebrates*. pp 395–437. Chapman and Hall, New York. [10.1, 10.2]

Crisp DT. (1990) Simplified methods of estimating daily mean stream water temperature. *Freshwater Biology* **23**: 457–62. [10.3]

Crowder LB. (1990) Community ecology. In: Schreck CB, Moyle PB (eds) *Methods for Fish Biology*, pp 609–32. American Fisheries Society, Bethesda, Maryland. [10.1]

Cummins KW. (1973) Trophic relations of aquatic insects. *Annual Review of Entomology* **18**: 183–206. [10.1]

Dawson FH. (1980) The origin, composition and downstream transport of plant material in a small chalk stream. *Freshwater Biology* **10**: 419–35. [10.1]

Dawson FH. (1988) Water flow and the vegetation of running waters. In: Symoens JJ (ed.) *Vegetation of Inland Waters. Handbook of Vegetation Science 15/1*, pp 283–309. Kluwer Academic Publishers, Dordrecht. [10.1]

Dennis WM, Isom BG (eds). (1984) *Ecological Assessment of Macrophyton: Collection, Use, and Meaning of Data*. American Society for Testing and Materials Special Technical Publication 843. American Society for Testing and Materials, Philadelphia. [10.1]

Denny FH. (1980) Solute movement in submerged angiosperms. *Biological Reviews* **55**: 65–92. [10.1]

Downing JA. (1979) Aggregation, transformation, and the design of benthos sampling programs. *Journal of the Fisheries Research Board of Canada* **36**: 1454–63. [10.3, 10.4]

Downing JA. (1989) Precision of the mean and the design of benthos sampling programmes: caution revised. *Marine Biology* **103**: 231–4. [10.4]

Downing JA, Anderson MR. (1986) Estimating the standing biomass of aquatic macrophytes. *Canadian Journal of Fisheries and Aquatic Sciences* **42**: 1094–104. [10.1, 10.3]

Eberhardt LL. (1963) Problems in ecological sampling. *Northwest Science* **37**: 144–54. [10.1, 10.2, 10.3]

Eberhardt LL. (1976) Quantitative Ecology and Impact Assessment. *Journal of Environmental Management* **4**: 213–17. [10.1, 10.2, 10.3]

Eberhardt LL. (1978) Appraising variability in population studies. *Journal of Wildlife Management* **42**: 207–38. [10.1, 10.2, 10.3, 10.4]

Elliott JM. (1977) *Some Methods for the Statistical Analysis of Samples of Benthic Invertebrates* 2nd edn. Freshwater Biological Association Scientific Publication No. 25, Cumbria, UK. [10.1, 10.3, 10.4]

Ellis JC, Lacey RF. (1980) Sampling: defining the task and planning the scheme. *Water Pollution Control* **79**: 699–718. [10.1]

El-Shaarawi AH. (1989) Inferences about the mean from censored water quality data. *Water Resources Research* **25**: 685–90. [10.3]

El-Shaarawi AH, Dolan DM. (1989) Maximum likelihood estimation of water quality concentrations from censored data. *Canadian Journal of Fisheries and Aquatic Sciences* **46**: 1033–9. [10.3]

Elwood JW, Newbold JD, O'Neill RV, van Winkle W. (1983) Resource spiraling: an operational paradigm for analyzing lotic ecosystems. In: Fontaine TD III, Bartell SM (eds) *Dynamics of Lotic Ecosystems*, pp 3–27. Ann Arbor Science Publishers, Ann Arbor. [10.1]

Frissell CA, Liss WJ, Warren CE, Hurley MD. (1986) A hierarchical framework for stream habitat classification: viewing streams in a watershed context. *Environmental Management* **10**: 199–214. [10.2]

Fry JC. (1988) Determination of biomass. In: Austin B (ed.) *Methods in Aquatic Bacteriology*, pp 27–82. John Wiley and Sons, New York. [10.1]

Fryer G. (1987) Quantitative and qualitative: numbers and reality in the study of living organisms. *Freshwater Biology* **17**: 447–55. [10.1]

Gale WF, Mohr HW Jr. (1978) Larval fish drift in a large river with a comparison of sampling methods. *Transactions of the American Fisheries Society* **107**: 46–55. [10.1]

Green RH. (1979) *Sampling Design and Statistical Methods for Environmental Biologists*. John Wiley and Sons, New York. [10.1, 10.3, 10.4]

Haedrich RL. (1983) Reference collections and faunal surveys. In: Nielsen LA, Johnson DL (eds) *Fisheries Techniques*, pp 275–82. The American Fisheries Society, Bethesda, Maryland. [10.1]

den Hartog C. (1982) Architecture of macrophyte-dominated aquatic communities. In: Symoens JJ, Hooper SS, Compere P (eds) *Studies on Aquatic Vascular Plants*, pp 222–34. Royal Botanical Society of Belgium, Brussels. [10.1]

den Hartog C, van der Velde G. (1988) Structural aspects of aquatic plant communities. In: Symoens JJ (ed.) *Vegetation of Inland Waters. Handbook of Vegetation Science 15/1*, pp 113–53. Kluwer Academic Publishers, Dordrecht. [10.1]

Harvey BC. (1987) Susceptibility of young-of-the-year fishes to downstream displacement by flooding. *Transactions of the American Fisheries Society* **116**: 851–5. [10.1]

Helsel DR. (1986) Estimation of distributional parameters for censored water quality data. In: El-Shaarawi AH, Kwiatkowski RE (eds) *Developments in Water Science*, pp 137–57. Elsevier, Amsterdam. [10.3]

Hill BH, Webster JR, Linkins AE. (1984) Problems in the use of closed chambers for measuring photosynthesis by a lotic macrophyte. In: Dennis WM, Isom BG (eds) *Ecological Assessment of Macrophyton: Collection, Use, and Meaning of Data*, pp 69–75. American Society for Testing and Materials Special Technical Publication 843. American Society for Testing and Materials, Philadelphia. [10.1]

Holder-Franklin MA. (1986) Ecological relationships of microbiota in water and soil as revealed by diversity measurements. In: Tate RL III (ed.) *Microbial Autecology*, pp 93–132. John Wiley and Sons, New York. [10.1]

Hurlbert SH. (1984) Pseudoreplication and the design of ecological field experiments. *Ecological Monographs* **45**: 1922–7. [10.1, 10.2]

Jordan C. (1989) The mean pH of mixed fresh waters. *Water Research* **23**: 1331–4. [10.3]

Karl DM. (1985) Determination of *in situ* microbial biomass, viability, metabolism and growth. In: Poindexter JS, Leadbetter ER (eds) *Bacteria in Nature, Vol 2. Methods and Special Applications in Bacterial Ecology*, pp 45–176. Plenum, New York. [10.1]

Kautsky L. (1988) Life stages of aquatic soft bottom macrophytes. *Oikos* **53**: 126–35. [10.1]

Kerekes J, Freedman B. (1989) Seasonal variations of water chemistry in oligotrophic streams and rivers in Kejimkujik-National-Park Novia Scotia. *Water Air and Soil Pollution* **46**: 131–44. [10.3]

Kogure K, Simidu U, Taga N. (1984) An improved direct viable count method for aquatic bacteria. *Archiv für Hydrobiologie* **102**: 117–22. [10.1]

Kosinski RJ. (1984) A comparison of the accuracy and precision of several openwater oxygen productivity techniques. *Hydrobiologia* **119**: 139–48. [10.3]

Lagler KF. (1978) Capture, sampling and examination of fishes. In: Bagenal T (ed.) *Methods for Assessment of Fish Production in Fresh Waters* 3rd edn, pp 7–47. IBP Handbook No. 3, Blackwell Scientific Publications, Oxford. [10.1]

Lamberti GA, Resh VH. (1987) Seasonal patterns of suspended bacteria and algae in two northern California streams. *Archiv für Hydrobiologie* **110**: 45–57. [10.1]

De Long EF, Wickham GS, Pace NR. (1989) Phylogenetic stairs: ribosomal RNA-based probes for identification of single cells. *Science* **243**: 1360–3. [10.1]

Lowe-McConnell RH. (1978) Identification of freshwater fishes. In: Bagenal T (ed.) *Methods for Assessment of Fish Production in Fresh Waters* 3rd edn, pp 48–83. IBP Handbook No. 3, Blackwell Scientific Publications, Oxford. [10.1]

Lundberg JG, McDade LA. (1990) Systematics. In: Schreck CB, Moyle PB (eds) *Methods for Fish Biology*, pp 65–108. American Fisheries Society, Bethesda, Maryland. [10.1]

Mahon R. (1980) Accuracy of catch effort methods for estimating fish density and biomass in streams. *Environmental Biology of Fishes* **5**: 343–64. [10.3]

Malone TC. (1980) Algal size. In: Morris T (ed.) *The Physiological Ecology of Phytoplankton*, pp 433–63. University of California Press, Berkeley. [10.1]

Merritt RW, Cummins KW (eds). (1984) *An Introduction to the Aquatic Insects of North America* 2nd edn. Kendall/Hunt Publishing Company, Dubuque. [10.1]

Merritt RW, Cummins KW, Resh VH. (1984) Collecting, sampling, and rearing methods for aquatic insects. In: Merritt RW, Cummins KW (eds) *An Introduction to the Aquatic Insects of North America* 2nd edn. pp 11–26. Kendall/Hunt Publishing Company, Dubuque. [10.1]

Millard SP, Lettenmaier DP. (1985) Optimal design of biological sampling programs using the analysis of variance. *Estuarine Coastal Shelf Science* **42**: 637–45. [10.1, 10.3]

Millard SP, Yearsley JR, Lettenmaier DP. (1985) Space–time correlation and its effects on methods for detecting aquatic ecological change. *Canadian Journal of Fisheries and Aquatic Sciences* **42**: 1391–400. [10.1]

Miller C. (1989) Down with numbers. *Biocycle* **30**: 20. [10.1]

Moriarty DJW. (1986) Measurement of bacterial growth rates in aquatic systems from rates of nucleic acid synthesis. *Advances in Microbial Ecology* **6**: 245–92. [10.1]

Morin A. (1985) Variability of density estimates and the optimization of sampling programs for stream benthos. *Canadian Journal of Fisheries and Aquatic Sciences* **42**: 1530–40. [10.3, 10.4]

Morin A, Mousseau TA, Roff DA. (1987) Accuracy and precision of secondary production estimates. *Limnology and Oceanography* **32**: 1342–52. [10.3]

Mulholland PJ, Steinman AD, Elwood JW. (1990) Measurement of phosphorus uptake length in streams: Comparison of radiotracer and stable PO_4 releases. *Canadian Journal of Fisheries and Aquatic Sciences* **32**: 817–20. [10.1]

Mundie JH. (1971) Sampling benthos and substrate materials, down to 50 microns in size, in shallow streams. *Journal of the Fisheries Research Board of Canada* **28**: 849–60. [10.1]

Needham PR, Usinger RL. (1956) Variability in the macrofauna of a single riffle in Prosser Creek, California, as indicated by the Surber sampler. *Hilgardia* **24**: 383–409. [10.4]

Niels AL, Bell PE. (1986) Determination of individual organisms and their activities *in situ*. In: Tate RL III (ed.) *Microbial Autecology*, pp 27–60. John Wiley and Sons, New York. [10.1]

Nielsen LA, Johnson DL (eds). (1983) *Fisheries Techniques*. The American Fisheries Society, Bethesda, Maryland. [10.1]

Norris RH, Georges A. (1985) *Importance of sample size for indicators of water quality*. International Symposium on Biological Monitoring of the State of the Environment (Bioindicators), New Delhi, 11–13 October 1984, pp 37–49. Indian Academy of Science. [10.3]

Norris RH, Georges A. (1986) Design and analysis for assessment of water quality. In: Deckker P, Williams WD (eds) *Limnology in Australia*, pp 555–72. Commonwealth Scientific and Industrial Research Organization, Melbourne, Australia and Dr W.Junk, Dordrecht, the Netherlands. [10.1, 10.3, 10.4]

Norris RH, Georges A. (1991) Analysis and interpretation of benthic macroinvertebrate surveys. In: Rosenberg DM, Resh VH (eds) *Freshwater Biomonitoring Using Macroinvertebrates*, pp 230–82. Chapman and Hall, New York [10.1, 10.3]

Peckarsky BL. (1984) Sampling the stream benthos. In: Downing JA, Rigler FH (eds) *A Manual on Methods for the Assessment of Secondary Productivity in Fresh Waters*, pp 131–60. IBP Handbook 17, Blackwell Scientific Publishers, Oxford. [10.1]

Peterson BJ, Hobbie JE, Corliss TL, Kriet K. (1983) A continuous-flow periphyton bioassay: tests of nutrient limitation in a tundra stream. *Limnology and Oceanography* **28**: 583–91. [10.1]

Power ME. (1990) Benthic turfs vs floating mats of algae in river food webs. *Oikos* **58**: 67–79. [10.1]

Power ME, Stewart AJ, Matthews WJ. (1988) Grazer control of algae in an Ozark mountain stream: effects of a short-term exclusion. *Ecology* **69**: 1894–8. [10.1]

Raleigh RF, Short C. (1981) Depletion sampling in stream ecosystems: assumptions and techniques. *Progress in Fish Culture* **43**: 115–20. [10.3]

Resh VH. (1979) Sampling variability and life history features: basic considerations in the design of aquatic insect studies. *Journal of the Fisheries Research Board of Canada* **36**: 290–311. [10.1, 10.2, 10.3, 10.4]

Resh VH. Jackson JK. (1992) Rapid assessment approaches to biomonitoring using benthic macroinvertebrates. In: Rosenberg DM, Resh VH (eds) *Freshwater Biomonitoring and Benthic Macroinvertebrates*, pp 192–229. Chapman and Hall, New York. [10.3]

Resh VH, McElravy EP. (1992) Contemporary quantitative approaches to biomonitoring using benthic macroinvertebrates. In: Rosenberg DM, Resh VH (eds) *Freshwater Biomonitoring and Benthic Macroinvertebrates*, pp 157–91. Chapman and Hall, New York. [10.1]

Resh VH, Rosenberg DM. (1989) Spatial–temporal variability and the study of aquatic insects. *Canadian Entomologist* **121**: 941–63. [10.3]

Resh VH, Brown AV, Covich AP, Gurtz ME, Li HW, Minshall GW *et al.* (1988) The role of disturbance in stream ecology. *Journal of the North American Benthological Society* **7**: 433–55. [10.2]

Richey JE, Perkins MA, Goldman CR. (1975) Effects of kokanee salmon (*Oncorhynchus nerka*) decomposition of the ecology of a subalpine stream. *Canadian Journal of Fisheries and Aquatic Sciences* **32**: 817–20. [10.1]

Riddle MJ. (1989) Precision of the mean and the design of benthos sampling programmes: caution advised. *Marine Biology* **103**: 225–30. [10.3, 10.4]

Ridley-Thomas CJ, Austin A, Lucey WP, Clark MJR. (1989) Variability in the determination of ash free dry weight for periphyton communities — a call for a standard method. *Water Research* **23**: 3–8. [10.3]

Rigler FH, Downing JA. (1984) The calculation of secondary productivity. In: Downing JA, Rigler FH (eds) *A Manual on Methods for the Assessment of Secondary Productivity in Fresh Waters*, pp 19–58. IBP Handbook No. 17. 2nd edn. Blackwell Scientific Publications, Oxford. [10.1]

Robson DS, Spangler GR. (1978) Estimation of population abundance and survival. In: Gerking SD (ed.) *Ecology of Freshwater Fish Production*, pp 26–51. John Wiley & Sons, New York. [10.1]

Rosenberg DM, Resh VH, Balling SS, Barnby MA, Collins JN, Durbin DV *et al.* (1981) Recent trends in environmental impact assessment. *Canadian Journal of Fisheries and Aquatic Sciences* **38**: 591–624. [10.1]

SAS Institute. (1987) *SAS/STAT Guide for Personal Computers.* Version 6. SAS Institute, Cary, North Carolina. [10.3]

Schwenneker BW, Hellenthal RA. (1984) Sampling considerations in using stream insects for monitoring water quality. *Environmental Entomology* **13**: 741–50. [10.3, 10.4]

Seber GAF. (1982) *Estimation of Animal Abundance.* Griffin, London. [10.2]

Snedecor GW, Cochran WG. (1967) *Statistical Methods* 6th edn. The Iowa State University Press, Ames, Iowa. [10.4]

Sokal RR, Rohlf FJ. (1981) *Biometry* 2nd edn. W.H. Freeman Company, London. [10.4]

Sprugel DG. (1983) Correcting for bias in log-transformed allometric equations. *Ecology* **64**: 209–10. [10.3]

Statzner B. (1978) The effects of flight behaviour on the larval abundance of Trichoptera in the Schierenseebrooks (North Germany). In: *Proceedings of the 2nd International Symposium on Trichoptera*, pp 121–34. Junk, The Hague. [10.3, 10.4]

Statzner B, Gore JA, Resh VH. (1988) Hydraulic stream ecology: observed patterns and potential applications. *Journal of the North American Benthological Society* **7**: 307–60. [10.2, 10.3, 10.4]

Steinman AD, McIntire CD. (1986) Effects of current velocity and light energy on the structure of periphyton assemblages in laboratory streams. *Journal of Phycology* **22**: 352–61. [10.1]

Stevenson RJ. (1983) Effects of current and conditions simulating autogenically changing microhabitats on benthic diatom immigration. *Ecology* **64**: 1514–24. [10.1]

Stewart-Oaten A, Murdoch WW, Parker KR. (1986) Environmental impact assessment: 'Pseudoreplication: in time?'. *Ecology* **67**: 929–40. [10.2]

Strauss RE, Bond CE. (1990) Taxonomic methods: morphology. In: Schreck CB, Moyle PB (eds) *Methods for Fish Biology*, pp 109–40. American Fisheries Society, Bethesda, Maryland. [10.1]

Taylor LR. (1961) Aggregation, variance and the mean. *Nature (London)* **189**: 732–5. [10.3]

Taylor LR. (1980) New light on the variance/mean view of aggregation and transformation: comment. *Canadian Journal of Fisheries and Aquatic Sciences* **37**: 1330–2. [10.3]

Tiku ML. (1967) Estimating the mean and standard deviation from a censored normal sample. *Biometrika* **54**: 155–8. [10.3]

Trotter DM, Hendricks AC. (1979) Attached, filamentous algal communities. In: Weitzel RL (ed.) *Methods and Measurements of Periphyton Communities: a Review*, pp 58–60. American Society for Testing and Materials Special Technical Publication 690. American Society for Testing and Materials, Philadelphia. [10.1]

Waters TF. (1972) The drift of stream insects. *Annual Review of Entomology* **17**: 253–72. [10.1]

Waters WE, Erman DC. (1990) Research methods: concept and design. In: Schreck CB, Moyle PB (eds) *Methods for Fish Biology*, pp 1–34. American Fisheries Society, Bethesda, Maryland. [10.1]

Waters WE, Resh VH. (1979) Ecological and statistical features of sampling insect populations in forest and aquatic environments. In: Patil GP, Rosenzweig ML (eds) *Contemporary Quantitative Ecology and Related Econometrics*, pp 569–617. International Co-operative Publishing House, Fairland, Maryland. [10.1]

Webster JR, Benfield EF. (1986) Vascular plant breakdown in freshwater ecosystems. *Annual Review of Ecology and Systematics* **17**: 567–94. [10.1]

Webster JR, Benfield EF, Golladay SW, Hill BH, Hornick LE, Kazmierczak RF, Berry WB. (1987) Experimental studies of physical factors affecting seston transport in streams. *Limnology and Oceanography* **32**: 848–63. [10.1]

Windell JT, Bowen SH. (1978) Methods for study of fish diets based on analysis of stomach contents. In: Bagenal T (ed.) *Methods for Assessment of Fish Production in Fresh Waters* 3rd edn, pp 219–26. IBP Handbook No. 3, Blackwell Scientific Publications, Oxford. [10.1]

Wright JF, Hiley PD, Ham SF, Berrie AD. (1981) Comparison of three mapping procedures developed for river macrophytes. *Freshwater Biology* **11**: 369–79. [10.1]

Zar JH. (1968) Calculation and miscalculation of the allometric equation as a model in biological data. *Bioscience* **18**: 1118–20. [10.3]

Zar JH. (1984) *Biostatistical Analysis* 2nd edn. Prentice Hall, Englewood Cliffs, New Jersey. [10.4]

11: Responses of Aquatic Biota to Hydrological Change

J. A. GORE

11.1 INTRODUCTION

The prevailing interactions of velocity, depth and substrate reflect the hydrological integrity of a river channel (Dingman 1984). It is not unreasonable to assume that the structure and function of most aquatic communities is tied to the stability or predictability of hydrological patterns and in-stream hydraulic conditions (Statzner & Higler 1986). Only recently have lotic ecologists begun to examine the potential impact of the change in hydrological and hydraulic conditions on the in-stream flora and fauna. The manner in which these hydraulic factors interact influences the distribution of aquatic biota along the length of a lotic system and within a given reach of that system. Although it might appear that adaptation and/or preference for certain ranges of hydraulic conditions are the physical challenges faced by lotic biota, it can be demonstrated (and considered from a management perspective) that increased genetic variability (and species vigour?) accompany a physically variable stream environment (Robinson *et al* 1992).

The basic processes by which hydrological and hydraulic conditions might influence biota have been reviewed by Newbury (1984) and Davis (1986). In introducing the concept of 'hydraulic stream ecology', Statzner *et al* (1988) emphasized the particular attention that must be paid to the complex hydraulic interactions which occur near the substrate or shear zone and their influence on benthic organisms. Although the interactions with complex hydraulic conditions may not be as well defined for those organisms which exist within the water column and can more actively escape turbulent conditions, Heede and Rinne (1990) also view the analysis of hydrological and hydraulic conditions to be critical for effective management of fisheries. In this chapter, the potential effects of hydrological change will be reviewed and it will be necessary to establish some basic hydraulic reference points to understand the magnitude of changes brought about by anthropomorphic alterations of runoff pattern, and discharge regime.

11.2 HYDROLOGICAL EFFECTS

Transition zones

As water flows from headwaters to mouth, changes in slope and channel shape, associated with changes of the sediment load, have the effect of altering the manner in which that flow body dissipates energy. Flow patterns, the distribution and frequency of occurrence of various depth, velocity and substrate combinations, can be expected to change with distance from the headwaters. In general, headwater streams have relatively shallow slopes and gentle flows, but relatively close to the source they undergo a shift to high-velocity flows as the slope dramatically increases. As the river nears the foothills or lowland areas another transitional zone is encountered in which flows are decelerated and the river begins to braid and deposit sediment loads. Farther downstream, with declining slope the river begins to form large sinuous meanders; this is a zone of diverse flow patterns as the river begins to cut backwater areas and form oxbow lakes. Statzner and Higler (1986) proposed that an index of hydraulic stress, laminar sublayer thickness and slope, could be used to describe a 'typical' pristine

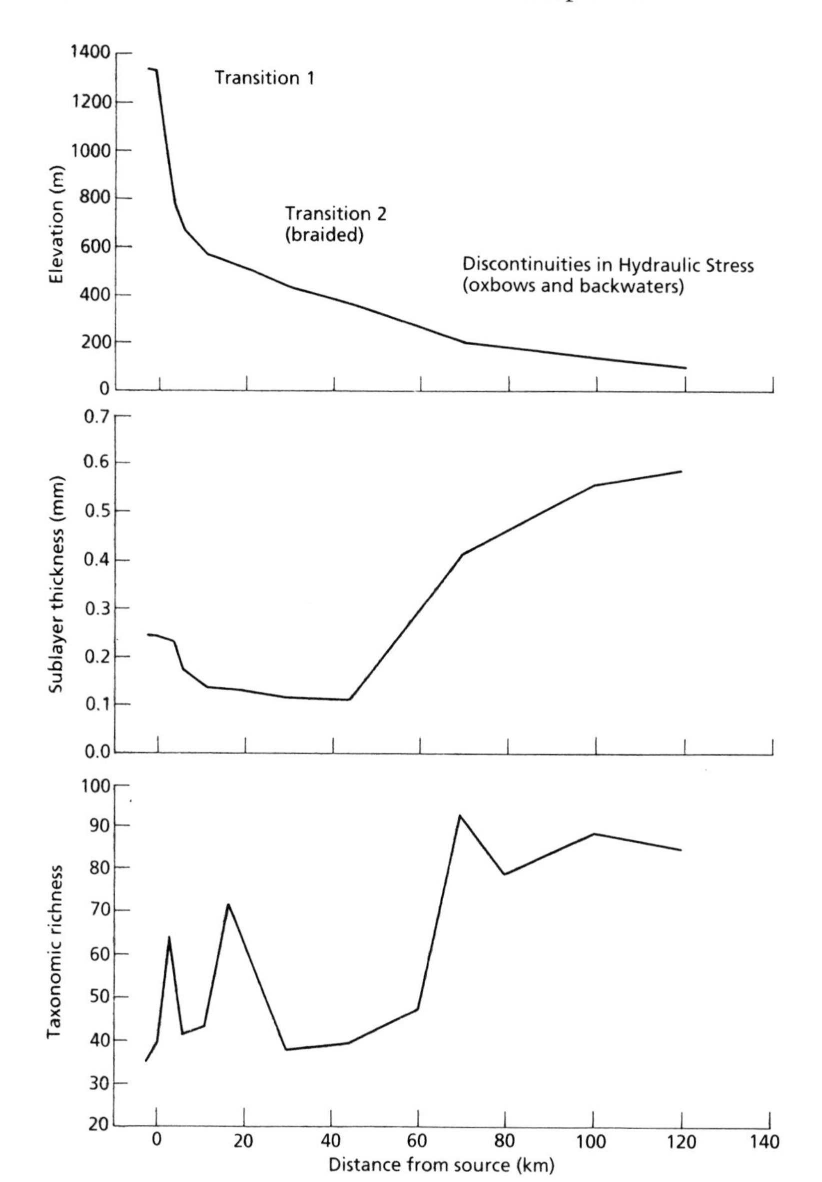

Fig. 11.1 Proposed relationship between hydraulic heterogeneity, complex hydraulics along the substrate (as quantified by sublayer thickness), and diversity of aquatic organisms (taxonomic richness) along the length of a typical river (after Statzner & Higler 1986).

flow for the purposes of comparison (Fig. 11.1). The source and headwaters, then, are characterized by low hydraulic stress while the first transitional zone (approaching the high slope zone) is characterized by elevated hydraulic stress. At the zone of transition to lowland streams, high hydraulic heterogeneity exists as the river begins to braid, but this zone is characterized by lower hydraulic stress. As the river meanders and forms backwater and oxbow lakes, numerous hydraulic discontinuities occur.

In an examination of 14 stream systems around the world, Statzner and Higler (1986) found a significant correlation between high benthic invertebrate diversity and the location of the transitional zones between low and high hydraulic stress. This is not surprising as the availability of substantially different hydraulic microhabitats would increase in these transitional zones and a diverse fauna, existing at the limits of their hydraulic tolerances, could be expected. Indeed, Statzner and Higler suggest that in the zones between the transitions, communities are more stable in their structure, and resilience to disturbance might be different from those communities at the transition zones. This general conclusion is echoed by other studies outside North America, where zonational patterns could not be attributed equally to patterns of particulate organic carbon use as in the river continuum concept (RCC; Vannote *et al* 1980). In Australian streams, for example, temporal and spatial flexibility of the fauna are more attributable to variability in streamflow (Lake *et al* 1986) while King (1981) found that changes in physico-chemical conditions (including water velocity) correlated well with changes in faunal distributions along the length of a short South African coastal river.

The relationship between changes in hydraulic condition and benthic invertebrates extends to the interstitial meiofauna as well. Ward and Voelz (1990) found that substrate condition and localized hydraulic conditions are more important than elevational factors in determining meiofaunal distributions. The areas of transition between meiofaunal communities coincided with the zones of hydraulic transition as predicted by Statzner and Higler (1986).

Longitudinal transition in fish communities is well documented (Sheldon 1968; Jenkins & Freeman 1972; Evans & Noble 1979; Platania 1991). Again, most of the patterns of species replacement or change in community structure are related to identifiable physical habitats related to site-specific hydraulic conditions. Chisholm and Hubert (1986) found that brook trout densities varied as a function of gradient and stream width-to-depth ratios with three identifiable zones of population density, varying from highest densities in low-gradient streams and lowest densities in highest-gradient areas. Platts (1979) also identified an abrupt transition in salmonid-dominated communities at the first zone of transition in hydraulic stress in low-order streams of Idaho. Statzner (1987b) has predicted that zonational changes occur at major hydrological breaks (usually at geomorphic changes in slope). In these areas where a high diversity of hydraulic conditions occurs, diversity of biota has been found to be at its highest.

The points of hydrological change along rivers usually coincide with optimum points of placement for hydraulic control structures, particularly impoundments. Thus, these ecotonal areas, which are reasonably fragile in structure (Naiman *et al* 1988), are often the first communities to be altered by anthropogenic changes which influence discharge regimes. Not only does the physico-chemical nature of release water from impounded waters in these areas influence the structure and function of downstream communities, but alterations in flow patterns influence the distribution of substrate particles and the pattern and distribution of flow which, in turn, create different patterns of microhabitat availability. These discontinuities can have the effect of 'resetting' community structure to mimic those of areas farther upstream where physico-chemical conditions are similar (Fig. 11.2) (Ward & Stanford 1983). The ability of the lotic biota to recover from these changes in physical environment is an index of the ultimate severity of the disturbance.

Superimposed upon these longitudinal and within-reach changes in hydraulic conditions are changes in seasonal hydrograph, short-term events such as floods, and potentially long-term 'events' such as the effects of human intervention upon the flow regime (i.e. afforestation, impoundments, diversions and abstractions) which have the potential to alter temperature patterns and the availability of particulate organic food for lotic biota.

Life-cycle requirements

The life cycles of lotic biota appear to be regulated by two major physical factors: temperature and hydraulic conditions. The impacts of temperature change on aquatic insects is well documented (Hynes 1970). Temperature and day-length seem to be the regulating features which synchronize hatching, maturation of larvae, emergence and mating of adults and, in some cases, egg-laying behaviour.

Not only are there certain temperature optima

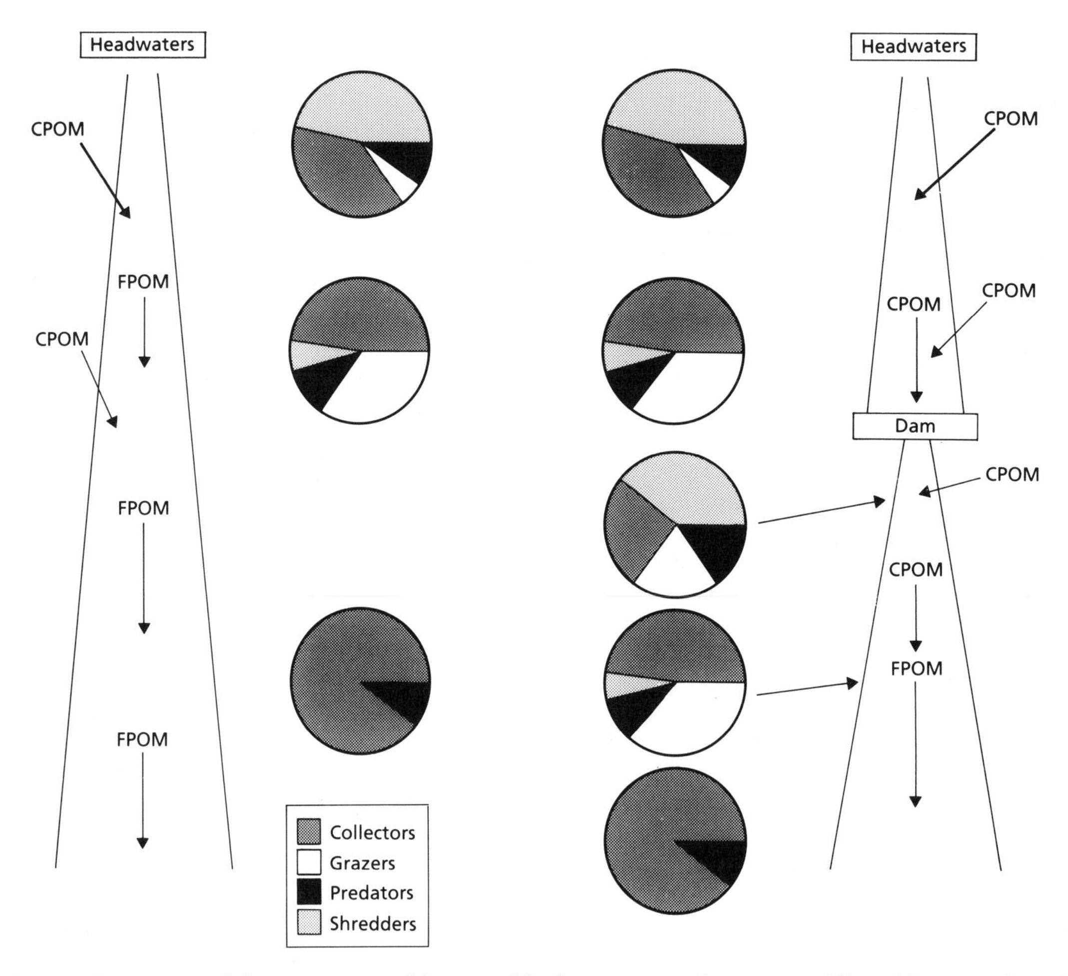

Fig. 11.2 Comparison of the composition of functional feeding groups at the upper, middle and lower reaches of a natural river system (left) and an impounded system (right) where the impacts of the dam include increased proportions of coarse particulate organic matter (CPOM) and less fine particulate organic matter (FPOM) in the tailwater. Increased clarity causes increased periphyton growth for grazers while hypolimnetic releases decrease the number of degree-days in the tailwater by as much as half that in the natural system. (After Stanford & Ward 1984.)

for the hatching of various eggs but a diel cycle also appears to enhance development (Humpesch 1978). Indeed, it appears that greater daily temperature changes reduce the degree-hours required for hatching of some species of aquatic insects (Sweeney & Schnack 1977). Although there is some debate about the importance of this diel temperature cycle, alteration of thermal regimes by change in hydrological condition could result in reduced success of many macroinvertebrate species.

Thermal cues also regulate the ability of aquatic invertebrates to enter a period of quiescence or rapid development prior to emergence as adults. In general, rising temperatures induce diapase and declining temperatures terminate that period of dormancy (Ward 1992). These responses preadapt a species to survive drought periods and to colonize temporary waters. In temperate-zone rivers, rising temperatures in the spring cue growth and maturation among larval and nymphal aquatic insects which have overwintered. In combination with day-length, the summation of degree-days synchronizes the emergence of adults. Declining autumnal temperatures may also cue the emergence of the more rapidly growing summer communities of benthic invertebrates (Ward 1992). Artificial disruption of thermal patterns and/or day-length cues has been shown to disrupt emergence patterns and reduce population success (Ward & Stanford 1982). Higher than normal discharge during snowmelt periods may depress downstream water temperatures and delay emergence (Canton & Ward 1981).

The combined influences of temperature and hydraulic preferences have also been demonstrated for lotic fish species, particularly benthic species, where distinct thermal tolerances correlate well to preferences for pool (higher temperature) or riffle (lower temperature) habitats (Hill & Matthews 1980; Ingersoll & Claussen 1984).

It can easily be demonstrated that any shift in hydrological pattern that leads to an alteration of the established thermal regime of a lotic ecosystem will ultimately lead to a dramatic change in the composition and survival of lotic biota. These hydrological alterations are discussed below.

Statzner *et al* (1988) have summarized the results of many examinations of the relationships between the distribution of lotic biota and the heterogeneity of flows within a stream reach. There is little doubt that a strong interaction between body morphology and hydrodynamic stresses controls the ability of lotic species to forage and maintain position within a preferred habitat. The relationship between flow conditions and life-cycle requirements is not as clear. However, there is recent evidence to indicate that flow patterns also control life-cycle success. Dudley and Anderson (1987) demonstrated that receding water level cued pupation for the tipulid, *Lipsothrix nigrilinea*. In this case, discharge was the governing factor rather than the hydraulic conditions. Pupation by the larvae could be delayed for a year or longer if water levels did not decline.

As growth leads to an alteration of both morphology and cross-sectional area of the organisms exposed to various hydrodynamic conditions, it could be expected that the organisms must migrate to areas of suitable hydraulic fitness in order to continue to forage and mature. Gersabeck and Merritt (1979) calculated distinct velocity preferenda for different instars of larval blackflies, while Gore *et al* (1986) demonstrated significantly different depth distributions for the final instars of the caddisfly, *Hydropsyche angustipennis*. Although insufficient physiological data exist to verify, Gore and Bryant (1990) have speculated that the apparent short-term preference of egg-bearing female crayfish for high-velocity riffles is a result of aeration requirements for successful incubation. Statzner *et al* (1988) concluded that the distribution of the various life stages of the hemipteran, *Aphelocheirus aestivalis*, was controlled by the ability of each instar to respond to changes in the distribution of hydraulic stresses across the substrate. Thus, early instars required low-stress areas (pool and moderate runs) while adults lived almost exclusively in high-velocity riffles. In stream reaches where a high diversity of depth and velocity combinations existed, there also existed a high diversity and density of instars and adults. Diversity of velocities and depths in stream reaches is directly related to discharge *and* to diversity of fish and macroinvertebrate species (Gore *et al* 1992b). Artificial discharge regimes which promote more homogeneous flow patterns might be expected to sustain lower diversities of lotic species and reduce or eliminate some species which are cued to certain hydraulic conditions.

11.3 DISRUPTION OF HYDROLOGICAL PATTERNS

Predictable hydrographic patterns are disrupted by human activity, which either impounds mainstem flow or diverts that flow for human consumption leaving a dramatically reduced flow and altered flow regime in the natural channel of the river. For the most part, significant alterations in physical and chemical conditions also result from impoundment for flood control and irrigation storage, impoundment for hydropower operation (especially hydropeaking), irrigation diversion or abstraction, channelization and afforestation.

Lakes as a hydrological influence

In comparison to the age of most lotic ecosystems, lakes represent a temporary (albeit lengthy) disruption to the flow patterns of riverine habitats. As standing bodies of water, lakes dramatically alter the physical, chemical and biological composition of the newly flowing water in the lake outlet. Since these disruptions constitute a natural geomorphological occurrence, the changes in hydrological and biological conditions can serve as a baseline for comparison to artificial disruptions (impoundments) which mimic lakes.

Standing water lacks the kinetic energy to maintain a large proportion of suspended solids. As a result, lakes act as settling basins for the suspended load carried by the input stream. The receiving stream which is fed primarily by epilimnetic water tends to be sediment poor. The mass and depth of the lake often lead to thermal and chemical stratification of the lake. With stratification, nutrients carried into the lake are concentrated in the hypolimnion (Wetzel 1983). The outlet stream, then, is nutrient poor with the exception of short periods of time when pulses of nutrients enter the receiving stream during mixis. Lentic systems are also characterized by a distinct community of planktonic organisms, which also appear as seston in the outlet streams. The pulses of seston occur on diel cycles as a result of vertical migration of zooplankton to the epilimnion, as seasonal pulses during particular phases of life cycles of planktonic and nectonic species, and as organic particulates generated by biotic activity within the epilimnion (Cushing 1964; Wotton 1982; Perry & Sheldon 1986). Epilimnetic waters are usually warmer than natural stream systems (Mackay 1979). These physical and biological changes are most often expressed in distinctive changes in the biota of the lake outlet ecosystem characterized by highly productive communities of filter-feeders (Richardson & Mackay 1991).

The containment of water within a lake has a tendency to ameliorate the effects of heavy precipitation and runoff events in the catchment. This buffering affect may stabilize discharges in the outlet stream and favour species which rely on constant hydraulic conditions for filter-feeding, particularly blackflies (Morin & Peters 1988). Finally, the mouth of the lake outlet is most often wide and shallow; this tends to reduce turbulence in the outflow water, making flow more laminar (Carlsson *et al* 1977) while making a greater portion of the water column and its sestonic load available to filtering macroinvertebrates (Morin *et al* 1988). Dense populations of filter-feeding blackflies and caddis flies are favoured.

The outflows of most lake ecosystems tend to have high gradients. The high gradient and the sediment-hungry nature of the outflow water combine to wash away fine particulates in the immediate outflow leaving larger substrate particles for attachment by filter-feeders. In addition, the sediment retention of the lake means that the outlet water does not carry a scouring suspended load and allows the increase of dense mats of attached periphyton (Simons 1979; Valett & Stanford 1987).

The more stable thermal regimes in outlet waters of lakes tend to result in fewer species in the community with dominance by filter feeders (Ward & Stanford 1982). Prolonged and elevated summer temperatures may promote higher growth rates while reduced ice-cover in temperate streams may also reduce mechanical damage to filtering species (Valett & Stanford 1987).

The abundance of filter-feeders displays a characteristic decline with distance from the lake outlet. Although increased predation may eliminate some of the filter-feeders (Sheldon & Oswood 1977), the reduction in density is apparently attributable to the rapid depletion of the seston by the very dense populations of filter-feeders near the outlet (Oswood 1979; Perry & Sheldon 1986). The recovery of downstream communities to resemble those of adjacent or upstream communities seems to be a function of the time for the stream to restore the physical heterogeneity that accompanies thermal variation, hydrographic variation from spates and variation in suspended load, emphasizing the importance of physical habitat structure on the structure and dynamics of lotic communities (Robinson & Minshall 1990).

Deep-release impoundments

Although reservoirs mimic the dynamics of most lakes in their abilities to trap suspended material, their stratification patterns and development of planktonic fauna, the structure which impounds the water differs significantly from the shallow and wide mouth of a lake outlet. Most impoundment structures are placed for the purpose of flood control, irrigation storage and/or hydroelectricity generation. In all of these cases, storage or development of a hydraulic head necessitates the controlled release of waters from the base of the dam structure. Thus, release waters are derived from the hypolimnion of the reservoir behind the dam. The effect of this hypolimnetic or deep-release can alter the composition of the tailwater fauna for a considerable distance downstream. The impacts of a cold yet thermally constant, nutrient-rich but sediment-hungry discharge are well documented (Ward & Stanford 1979; Lillehammer & Saltveit 1984; Petts 1984; Craig & Kemper 1987).

In general, the effects of hypolimnetic releases

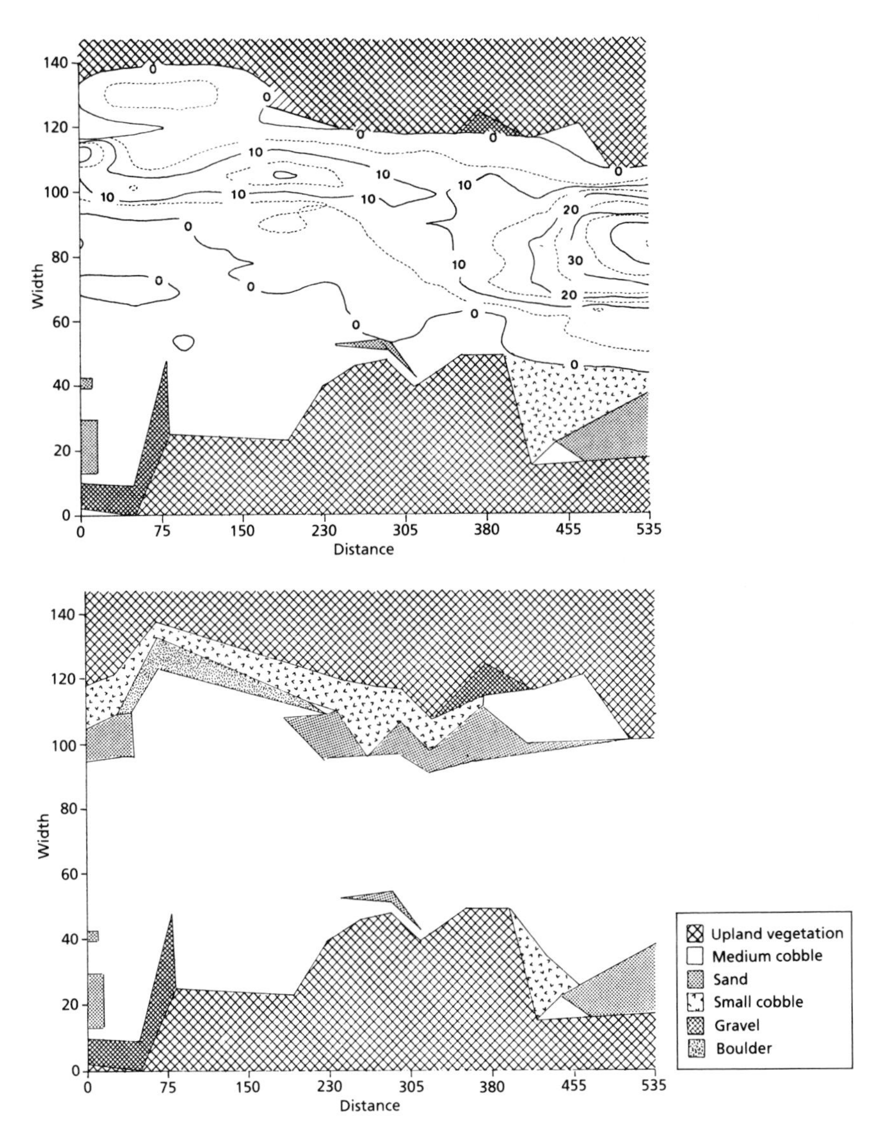

Fig. 11.3 Substrate and flow maps of a site on the Caney Fork River (Tennessee, USA) approximately 500 m downstream of a peaking hydropower facility, showing the degraded channel shape with a straight *thalweg* and substrate dominated by medium cobble (after Gore *et al* 1990).

are to improverish the downstream macroinvertebrate communities by the action of thermal constancy which eliminates the thermal cues needed to complete life cycles (as described above). Nutrient-rich and sediment-poor waters allow the abundant growth of periphyton mats, which further eliminate macroinvertebrate habitats. Fish communities, again due to altered thermal regime and reduced food availability, are also impoverished. The zone of recovery, i.e. the length of stream in which restoration of thermal variability occurs, can extend to distances of over 25 km.

Because the reservoir traps most of the inflowing sediment the released water entrains bed sediments and erodes the river channel, resulting in a channelized system with relatively homogeneous flow patterns (Fig. 11.3). Degradation of

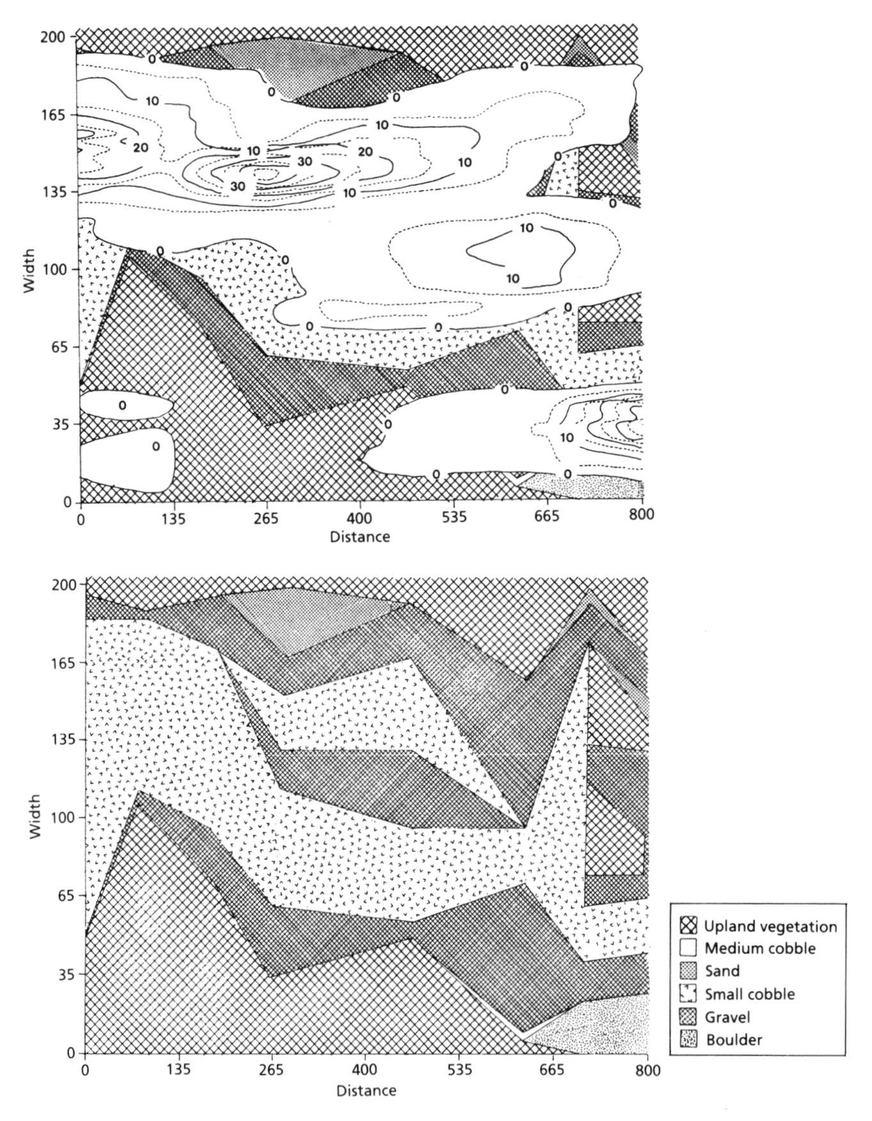

Fig. 11.4 Substrate and flow maps of a site on the Caney Fork River (Tennessee, USA) proximately 5 km downstream of a peaking hydropower facility, showing the aggraded channel with a braided appearance, split *thalweg*, and substrate dominated by sand and gravel (after Gore *et al* 1990).

the channel continues until armouring by medium and large cobbles occurs.

In many instances, degradation reduces the water table in the floodplain. Floodplains themselves serve as refugia for some aquatic organisms and as a critical habitat for the completion of life cycles of many micro-organisms which, in turn, serve as food for the lotic communities. Reduced floodplain inundation could reduce this reserve of food for the lotic community, especially juvenile fish, which often spend the first few weeks of their lives in these floodplain areas (Boulton & Lloyd 1992). Combined with water-table reductions, decreased bank stability often results in the alteration of the composition and use of the riparian community while the clear water itself induces the increase in growth of algae (Gore 1977).

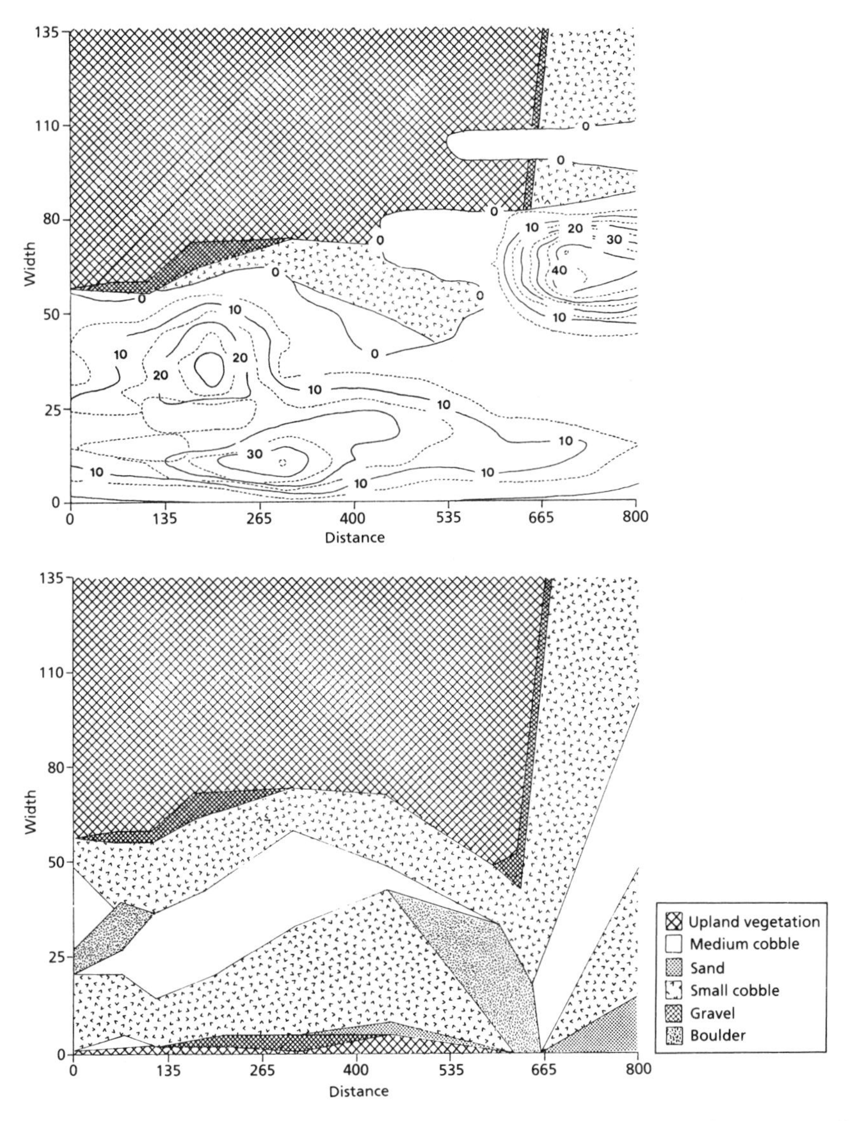

Fig. 11.5 Substrate and flow maps of a site on the Caney Fork River (Tennessee, USA) approximately 15 km downstream of a peaking hydropower facility, showing a minimally impacted section of channel with high flow diversity, heterogeneous substrate and sinuous internal meander (after Gore *et al* 1990).

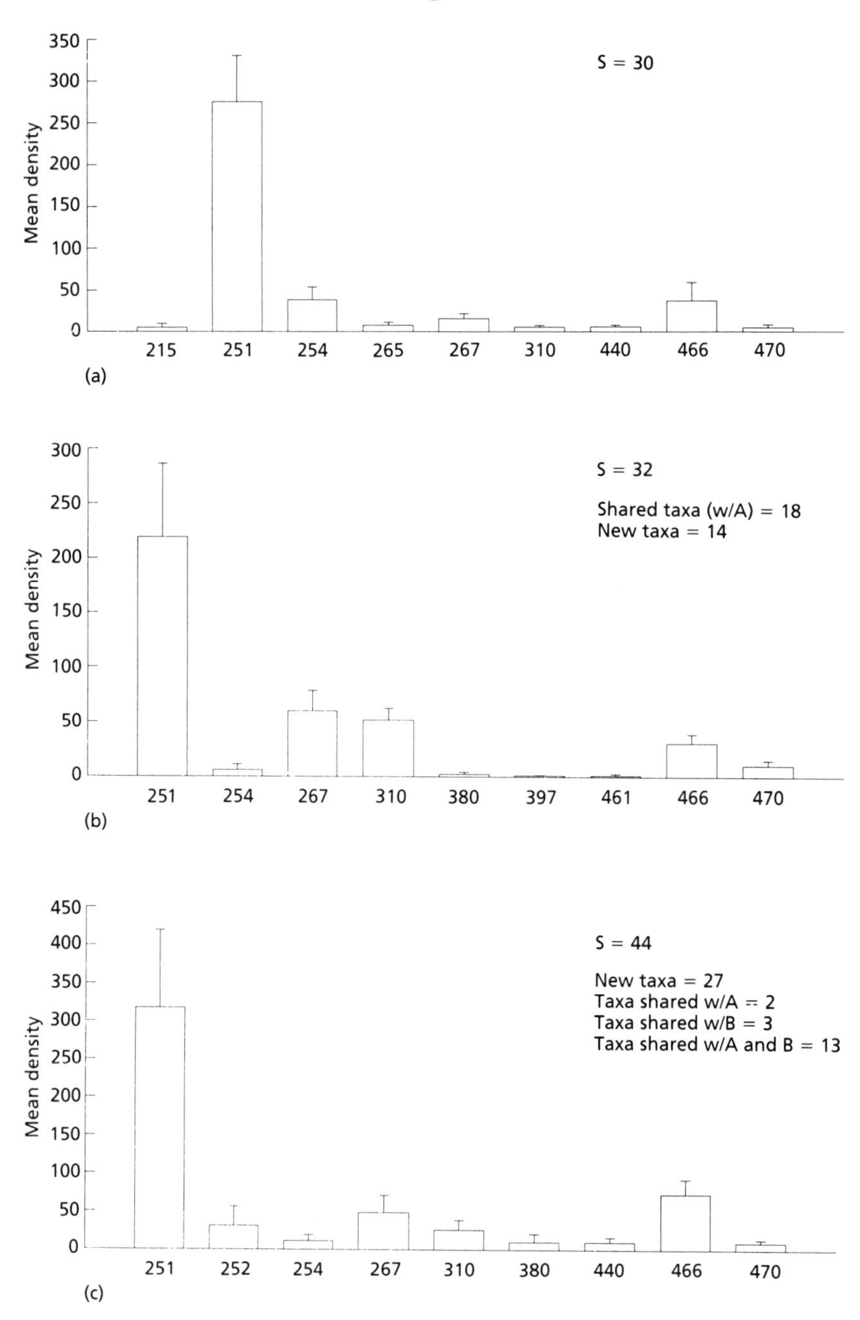

Fig. 11.6 Distribution and density of the 10 most common/dominant midge (Chironomidae) taxa at each of three stations downstream of a hydropower facility (Caney Fork, Tennessee, USA). (a) Zone of degradation. (b) Zone of aggradation/deposition. (c) Historical channel. Taxa listed are: 215 = *Cardiocladius obscurus*; 251 = *Cricotopus tremulus*; 252 = *Cricotopus trifascia*; 254 = *Cricotopus bicinctus*; 265 = *Dicrotendipes* sp.; 267 = *Dicrotendipes nervosa*; 310 = *Microtendipes* sp.; 380 = *Phaenopsectra* sp.; 397 = *Potthastia longimanus*; 440 = *Rheotanytarsus* sp.; 461 = *Stictochironomus devinctus*; 466 = *Synorthocladius semivirens*; 470 = *Tanytarsus* sp. (Data from Gore & Pennington, personal communication, with permission.)

Farther downstream, a zone of aggradation can occur where finer sediments from the degraded tailwaters are redeposited, forming many small islands (Figs 11.4 & 11.5). Fuller discussion of geomorphological changes is presented by Brookes (1994).

Aside from the variety of thermal and chemical changes which may alter the composition of downstream communities – see reviews by Ward and Stanford (1979) and Petts (1984) – the changes in flow patterns and substrate composition will alter the amounts and kinds of physical habitat available. Gore and Pennington (1988) reported significant changes in chironomid species in the zones of degradation and aggradation related, in part, to the dramatic changes in dominant substrate when compared to the historical channel (Fig. 11.6). More recently, Grzybkowska *et al* (1990) found that densities of dominant macroinvertebrates, mostly chironomids, were highest along the banks of tailwaters and declined abruptly towards mid-river. In general, higher species richness occurred in depositional areas (Gaschignard 1984). Although not mentioned by these authors, it is not surprising that densities would be reduced near mid-river, the much higher velocities resulting in increased shear across the substrate and decreased suitability of hydraulic habitat.

Release schedules

Although it is difficult to separate the effects of changes in physico-chemical conditions from those of changes in flow pattern, it can be assumed that loss of preferred hydraulic habitat is partially responsible for the impoverished lotic communities which exist in the tailwaters of most impoundments. Of particular importance is the increased shear that peaking hydropower imparts on downstream communities. Release schedules have a dramatic effect on community composition. Unless releases of water are constant, the wetted perimeter of the channel is continually changing. Depending upon the habitat preferences of the species and the duration of time that the substrate is wetted, there is a continual cycle of invasion and stranding by various species of benthic invertebrates and fish (Corrarino & Brusven 1983; Perry & Perry 1986). The effect of reducing discharges also imparts on community composition by increasing the amount of downstream drift of benthic species. Most often increased drift has been reported to accompany flow reductions (Perry & Perry 1986). Gore (1977) and Corrarino and Brusven (1983) reported catastrophic drift of invertebrates when discharges declined rapidly. Gore attributed this to elimination of hydraulic habitat requirements of nearly all benthic species as discharge declined. Peaking hydropower further exacerbates the problems of adjusting to a predictable flow regime. Peaking releases occur as human demand dictates (when electricity consumption is higher than normal distribution loads). As a result the hydrograph of peaking tailwaters is flashy and significantly different from the input stream (Gore *et al* 1989). Although these tailwaters may have relatively stable beds and channel morphology after years of operation, analysis has revealed that habitat availability varies according to the release schedule, with less hydraulic habitat available as discharge increases (Gore *et al* 1989). The response of lotic biota to these changes in discharge is not clearly understood and it is not known if short-term loss of the preferred hydraulic conditions in the habitat is sufficient to alter population success or community dynamics. Gore *et al* (1993) have predicted that shear forces decline rapidly as water surface elevation increases and that the habitat disturbance occurs only for a brief time during the rising limb of each peaking wave. However, a considerable amount of behavioural observation is needed to verify this prediction.

The sudden increase in discharge in tailwater areas can be equivalent to floods with yearly recurrence in the historical channel. This often leads to daily increases in drift of macroinvertebrates as shear dislodges foraging and resting individuals. Increases in drift are usually associated with initial discharge surges (Matter *et al* 1983; Irvine 1985). Although drift has been reported to decline within an hour of the initial surge (Layzer *et al* 1989), Matter *et al* (1983) calculated that as much as 14% of standing crop in tailwaters can be eliminated each month by the drift initiated through the increased shear resulting from peaking operations. The affect of sudden flow fluctuations on tailwater fish communities is not as clear; however, DeJalon *et al* (1988) reported reductions in trout production as a result of the decline in the food source, the benthic macroinvertebrates lost to drift and other reservoir-related phenomena. Both fish and macroinvertebrates shift their physical habitat requirements as spawning or changes in water temperature dominate fish life-histories or changes in size and state of maturity alter habitat preferences of macroinvertebrates. Thus, the magnitude of drift and the success of recruitment to fish populations can also vary in tailwaters affected by hydropeaking operations (Campbell & Neuner 1985; Layzer *et al* 1989). The loss of habitat heterogeneity (i.e. decline in usable habitat) and the effects of unpredictable shear across the substrate serve to aid the impoverishment of tailwater fauna, regardless of the physicochemical conditions which arise as a result of the release structure.

Land use practices

In many cases, demand for water to irrigate croplands or the removal of forest or riparian vegetation can significantly alter the hydraulic conditions within the channel. Irrigation withdrawal, on the one hand, reduces mean discharge within the channel, while return water from poorly managed croplands, in addition to being chemically altered, may result in unusually higher flows in downstream receiving stream channels.

In the case of logging practices such as clearcutting, removal of the dominant vegetation results in reduced water retention by the watershed itself and dramatic increases in surface runoff and the volume of water being transported by the river channel. Although the majority of studies on the effects of clearcutting and other logging practices have concentrated on the consequences of the alteration of sediment transport, thermal regime and availability of particulate organic material as a food source for biota in the effected lotic ecosystem (see, for example, Wallace & Gurtz 1986; Webster *et al* 1990), a few studies have examined concurrent alterations in the hydraulic and hydrological conditions of the river which alter hydraulic habitat for the biota.

The availability of woody debris in low-order streams determines not only the quantity of coarse particulate organic matter available to drive community dynamics, but also the hydraulic characteristics of the channel. Bisson *et al* (1987) have shown that removal of trees and deprivation of the channel of the woody debris contributed by downed trees alters the ability of the channel to maintain pool habitats. In high-gradient streams (10–15%), an increase in average discharge resulting from clearcutting can cause scouring of the channel bed and a coarsening of the substrate, while low-gradient streams tend to be accumulators of sediment. Reaches with coarser substrate provide more diverse microhabitats and higher densities and diversities of macroinvertebrates and fish (Murphy & Hall 1981). Foraging efficiency of trout predators can increase in pools in logged catchments since sedimentation tends to occlude refugia on the substrate and higher overland flow results in increased drift densities of food sources (Wilzbach *et al* 1986). These relationships seem to be related to the amount of logging done in the watershed and the ability of the remaining vegetation to alter surface flow rates. With at least 25% of the forest remaining after logging activities, little significant impact on invertebrate communities could be ascertained in a series of low-order streams in Oregon (Carlson *et al* 1990).

Abstractions and diversions to provide water for irrigation, hydropower and public consumption usually result in significant decreases in water volume flowing through the river channel. Armitage and Petts (1992) have demonstrated that combinations of change in discharge, baseflow and substrate are the factors which govern numbers of species, numbers of families and total abundance of benthic species in stream channels affected by abstraction. Although it is possible that the heterogeneity of flow conditions may remain the same as in prewithdrawal conditions, hydraulic patterns shift to lower velocities and depths. Thus, one might expect that the preferred habitat conditions for high-velocity, riffle-dwelling species and deep-pool-dwelling species might be substantially reduced. In addition, the wetted area available to provide hydraulic cover and overhead cover for edge-dwelling species will also be reduced and, despite the fact that the same amount of edge area exists under reduced flows, the amount of habitat for edge dwellers will also be reduced. Alteration of the benthic community begins as soon as abstraction causes flows to fall below the natural low-stage conditions (Jäger *et al* 1985). Densities and biomass of macroinvertebrates have been reported to be reduced by as much as 75% (Heger & Moog 1986). Initial loss of invertebrates is immediate and attributable to drift, as an escape response from less than favourable physical habitat conditions (Gore 1977). However, species which are adapted to life on intermittently wetted substrates may persist (Bickerton *et al* 1993). Continued low flows do not provide habitat criteria favourable to recolonization (Gore & Milner 1990). Thus, the food support for fish communities does not exist and fish numbers, biomass and diversity also decline. Survival is dependent upon the number of refugia available and contribution of colonists from upstream drift (Armitage 1987).

The effects of reduced flows in a channel reach beyond the biota of the stream channel itself as the flow provides the supply for bank storage, hence riparian growth. In areas where reduced flows prevailed (downstream of hydropower facilities), reduced soil moisture caused lower growth rates and declining abundance of riparian plants, especially of juveniles (Stromberg & Patten 1990). Smith *et al* (1991) predicted that the elimination of floods and high flows could lead to the selective mortality of juvenile plants and the reduction of riparian strip width.

11.4 HYDRAULIC REFERENCE POINTS

In general, there are five major hydraulic conditions which most affect the distribution and ecological success of lotic biota. These are velocity profile, suspended load, bedload movement, water column effects such as turbulence, and substratum interactions (near-bed hydraulics). Singly, or in combination, the changes in these instream conditions can alter the distribution of biota and disrupt community structure. Within a stream reach the distribution of aquatic biota is governed by the interactions of these hydraulic conditions upon the morphology and behaviour of the individual organisms. Full discussions of suspended sediment transport and instream hydraulics and sediment transport are discussed in Calow and Petts (1994).

Complex hydraulics

For any organisms living upon or near the substrate of a river or stream, the interactions of depth and velocity with the profile of the substrate particles is of critical importance to the range of potential microhabitats available (Statzner *et al* 1988). Table 11.1 summarizes some of the critical hydraulic conditions which a benthic organism must encounter. Although studies of flow in flumes have indicated the existence of a non-moving boundary layer (up to 2 mm in thickness) (Nowell & Jumars 1984) and lotic scientists have indicated that morphological and behavioural

Table 11.1 Instream hydraulic characteristics which have been shown to influence the distribution of riverine biota

Simple hydraulic characteristics:	
D:	mean depth (cm)
k:	substrate characteristic (cm)
S:	slope of water surface
w:	stream width (cm)
ρ:	density of water ($g\,cm^{-3}$)
g:	acceleration due to gravity ($cm\,s^{-2}$)
Q:	discharge ($m^3\,s^{-1}$)
U:	mean current velocity ($cm\,s^{-1}$)
ν:	kinematic viscosity ($cm^2\,s^{-1}$)
Complex hydraulic characteristics relevant to describe flow for an area within a stream reach or for an entire stream reach:	
Fr:	Froude number (dimensionless); $Fr > 1$ (shooting) $Fr < 1$ (tranquil)
Re:	Reynolds number (dimensionless); drag characteristics
Re^*:	boundary Reynolds number (dimensionless)
U^*:	shear velocity ($cm\,s^{-1}$)
δ':	thickness of viscous sublayer (cm)
τ:	shear stress ($dyne\,cm^{-2}$)

Formulae:
$Fr = U/(gD)^{0.5}$, $Re = UD/\nu$, $Re^*_1 = U^*_1 k/\nu$, $Re^*_2 = U^*_2 k/\nu$, $U^*_1 = (\tau/\rho)^{0.5}$, $U^*_2 = U/5.75.\log_{10}(12\,D/k)$, $\delta'_1 = 11.5\nu/U^*_1$, $\delta'_2 = 11.5\nu/U^*_2$, $\tau = gSD$.

adaptations are directed towards existence within this boundary layer (Hynes 1970), recent investigations have determined that very few benthic organisms are sufficiently streamlined to take advantage of this boundary layer (see discussions below and Statzner 1987a). Indeed, a true boundary layer rarely exists under natural flow conditions because of the complex profile of the substrate. Thus, aquatic organisms are restricted to those combinations of velocity, depth and substrate which allow morphological and/or behavioural resistance to flow to be exceeded by energetic gains from foraging in these areas. That is, conditions such as shear stress and the thickness of the slower-moving laminations of water which constitute the viscous sublayer are the major determinants of the distribution of most benthic organisms within a given stream reach. Davis and Barmuta (1989) classify near-bed flows as hydraulically smooth or hydraulically rough – chaotic, wake interference, isolated roughness or skimming flows. Each of these flow types is associated with recognizable assemblages of benthic biota. Hydraulic fitness of swimming organisms, and by implication, fish community structure, is also affected by the distribution of turbulent zones (Scarnecchia 1988).

Any hydrological change which leads to increases in shear velocities or shear stresses or reduction in the thickness of the viscous sublayer will reduce the availability of adequate microhabitats to some species while increasing it for others and, presumably, alter the abundances of individuals in the community. For example, Ciborowski and Craig (1989) found that increases in velocity resulted in increased drift and significant changes in aggregations of larval blackflies. Indeed, the position of the nearest upstream organisms influenced the hydraulic conditions of the downstream individual and its ability to feed. In turn, community functioning and food web dynamics may be changed to the extent that some species will be locally eliminated from the community.

Near-bed conditions are further changed with changes in substrate composition or distribution since increased hydraulic roughness increases the rate of sediment deposition (Harvey & Watson 1986). Increased sedimentation rates have been shown to significantly alter the composition of benthic communities as interstitial microhabitats are eliminated (increased predation success; Brusven & Rose 1981) and scour or deposition eliminates primary production (Brusven & Prather 1974; Chutter 1969).

11.5 INTERACTIONS OF FLOW AND MORPHOLOGY

Ambühl (1959) indicated that most benthic invertebrates were sufficiently streamlined to avoid the shear conditions along the substrate and could, therefore, forage across the substratum with relative impunity. This notion was generally accepted by most lotic ecologists and contributed to the search for other land–water interactions which might serve as a major template to the distribution of aquatic biota (both within reaches and along the length of river ecosystems). However, in the past decade a number of scientists have examined the interactions of complex hydraulics and the responses of biota to changes in these conditions. Gore (1983) demonstrated a significant correlation between shell length (as an indicator of exposure to sheer velocities) and velocity distributions among pleurocerid snails. Using laser doppler anemometry, Statzner and Holm (1982, 1989) and Statzner (1987a, 1988) demonstrated that benthic invertebrates do not exhibit the streamlining conditions which Ambühl reported. Indeed, Reynolds numbers (the interaction between frontal area exposed to current, the length of the body, and the velocity and viscosity of the water) were sufficiently high to predict that in order to compensate for the forces of pressure drag and friction drag, and to maintain an energetic gain from foraging activity along the substrate, most benthic organisms were quite limited in the range of conditions in which they could exist. High drag coefficients may be of some advantage to organisms swimming in still waters, where the edges of the prothorax or elytra may aid predators in making fast turns and stops (Nachtigall & Bilo 1965). However, resistance to flow in lotic ecosystems implies that body shape (a compromise between hemispherical and ovoid) and behavioural responses to changes in flow restrict the ability of lotic organisms to move freely within and between stream reaches by limiting energy gains to occupancy of a narrow range of flow microhabitats. Indeed, because of the changes in surface area : volume ratios during the growth of an individual, many species (mayflies, for example) must move to higher-velocity microhabitats as they grow in order to satisfy oxygen requirements (Kovalak 1978). Resistance to entry into the drift increases as much as 1000-fold during the growth of benthic invertebrates (Waringer 1989) and is a function of increase in size and alteration in cross-section profile presented to the direction of flow (Dussart 1987). Gore (1983) suggested that exposure of shells to higher-velocity lamina flow may result in the erosion or decollation of the top spires of lotic pleurocerid snails. This decollation can result in further bacterial and fungal deterioration of the shells (Burch 1982). Thus, increases in average velocity in a channel might be expected to differentially affect cohorts of the same benthic population. Statzner *et al* (1988) used these hydraulic parameters to demonstrate that the density and distribution of specific cohorts of benthic organisms could be predicted if complex hydraulic conditions such as Reynolds velocity, laminar sublayer thickness, and even Froude number were known at the point of sampling.

The effect of flow is not limited to the interactions of complex hydraulics at the level of the substrate upon benthic invertebrates. Brewer and Parker (1990) determined that the upslope distribution of macrophyte species was determined by the break force upon the stems. The tensile strength was different for all species and the force required to break the stem varied as a power function of the cross-sectional area. Thus, thick-stemmed species are the only species occurring in high-gradient streams, while lowland rivers and large-order rivers tend to present more flow microhabitats for a variety of macrophyte species. Similarly, Scarnecchia (1988) used the fineness ratio (length/maximum diameter) of Webb (1975) to demonstrate that homogeneity of flows in a channelized river tended to reduce the diversity of fish communities. Where an *Fr* value of 4.5 gives minimum drag for fish, stream reaches of high uniform flows are characterized by fish with this more highly streamlined characteristic while species with *Fr* values less than 4 are virtually eliminated from these areas.

11.6 EFFECTS OF FLOW ON BEHAVIOUR AND PHYSIOLOGY

As hydraulic conditions change within a stream channel, the effects upon stream biota are either direct mechanical damage and removal of the organisms from the preferred microhabitat or changes in the behaviour and physiological conditions of the biota. These changes in behaviour and physiology may mean a net loss or reduction in energy for growth and reproduction or the inability to breed or forage in appropriate habitats.

Although Pfeifer and McDiffett (1975) indicated the potential for large increases in nutrient uptake with increased velocity, a minimum current of about $5\,\mathrm{cm\,s^{-1}}$ across the surface of periphyton beds has been shown to stimulate increased primary production in lotic ecosystems (Lock & John 1979). The influence of changes in current velocity extend to composition of periphyton communities themselves. Peterson (1986, 1987) found that reduced mean daily flows supported higher diatom biomass in partially exposed areas, with a higher diversity of taxa including unattached colonial species. Those species inhabiting slow current areas were less resistant to desiccation, probably due to slow nutrient renewal.

Among fish species, changes in hydraulic conditions may differentially affect certain cohorts or functional groups within the community. These changes are related to energetic balances tuned to body morphology (see above) and physiological abilities. Among fish, morphological adaptations which minimize exposure to shear and enhance preference for low velocities (and low oxygen consumption) can be predicted by an energy-cost hypothesis (Facey & Grossman 1992). Although it is well known that fish can rapidly adjust their swimming speeds and positions in streams with changes in current velocity (Jones 1968; Hanson & Li 1978), Trump and Leggett (1980) were able to demonstrate that the most efficient swimming conditions (particularly during migration) were conditions in which current velocities can be rapidly detected and are not rapidly varying. Under suboptimal swimming conditions, energetic losses may lead to decreased spawning success or condition factor. In the same manner Gibson (1983) has noted that the change from aggressive to schooling behaviour for migration is not only triggered by changes in temperature in the stream ecosystem but is also tied to changes in buoyancy associated with changes in current velocity.

Among benthic species, shifts in behaviour and/or physiological demand according to hydraulic changes are also common. At minimum, many species of lotic insects have been shown to increase respiratory activity with decreases in current velocity (Feldmeth 1970). Kovalak (1976) found larvae of the trichopteran, *Glossosoma*, to occupy up-stream faces of rocks during times of higher temperatures and lower dissolved oxygen concentrations. Similarly, mayfly nymphs change position on the surface of substrate particles in relation to current velocity and dissolved oxygen (DO) concentrations. As DO decreases, nymphs take positions in higher current velocities, usually exposing themselves to greater shear and mechanical removal (as drift) or greater predation (Wiley & Kohler 1980). At least some species of crayfish exhibit the same apparent high-risk activity (exposure to predation) in order to oxygenate egg clusters (Gore & Bryant 1990). With increases in stream velocity, many crayfish species can alter body posture to counteract the effects of drag, maintain position, and take advantage of increased oxygen in these areas (Maude & Williams 1983).

Drift

The phenomenon of downstream drift of benthic species has been attributed to a variety of factors including negative phototactic responses, escape from introduced toxic substances or increased sediment load, escape from predation, interference competition as density increases, and as a response to a deteriorating microhabitat. An excellent review of the phenomenon of drift is provided by Brittain and Eikeland (1988). Although the influence of some of the factors is controversial, there is little doubt that changes in velocity pattern result in increased drift from areas in which the hydraulic habitat is no longer suitable. Varying discharges, particularly as a result of hydropower operation, are known to initiate catastrophic drift among stream invertebrates and larval fish (Cushman 1985). However, it is not certain if the rate of change of discharge is the primary influencing factor; releases which simulate large floods do not seem to initiate greater drift densities than gradual increases in flow (Irvine & Henriques 1984). However, it does appear that initial changes in flow initiate more drift than subsequent similar flow changes (Irvine 1985), which seems to indicate that areas with unpredictable flow patterns are continually occupied by invertebrates at lower densities than the hydraulic habitat might support under more predictable or stable flow conditions (Ciborowski & Clifford 1983). The addition of sediment to increased flows further increases drift density as scour tends to remove some individuals maintaining position in adequate flow habitats (Ciborowski *et al* 1977). The importance of the hydraulic habitat to the success of benthic species is underlined by the observations that drift has also been seen to increase during conditions of decreasing flow. Catastrophic drift is usually initiated at some discharge greater than conditions in which the majority of the substrate is dewatered indicating that lower velocity tolerances or preferences can also be exceeded (Gore 1977; Corrarino & Brusven 1983). In either case, increased or decreased discharges which promote hydraulic conditions exceeding the preferences of the benthic species (and/or increase sediment loads to the flow; see Culp *et al* 1986) can result in a net loss of benthic species from a given stream reach.

11.7 MANAGEMENT/MITIGATION IMPLICATIONS

It should be apparent that maintenance of a functioning stream ecosystem is, in part or whole, a function of the hydraulic heterogeneity of the river channel. Statzner *et al* (1988) and Heede and Rinne (1990) have demonstrated the close relationship between stream biota and the hydraulic and hydrological conditions which define the physical habitat of most lotic biota. Stream managers should, then, focus attention on the development of techniques and release schemes which provide for currently existing hydraulic conditions within the channel or, in the case of degraded systems, provide new conditions which increase hydraulic heterogeneity to allow the recolonization of the greatest numbers and kinds of biota. Obviously, this does not imply that high diversity will yield a stable stream community, but the greatest variety of physical conditions should allow the fastest rate of colonization and recovery from disturbance as at least *some* habitat will be available to local species as colonizers. A period of dynamic adjustment between competitors, predators and prey will likely adjust the community structure to the proportions and conditions of available hydraulic habitat.

Two management techniques are currently available to ensure maintenance of adequate hydraulic habitats. One method, instream flow prediction and managed discharge regimes, is an *a priori* approach; the other method, channel engineering and restoration, provides an *a posteriori* way of increasing or altering the available hydraulic habitat.

Instream flow techniques encompass a body of field measurements and computer techniques which combine hydrological and hydraulic conditions in the channel with biological criteria for those same parameters in order to predict gain or loss of habitat under new flow regimes or to formulate recommendations to ensure that adequate minimum flows be maintained for retention of the target organisms of management interest. These techniques are summarized in Wesche and Rechard (1980), Gustard *et al* (1987), and Calaw & Petts (1994). Gore and Nestler (1988) point out that the body of instream flow models currently used for flow recommendations in North America and Europe should be viewed as management models which allow the assessment of economic and biological benefits when designing release schedules rather than an exacting model to predict the ecological consequences of changes in flows. Despite the lack of ecological prediction, these techniques provide a useful tool to stream managers to ascertain which flows will substantially alter habitat availability and provide a 'window' of parameters in which hydrological conditions for maintaining a reasonably stable downstream ecosystem can be produced.

Channel engineering is the construction, after channel disturbance such as diversion and/or channelization, of artificial meanders, embankments and obstacles to increase hydraulic heterogeneity in the channel by reconstructing pool/riffle sequences and providing a high diversity of substrate particles. The construction of these structures is examined in detail by Gore (1985), Brookes, (1988) Gore and Petts (1989), and Calow & Petts (1994). Gore (1985), has suggested that combining instream flow techniques with channel engineering may improve the design of restoration projects and allow a cost/benefit analysis prior to construction of various structures such as weirs and wings. Traditionally, the use of habitat restoration features has been restricted to low-order, high-gradient streams dominated by salmonid fisheries or forage fish. However, recent instream flow analyses suggest that reregulation weirs in large rivers can provide increases in hydraulic heterogeneity and increased habitat availability for target fish (Martin *et al* 1985). Indeed, Gore *et al* (1992b) provide a list of techniques and restoration structures aimed at increasing top width, restoring internal meander, and providing dampening of peaking hydropower waves as methods to establish predictably available habitat for fish and benthos in the receiving streams of large-order rivers.

In both cases, the result is some amount of control of the hydrological and hydraulic conditions which influence the availability of physical habitats for lotic biota and a means to conserve the integrity of lotic ecosystems.

REFERENCES

Ambühl H. (1959) Die Bedeutung der Strömung als ökologischer Faktor. *Schweizerische Zeitschrift für Hydrobiologie* **21**: 133–264. [11.5]

Armitage PD. (1987) The classification of tailwater sites receiving residual flows from upland reservoirs in Great Britain, using macroinvertebrate data. In: Craig JF, Kemper JB (eds) *Regulated Streams. Advances in Ecology*, pp. 131–44. Plenum Press, New York. [11.3]

Armitage PD, Petts GE. (1992) Biotic score and prediction to assess the effects of water abstractions on river macroinvertebrates for conservation purposes.

Aquatic Conservation **2**: 1–17. [11.3]

Bickerton M, Petts G, Armitage P, Castella E. (1993) Assessing the ecological effects of groundwater abstraction on chalk streams: three examples from eastern England. *Regulated Rivers* **2**: 121–34. [11.3]

Bisson PA, Bilby RE, Bryant MD, Andrew CD, Grette GB, House RA, Murphy ML, Koski KV, Sedell JR. (1987) Large woody debris in forested streams in the Pacific Northwest: past, present, and future. In: Salo EO, Cundy TW (eds) *Proceedings of the Symposium on Streamside Management: Forestry and Fishery Interactions*, pp. 143–231. University of Washington Press, Seattle, WA. [11.3]

Boulton AJ, Lloyd LN. (1992) Flooding frequency and invertebrate emergence from dry floodplain sediments on the River Murray, Australia. *Regulated Rivers* **7**: 137–51. [11.3]

Brewer CA, Parker M. (1990) Adaptations of macrophytes to life in moving water: upslope limits and mechanical properties of stems. *Hydrobiologia* **194**: 133–42. [11.5]

Brittain JE, Eikeland TJ. (1988) Invertebrate drift – a review. *Hydrobiologia* **166**: 77–93. [11.6]

Brookes A. (1988) *Channelized Rivers. Perspectives for Environmental Management*. John Wiley & Sons, Chichester. [11.7]

Brookes A. (1994) River channel change. In: Calow P, Petts GE (eds) *The Rivers Handbook*, vol. 2, pp. 55–75. Blackwell Scientific Publications, Oxford. [11.3]

Brusven MA, Prather KV. (1974) Influence of stream sediments on distribution of macrobenthos. *Journal of the Entomological Society of British Columbia* **71**: 24–32. [11.4]

Brusven MA, Rose ST. (1981) Influence of substrate composition and suspended sediment on insect predation by the torrent sculpin, *Cottus rhotheus*. *Canadian Journal of Fisheries and Aquatic Sciences* **38**: 1444–8. [11.4]

Burch JB. (1982) *Freshwater Snails (Mollusca: Gastropoda) of North America*. EPA-600/3-82-026, US Environmental Protection Agency, Cincinnati, Ohio. [11.5]

Calow P, Petts GE. (eds) (1994) *The Rivers Handbook*, vol. 1. Blackwell Scientific Publications, Oxford. [11.4]

Campbell RF, Neuner JH. (1985) Seasonal and diurnal shifts in habitat utilized by resident rainbow trout in western Washington Cascade Mountain streams. In: Olson FW, White RG, Hamre RH (eds) *Symposium on Small Hydropower and Fisheries*, pp. 39–48. American Fisheries Society, Bethesda, MD. [11.3]

Canton SP, Ward JV. (1981) Emergence of Trichoptera from Trout Creek, Colorado, USA. In: Moretti GP (ed) *Proceedings of the Third International Symposium on Trichoptera*, pp. 39–45. W. Junk, The Hague, Netherlands. [11.2]

Carlson JY, Andrus CW, Froehlich HA. (1990) Woody debris, channel features, and macroinvertebrates of streams with logged and undisturbed riparian timber in northeastern Oregon, USA. *Canadian Journal of Fisheries and Aquatic Sciences* **47**: 1103–11. [11.3]

Carlsson M, Nilsson LM, Svensson B, Ulfstrand S, Wotton RS. (1977) Lacustrine seston and other factors influencing the blackflies (Diptera: Simuliidae) inhabiting lake outlets in Swedish Lapland. *Oikos* **29**: 229–38. [11.3]

Chisholm IM, Hubert WA. (1986) Influence of stream gradient on standing stock of brook trout in the Snowy Range, Wyoming. *Northwest Science* **60**: 137–9. [11.2]

Chutter FM. (1969) The effects of silt and sand on the invertebrate fauna of streams and rivers. *Hydrobiologia* **34**: 57–76. [11.4]

Ciborowski JJH, Clifford HF. (1983) Life histories, microdistribution and drift of two mayfly (Ephemeroptera) species in the Pembina River, Alberta, Canada. *Holarctic Ecology* **6**: 3–10. [11.6]

Ciborowski JJH, Craig DA. (1989) Factors influencing dispersion of larval black flies (Diptera: Simuliidae): effects of current velocity and food concentration. *Canadian Journal of Fisheries and Aquatic Sciences* **46**: 1329–41. [11.4]

Ciborowski JJH, Pointing PJ, Corkum LD. (1977) The effect of current velocity and sediment on the drift of the mayfly *Ephemerella subvaria* McDunnough. *Freshwater Biology* **7**: 567–72. [11.6]

Corrarino CA, Brusven MA. (1983) The effects of reduced stream discharge on insect drift and stranding of near shore insects. *Freshwater Invertebrate Biology* **2**: 88–98. [11.3, 11.6]

Craig JF, Kemper JB. (eds) (1987) *Regulated Streams. Advances in Ecology*. Plenum Press, New York. [11.3]

Culp JM, Wrona FJ, Davies RW. (1986) Response of stream benthos and drift to fine sediment deposition versus transport. *Canadian Journal of Zoology* **64**: 1345–51. [11.6]

Cushing CE. (1964) Plankton and water chemistry in the Montreal River lake–stream system, Saskatchewan. *Ecology* **45**: 306–13. [11.3]

Cushman RM. (1985) Review of ecological effects of rapidly varying flows downstream from hydroelectric facilities. *North American Journal of Fisheries Management* **5**: 330–9. [11.6]

Davis JA. (1986) Boundary layers, flow microenvironments and stream Benthos. In: De Deckker P, Williams WD (eds) *Limnology in Australia*, pp. 293–312. CSIRO, Melbourne. [11.1]

Davis JA, Barmuta LA. (1989) An ecologically useful classification of mean and near-bed flows in streams and rivers. *Freshwater Biology* **21**: 271–82. [11.4]

DeJalon DG, Montes C, Barcelo E, Casado C, Menes F. (1988) Effects of hydroelectric scheme on fluvial ecosystems within the Spanish Pyrenees. *Regulated Rivers* **2**: 479–91. [11.3]

Dingman SL. (1984) *Fluvial Hydrology*. W.H. Freeman, San Francisco. [11.1]

Dudley TL, Anderson HH. (1987) The biology and life cycles of *Lipsothrix* spp. (Diptera: Tipulidae) inhabiting wood in western Oregon streams. *Freshwater Biology* **17**: 437–51. [11.2]

Dussart GBJ. (1987) Effects of water flow on the detachment of some aquatic pulmonate gastropods. *American Malacological Bulletin* **5**: 65–72. [11.5]

Evans JW, Noble BL. (1979) The longitudinal distribution of fishes in an east Texas stream. *American Midland Naturalist* **101**: 333–43. [11.2]

Facey DE, Grossman GD. (1992) The relationship between water velocity, energetic costs, and microhabitat in four North American stream fishes. *Hydrobiologia* **239**: 1–6. [11.6]

Feldmeth CR. (1970) The respiratory energetics of two species of stream caddis fly larvae in relation to water flow. *Comparative Biochemistry and Physiology* **32**: 193–202. [11.6]

Gaschignard O. (1984) Impact d'un crue sur les macroinvertébrés benthiques d'un bras du Rhône. *Verhandlungen Internationalen Vereinigung für theoretische und angewandte Limnologie* **22**: 1997–2001. [11.3]

Gersabeck EF Jr, Merritt RW. (1979) The effect of physical factors on the colonization and relocation behavior of immature black flies (Diptera: Simuliidae). *Environmental Entomology* **8**: 34–9. [11.2]

Gibson RJ. (1983) Water velocity as a factor in the change from aggressive to schooling behavior and subsequent migration of Atlantic salmon smolt (*Salmo salar*). *Naturaliste Canadien* **110**: 143–8. [11.6]

Gore JA. (1977) Reservoir manipulations and benthic macroinvertebrates in a prairie river. *Hydrobiologia* **55**: 113–23. [11.3, 11.6]

Gore JA. (1983) Considerations of size related flow preferences among macroinvertebrates used in instream flow studies. In: Shuval HI (ed) *Developments in Ecology and Environmental Quality*, vol. II, pp. 389–98. Balaban Int. Publ., Jerusalem. [11.5]

Gore JA. (ed) (1985) *The Restoration of Rivers and Streams*. Butterworth, Boston. [11.7]

Gore JA, Bryant RM Jr. (1990) Temporal shifts in physical habitat of the crayfish, *Orconectes neglectus* (Faxon). *Hydrobiologia* **199**: 131–42. [11.2, 11.6]

Gore JA, Milner AM. (1990) Island biogeographic theory: can it be used to predict lotic recovery rates? *Environmental Management* **14**: 737–53. [11.3]

Gore JA, Nestler JM. (1988) Instream flow studies in perspective. *Regulated Rivers* **2**: 93–101. [11.7]

Gore JA, Pennington W. (1988) Changes in larval chironomid habitat with distance from peaking hydropower operations. Annual Meeting of the North American Benthological Society, Tuscaloosa, AL. [11.3]

Gore JA, Petts GE. (eds) (1989) *Alternatives in Regulated River Management*. CRC Press, Boca Raton, FL. [11.7]

Gore JA, Statzner B, Resh VH. (1986) Physical habitat characteristics and microdistribution of final instars of *Hydropsyche angustipennis* (Curtis). *Fifth International Symposium on Trichoptera, Lyon, France*. W. Junk, The Hague, Netherlands. [11.2]

Gore JA, Nestler JM, Layzer JB. (1989) Instream flow predictions and management options for biota affected by peaking-power hydroelectric operations. *Regulated Rivers* **3**: 35–48. [11.3]

Gore JA, Nestler JM, Layzer JB. (1990) Habitat factors in tailwaters with emphasis on peaking hydropower. Technical Report EL-90-2, US Army Corps of Engineers, Waterways Experiment Station, Vicksburg, MS. [11.3]

Gore JA, Crawford DJ, Nestler JM. (1992a) Habitat restoration and enhancement techniques associated with large order, warmwater rivers influenced by peaking hydropower. Technical Report, US Army Engineers, Waterways Experiment Station, Vicksburg, MS (in press).

Gore JA, Layzer JB, Russell IA. (1992b) Non-traditional applications of instream flow techniques for conserving habitat of biota in the Sabie River of southern Africa. In: Boon PJ, Petts GE, Calow P (eds) *River Conservation and Management*, pp. 161–77. John Wiley, New York. [11.2, 11.7]

Gore JA, Niemela S, Resh VH, Statzner B. (1993) Near-substrate hydraulic conditions under artificial floods from peaking hydropower operation: disturbance intensity and duration. *Regulated Rivers* (in press) [11.3]

Grzybkowska M, Hejduk J, Zieliński P. (1990) Seasonal dynamics and production of Chironomidae in a large lowland river upstream and downstream from a new reservoir in Central Poland. *Archiv für Hydrobiologie* **119**: 439–55. [11.3]

Gustard A, Cole G, Marshall D, Bayliss A. (1987) A study of compensation flows in the U.K. Report No. 99, Institute of Hydrology, Oxfordshire, UK. [11.7]

Hanson CH, Li HW. (1978) A research program to examine fish behavior in response to hydraulic flow fields – development of biological design criteria for proposed water diversions. Completion Report, Project C-7679, OWRT, US Department of Interior. [11.6]

Harvey MD, Watson CC. (1986) Fluvial processes and morphological thresholds in incised channel restoration. *Water Resources Bulletin* **22**: 359–68. [11.4]

Heede BH, Rinne JN. (1990) Hydrodynamic and fluvial morphologic processes: implications for fisheries management and research. *North American Journal of Fisheries Management* **10**: 249–68. [11.1, 11.7]

Heger H, Moog O. (1986) Der Einflu von Wasserableitungen auf das Benthos des Landeckbaches in Osttirol (Österreich). *Berichte Naturwissenschaftlich-*

Medizinischen Vereins in Innsbruck **73**: 199–214. [11.3]

Hill LG, Matthews WJ. (1980) Temperature selection by the darters *Etheostoma spectabile* and *Etheostoma radiosum* (Pisces: Percidae). *American Midland Naturalist* **104**: 412–15. [11.2]

Humpesch U. (1978) Preliminary notes on the effect of temperature and light-condition on the time of hatching in some Heptageniidae (Ephemeroptera). *Verhandlungen der Internationalen Vereinigung für theoretische und angewandte Limnologie* **20**: 2605–11. [11.2]

Hynes HBN. (1970) *The Ecology of Running Waters.* University of Toronto Press, Toronto. [11.2, 11.4]

Ingersoll CG, Claussen DJ. (1984) Temperature selection and critical thermal maxima of the fantial darter, *Etheostoma flabellare*, and johnny darter, *E. nigrum*, related to habitat and season. *Environmental Biology of Fishes* **11**: 131–8. [11.2]

Irvine JR. (1985) Effects of successive flow perturbations on stream invertebrates. *Canadian Journal of Fisheries and Aquatic Sciences* **42**: 1922–7. [11.3, 11.6]

Irvine JR, Henriques PR. (1984) A preliminary investigation on effects of fluctuating flows on invertebrates of the Hawea River, a large regulated river in New Zealand. *New Zealand Journal of Marine and Freshwater Research* **18**: 283–90. [11.6]

Jäger P, Kawecka B, Margreiter-Kownacka M. (1985) Zur Methodik der Untersuchungen der Auswirkungen des Wasserentzuges in Restwasserstrecken auf die Benthosbiozönosen (Fallbeispiel: Radurschlbach): *Österreichische Wasserwirtschaft* (Jahrgang 37) 7/8: 190–202. [11.3]

Jenkins RE, Freeman CA. (1972) Longitudinal distribution and habitat of the fishes of Mason Creek, an upper Roanoke River drainage tributary, Virginia. *Virginia Journal of Science* **23**: 194–202. [11.2]

Jones FRH. (1968) *Fish Migration.* St. Martins, New York. [11.6]

King JM. (1981) The distribution of invertebrate communities in a small South African river. *Hydrobiologia* **83**: 43–65. [11.2]

Kovalak WP. (1976) Seasonal and diel changes in the positioning of *Glossosoma nigrior* Banks (Trichoptera: Glossosomatidae) on artificial substrates. *Canadian Journal of Zoology* **54**: 1585–94. [11.6]

Kovalak WP. (1978) Relationships between size of stream insects and current velocity. *Canadian Journal of Zoology* **56**: 178–86. [11.5]

Lake PS, Barmuta LA, Boulton AJ, Campbell IC, St. Clair RM. (1986) Australian streams and Northern Hemisphere stream ecology: comparisons and problems. *Proceedings of the Ecological Society of Australia* **14**: 61–82. [11.2]

Layzer JB, Nehus TJ, Pennington W, Gore JA, Nestler JM. (1989) Seasonal variation in the composition of drift below a peaking hydroelectric project. *Regulated Rivers* **3**: 29–34. [11.3]

Lillehammer A, Saltveit SJ. (eds) (1984) *Regulated Rivers.* Universitetsforlaget AS, Oslo. [11.3]

Lock MA, John PH. (1979) The effect of flow patterns on uptake of phosphorus by river periphyton. *Limnology and Oceanography* **24**: 376–83. [11.6]

Mackay RJ. (1979) Life history patterns of some species of *Hydropsyche* (Trichoptera: Hydropsychidae) in southern Ontario. *Canadian Journal of Zoology* **57**: 963–75. [11.3]

Martin JL, Curtis LT, Nestler JM. (1985) Effects of flow alterations on trout habitat in the Cumberland River below Wolf Creek Dam. Misc. Paper E-86-11. U.S. Army Engineer, Waterways Experiment Station, Vicksburg, MS. [11.7]

Matter WJ, Hudson PL, Saul GE. (1983) Invertebrate drift and particulate organic material transport in the Savannah River below Lake Hartwell during a peak power generation cycle. In: Fontaine TD, Bartell SM (eds) *Dynamics of Lotic Ecosystems*, pp. 357–69. Ann Arbor Press, Michigan. [11.3]

Maude SH, Williams DD. (1983) Behavior of crayfish in water currents: hydrodynamics of eight species with reference to their distribution patterns in southern Ontario. *Canadian Journal of Fisheries and Aquatic Sciences* **40**: 68–77. [11.6]

Morin A, Peters RH. (1988) Effect of microhabitat features, seston quality, and periphyton abundance of over-wintering black fly larvae in southern Québec. *Limnology and Oceanography* **33**: 431–46. [11.3]

Morin A, Back C, Chalifour A, Boisvert J, Peters RH. (1988) Effect of black fly ingestion and assimilation on seston transport in a Quebec lake outlet, *Canadian Journal of Fisheries and Aquatic Sciences* **45**: 705–14. [11.3]

Murphy ML, Hall JD. (1981) Varied effects of clear-cut logging on predators and their habitat in small streams of the Cascade Mountains, Oregon. *Canadian Journal of Fisheries and Aquatic Sciences* **38**: 137–45. [11.3]

Nachtigall W, Bilo D. (1965) Die Strömungsmechanik des *Dytiscus*-Rumpfes. *Zeitschrift fur Vergleichende Physiologie* **50**: 371–401. [11.5]

Naiman RJ, Décamps H, Pastor J, Johnston CA. (1988) The potential importance of boundaries to fluvial ecosystems. *Journal of the North American Benthological Society* **7**: 289–306. [11.2]

Newbury RW. (1984) Hydrologic determinants of aquatic insect habitats. In: Resh VH, Rosenberg DM (eds) *The Ecology of Aquatic Insects*, pp. 323–75. Praeger, New York. [11.1]

Nowell ARM, Jumars PA. (1984) Flow environments of aquatic benthos. *Annual Review of Ecology and Systematics* **15**: 303–28. [11.4]

Oswood MW. (1979) Abundance patterns of filter-feeding caddisflies (Trichoptera: Hydropsychidae) and seston in a Montana (U.S.A.) lake outlet. *Hydrobiologia* **63**: 177–83. [11.3]

Perry SA, Perry WB. (1986) Effects of experimental flow regulation on invertebrate drift and stranding in the Flathead and Kootenai Rivers, Montana, USA. *Hydrobiologia* **134**: 171–82. [11.3]

Perry SA, Sheldon AL. (1986) Effects of exported seston on aquatic insect faunal similarity and species richness in lake outlet streams in Montana, USA. *Hydrobiologia* **137**: 65–77. [11.3]

Peterson CG. (1986) Effects of discharge reduction on diatom colonization below a large hydroelectric dam. *Journal of the North American Benthological Society* **5**: 278–89. [11.6]

Peterson CG. (1987) Influences of flow regime on development and desiccation response of lotic diatom communities. *Ecology* **68**: 946–54. [11.6]

Petts GE. (1984) *Impounded Rivers. Perspectives for Ecological Management*. John Wiley & Sons, Chichester. [11.3]

Pfeifer RF, McDiffett WF. (1975) Some factors affecting primary productivity of stream riffle communities. *Archive für Hydrobiologie* **75**: 306–17. [11.6]

Platania SP. (1991) Fishes of the Rio Chama and upper Rio Grande, New Mexico, with preliminary comments on their longitudinal distribution. *Southwestern Naturalist* **36**: 186–93. [11.2]

Platts WS. (1979) Relationships among stream order, fish populations, and aquatic geomorphology in an Idaho river drainage. *Fisheries* **4**: 5–9. [11.2]

Richardson JS, Mackay RJ. (1991) Lake outlets and the distribution of filter feeders: an assessment of hypotheses. *Oikos* **62**: 370–80. [11.3]

Robinson CT, Minshall GW. (1990) Longitudinal development of macroinvertebrate communities below oligotrophic lake outlets. *Great Basin Naturalist* **50**: 303–11. [11.3]

Robinson CT, Reed LM, Minshall GW. (1992) Influence of flow regime on life history, production, and genetic structure of *Baetis tricaudatus* (Ephemeroptera) and *Hesperoperla pacifica* (Plecoptera). *Journal of the North American Benthological Society* **11**: 278–89. [11.1]

Scarnecchia DL. (1988) The importance of streamlining in influencing fish community structure in channelized and unchannelized reaches of a prairie stream. *Regulated Rivers* **5**: 155–66. [11.4, 11.5]

Sheldon AL. (1968) Species diversity and longitudinal succession in stream fishes. *Ecology* **49**: 193–8. [11.2]

Sheldon AL, Oswood MW. (1977) Blackfly (Diptera: Simuliidae) abundance in a lake outlet: test of a predictive model. *Hydrobiologia* **56**: 113–20. [11.3]

Simons DB. (1979) Effects of stream regulation on channel morphology. In: Ward JV, Stanford JA (eds) *The Ecology of Regulated Streams*, pp. 95–111. Plenum Press, New York. [11.3]

Smith SD, Wellington AB, Nachlinger JL, Fox CA. (1991) Functional responses of riparian vegetation to streamflow diversion in the eastern Sierra Nevada. *Ecological Applications* **1**: 89–97. [11.3]

Stanford JA, Ward JV. (1984) The effects of regulation on the limnology of the Gunnison River: A North American case history. In: Lillehammer A, Saltveit SJ (eds) *Regulated Rivers*, pp. 467–80. Universitetsforlaget AS, Oslo. [11.2]

Statzner B. (1987a) Ökologische Bedeutung der sohlennahen Strömungsgeschwindigkeit für benthische Wirbellose in Flie gewässern. Habil. Thesis, University of Karlsruhe. [11.4, 11.5]

Statzner (1987b) Characteristics of lotic ecosystems and consequences for future research direction. In: Schulze E-D, Zwölfer H (eds) *Potentials and Limitations of Ecosystem Analysis*, pp. 365–90. Ecol. Stud. 61, Springer-Verlag, Berlin. [11.2]

Statzner B. (1988) Growth and Reynolds number of lotic macroinvertebrates: a problem for adaptation of shape to drag. *Oikos* **51**: 84–7. [11.5]

Statzner B, Higler B. (1986) Stream hydraulics as a major determinant of benthic invertebrate zonation patterns. *Freshwater Biology* **16**: 127–39. [11.1, 11.2]

Statzner B, Holm TF. (1982) Morphological adaptations of benthic invertebrates to stream flow – an old question studied by means of a new technique (Laser Doppler Anemometry). *Oecologia* **53**: 290–2. [11.5]

Statzner B, Holm TF. (1989) Morphological adaptation of shape to flow: Microcurrents around lotic macroinvertebrates with known Reynolds numbers at quasi-natural flow conditions. *Oecologia* **78**: 145–57. [11.5]

Statzner B, Gore JA, Resh VH. (1988) Hydraulic stream ecology: observed patterns and potential applications. *Journal of the North American Benthological Society* **7**: 307–60. [11.1, 11.2, 11.4, 11.5, 11.7]

Stromberg JC, Patten DC. (1990) Riparian vegetation instream flow requirements: a case study from a diverted stream in the eastern Sierra Nevada, California, USA. *Environmental Management* **14**: 185–94. [11.3]

Sweeney BW, Schnack JA. (1977) Egg development, growth, and metabolism of *Sigara alternata* (Say) (Hemiptera: Corixidae) in fluctuating thermal environments. *Ecology* **58**: 265–77. [11.2]

Trump CL, Leggett WC. (1980) Optimum swimming speeds in fish: the problem of currents. *Canadian Journal of Fisheries and Aquatic Sciences* **37**: 1086–92. [11.6]

Valett HM, Stanford JA. (1987) Food quality and hydropsychid caddisfly density in a lake outlet stream in Glacier National Park, Montana, USA. *Canadian Journal of Fisheries and Aquatic Sciences* **44**: 77–82. [11.3]

Vannote RL, Minshall GW, Cummins KW, Sedell JR,

Cushing CE. (1980) The river continuum concept. *Canadian Journal of Fisheries and Aquatic Sciences* **37**: 130–7. [11.2]

Wallace JB, Gurtz ME. (1986) Response of *Baetis* mayflies (Ephemeroptera) to catchment logging. *American Midland Naturalist* **115**: 25–41. [11.3]

Ward JV. (1992) *Aquatic Insect Ecology. 1. Biology and Habitat.* John Wiley, New York. [11.2]

Ward JV, Stanford JA. (1979) Ecological factors controlling stream zoobenthos with emphasis on thermal modification of regulated streams. In: Ward JV, Stanford JA (eds) *The Ecology of Regulated Streams*, pp. 35–56. Plenum Press, New York. [11.3]

Ward JV, Stanford JA. (1982) Thermal responses in the evolutionary ecology of aquatic insects. *Annual Review of Entomology* **27**: 97–117. [11.2, 11.3]

Ward JV, Stanford JA. (1983) The serial discontinuity concept of lotic ecosystems. In: Fontaine TD III, Bartell SM (eds) *Dynamics of Lotic Ecosystems*, pp. 29–42. Ann Arbor Press, Ann Arbor, Michigan. [11.2]

Ward JV, Voelz NJ. (1990) Gradient analysis of interstitial meiofauna along a longitudinal stream profile. *Stygologia* **5**: 93–9. [11.2]

Waringer JA. (1989) Resistance of cased caddis larva to accidental entry into the drift: the contribution of active and passive elements. *Freshwater Biology* **21**: 411–20. [11.5]

Webb PW. (1975) Hydrodynamics and energetics of fish populations. *Fisheries Research Board of Canada Bulletin* No. 190. [11.5]

Webster JR, Golladay SW, Benfield EF, D'Angelo DJ, Peters GT. (1990) Effects of forest disturbance on particulate organic matter budgets of small streams. *Journal of the North American Benthological Society* **9**: 120–40. [11.3]

Wesche TA, Rechard PA. (1980) A summary of instream flow methods for fisheries and related research needs. USDA, Forest Service, Eisenhower Consortium Bulletin No. 9. [11.7]

Wetzel RG. (1983) *Limnology*, 2nd edn. Saunders College Publ., Philadelphia. [3.3]

Wiley MJ, Kohler SL. (1980) Positioning changes of mayfly nymphs due to behavioral regulation of oxygen consumption. *Canadian Journal of Zoology* **58**: 618–22. [11.6]

Wilzbach MA, Cummins KW, Hall JD. (1986) Influence of habitat manipulations on interactions between cutthroat trout and invertebrate drift. *Ecology* **67**: 898–911. [11.3]

Wotton RS. (1982) Does the surface film of a lake provide a source of food for animals living in lake outlets? *Limnology and Oceanography* **27**: 959–60. [11.3]

12: Prediction of Biological Responses

P. D. ARMITAGE

12.1 INTRODUCTION

The first aim of most scientific investigations is to observe and describe phenomena. Once the descriptive data have been collected attempts can be made to organize and clarify what has been observed. Then through the use of experimentation, the mechanisms which determine these observations can be investigated. These three steps of study provide the basis for developing models to predict the response of the components of the system under investigation to changed conditions. It is this latter aspect which will be addressed in this chapter.

Increasing demands on river systems for water supply and amenity have put pressure on scientists to obtain and use basic data to develop guides for the optimal management of aquatic resources and maintenance of environmental integrity. These basic data may concern different levels of organization, for example system, community, population, organism, cell or molecule, and relate to such factors as pollution, fisheries, flow regulation (supply, flood control and hydropower) and whole catchment management.

These data obviously encompass the whole gamut of lotic studies and it is beyond the remit of this chapter to cover all aspects. Instead emphasis will be placed on reviewing current methodologies and providing selected examples of their application.

Stimulus and response

In animals a stimulus, such as a variation in light levels or the introduction of a pollutant, will produce a change in part of an organism, usually a sense receptor. The animal reacts to this change through the action of muscles and glands mediated to a greater or lesser extent by the nervous and endocrine systems. The sum of these effects constitutes the behaviour of the individual. In this chapter the term 'response' will be considered to be the overall reaction of an individual or community to a stimulus or range of stimuli.

In most studies in lotic environments it is the community, population and organism levels of organization which are most studied. However, recent studies (Guyomard 1989, and references therein) have started to examine the responses of biota to environmental variables at the genetic level, and this is a developing field of research.

In a natural river system the biotic communities have evolved in response to the prevailing physico-chemical conditions (temperature, light, flow conditions, substratum, oxygen concentration, ionic composition) and biotic interactions such as competition and predation to which a biotic unit can respond. It is important to note that the stimuli are not mutually exclusive and that a biological response is more often than not a result of the various combinations of primary stimuli. Only in extreme cases, for example severe heavy metal pollution or physical disturbance, is there one main factor affecting the response.

The need for predictive models

Biological monitoring in lotic systems provides information on the quality status of the river environment and is an essential feature for the maintenance of good-quality water-courses. It does not, however, provide information on the probable impacts of changes in the environment. This type of information is essential for planning and management in the water industry. It is here that the ability to predict the biological responses of river biota or whole river ecosystems to environmental change has its main use.

Biological responses are predicted from knowledge based on field observation and experimental work. Given sufficient data on the effects of physico-chemical changes on biological responses it would be possible to develop mathematical models. These may be simple or complex, pragmatic or conceptual, or combinations of all of these. At one extreme there are explanatory or system models whose behaviour is considered to more or less duplicate the true behaviour of the modelled population; at the other extreme there are predictive models whose primary purpose is not to explain population behaviour but rather predict it in the future. The classification and description of model types is beyond the scope of this chapter and the reader is referred to Jeffers (1982) for detailed information on this subject.

Aspects of the water industry which affect the river environment include water supply, flood control, effluent purification, waste disposal and recreation. In addition, changes in river catchments such as land drainage, urban development and large-scale changes in agricultural land use have major impacts on the water courses which drain them.

The water industry has a need to predict the likely responses of river biota to all these aspects of water management. There is also a need for information on ecotoxicological aspects in order to assess the effects of toxic compounds on river biota and predict the possible consequences on the environment as a whole.

Fisheries are a major recreational use of rivers and it is important to be able to predict yields and angling success, assess stocks and predict responses of fish to changes in food and habitat availability. This latter aspect has been the subject of much research and one particular approach, the Instream Flow Incremental Methodology (IFIM: Bovee 1982) is described by Stalnaker (1994).

As background to these impacts and aspects of management there is an overwhelming need for careful studies which describe responses of river biota to changes in the physical and chemical environment and to other biota both in natural and disturbed systems. This basic information is the substance of all predictive and conceptual models.

It is important to recognize the central role of predictive models in biomonitoring. The concept of monitoring has no meaning unless the measurements are tested against some model of system performance (Radford 1988). In the simplest case the model may be that no change is expected with time. This is the assumption made by many monitoring techniques. In more complicated situations the results of monitoring depend on comparisons with a standard or conceptual model based on real observations. Figure 12.1 illustrates the relation between monitoring, prediction and the development of prediction into simulation procedures which can be used to forecast responses at system or lower levels of organization.

Hydrology is a crucial aspect of the lotic system and knowledge of the behaviour of hydrological variables has many applications in the prediction of biological responses. Hydrological modelling is a well-developed field of research (Reed 1987; Trapp *et al* 1990). Similarly the concentration and distribution of chemical parameters in lotic systems will have a major influence on water quality (Tipping 1989). Neither chemical nor hydrological modelling are discussed directly in this chapter but their role in predicting biological responses is noted where appropriate.

Prediction is not only needed in monitoring activities. Predictive methodologies can offer the

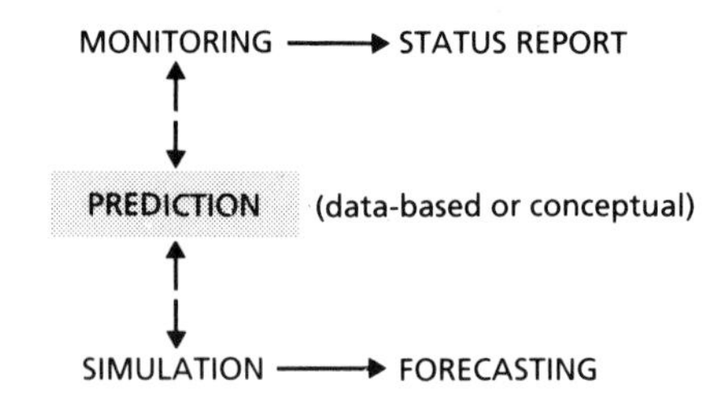

Fig. 12.1 The relationship between prediction, monitoring and simulation.

opportunity for rapid scientific development by placing individual work in context, by providing unambiguous criteria for the evaluation of theories, and by elaborating general, quantitative theories which can then be combined to yield sometimes unexpected results (Peters 1986).

12.2 PREDICTION OF BIOLOGICAL RESPONSES

General biological monitoring

A river is prone to a wide range of environmental disturbances. The river biota will respond to both natural and anthropogenic influences whether these be chemical, physical or combinations of the two. It is the role of biomonitoring to assess the extent and direction of these responses. These assessments can take place in the field, in the laboratory or in both.

Many groups of organisms have been employed to assess both environmental quality and ecotoxicity. Macroinvertebrates appear to have been used most often, for their role in biological assessment, see Metcalfe-Smith (1994). For an exhaustive treatment of biological indicators of pollution and environmental management see Hellawell (1989).

Environmental assessment

In the area of environmental assessment the prime object is to identify change, the assumption generally being that no change is expected. For example, a biotic index based on macroinvertebrates could be used to type or classify a site. When this site is revisited (provided that either the period between samples is within about 2 weeks or in the same season) a similar value of the index might be expected. Divergence from this indicates change. All biotic indices, whatever group they are based on, operate largely in this way. If previous samples are available for comparison it is possible to assess the amount of change. However, such techniques are not helpful when an assessment of a new site is required; neither can they provide a measure of any perturbed site's divergence from the 'norm'.

Workers in the UK (Moss *et al* 1987) have used multivariate techniques to develop a software package, RIVPACS (River In Vertebrate Prediction And Classification System), which can predict the probability of macroinvertebrate species occurrence at new sites from known environmental features. The database for RIVPACS consists of 438 sites on 81 unpolluted river systems covering all major types of geology and topography in Great Britain. The database comprises species lists from all sites for spring, summer and autumn samples, together with associated physical and chemical data for each location.

The various stages in the development of this package have been described elsewhere (Moss *et al* 1987; Furse *et al* 1987; Wright *et al* 1989) and will not be dealt with here. The predicted probabilities of taxon occurrence (at family or species level) can be converted into predicted biotic index values for the site. If required, RIVPACS can be adapted to provide target values for almost any macroinvertebrate-based index in common use. The ratio of observed index value over predicted value provides an index of environmental quality.

This combination of attributes makes RIVPACS a powerful tool for environmental assessment. It has been used to assess the effects of river regulation on faunal communities (Armitage 1989) and to describe the extent of pollution in rivers (Armitage *et al* 1990). A recent study explores its use in assessing physical disturbance, heavy metal pollution and organic pollution (Armitage *et al* 1992) (Fig. 12.2). The figure clearly shows the marked effects of pollution and more importantly compares this with a predicted standard.

RIVPACS can also be used to simulate the response of individual taxa or groups of taxa to single environmental variables or groups of variables (Fig. 12.3). This can provide insight into the critical values of particular variables (Armitage 1989).

Winget (1985) in the USA describes a biotic index which is sensitive to all types of environmental stress. Four variables – gradient, substrate roughness, total alkalinity and sulphate – are used to derive 'tolerance quotients' (TQs) for benthic species. The predicted community TQ is the mean of the TQs for a predicted macroinvertebrate community. Management success in

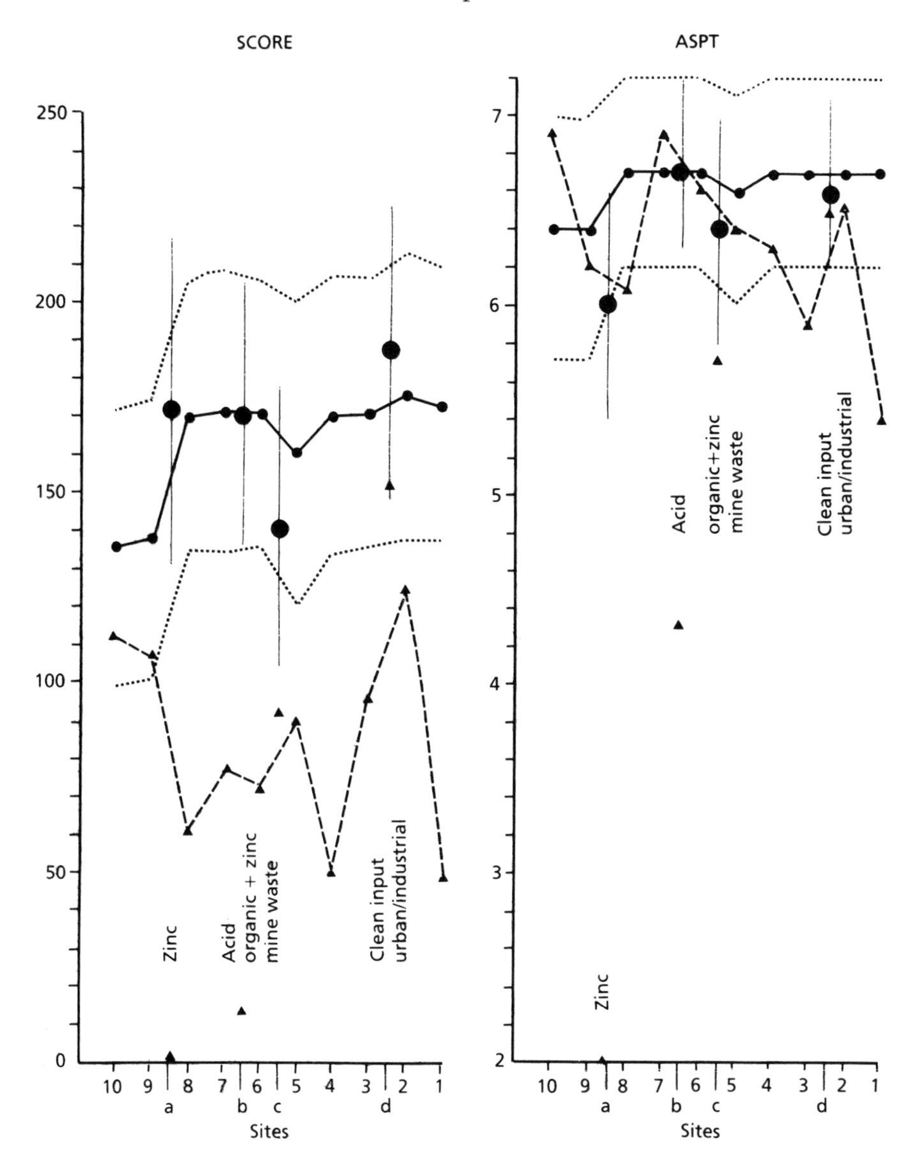

Fig. 12.2 The variation in Biological Monitoring Working Party (BMWP) Score (SCORE) and average score per taxon (ASPT) in the River Nent system. Predicted values in main stream (sites 10–1): small closed circles; in tributaries (sites a–d): large closed circles. 95% confidence limits indicated with dotted line for main stream and with fine vertical lines for tributaries. Observed values: closed triangles. Inputs into the main stream are indicated in the figure (from Armitage *et al* 1992).

reclamation/restoration works is judged by physical habitat and return of actual macroinvertebrate community condition to the predicted potential condition.

Both RIVPACS and the biotic condition scheme employ the same basic philosophy and use macroinvertebrates to assess environmental quality. Of course many other groups have been used for environmental assessments (Hellawell 1989) but few of these have been able to predict

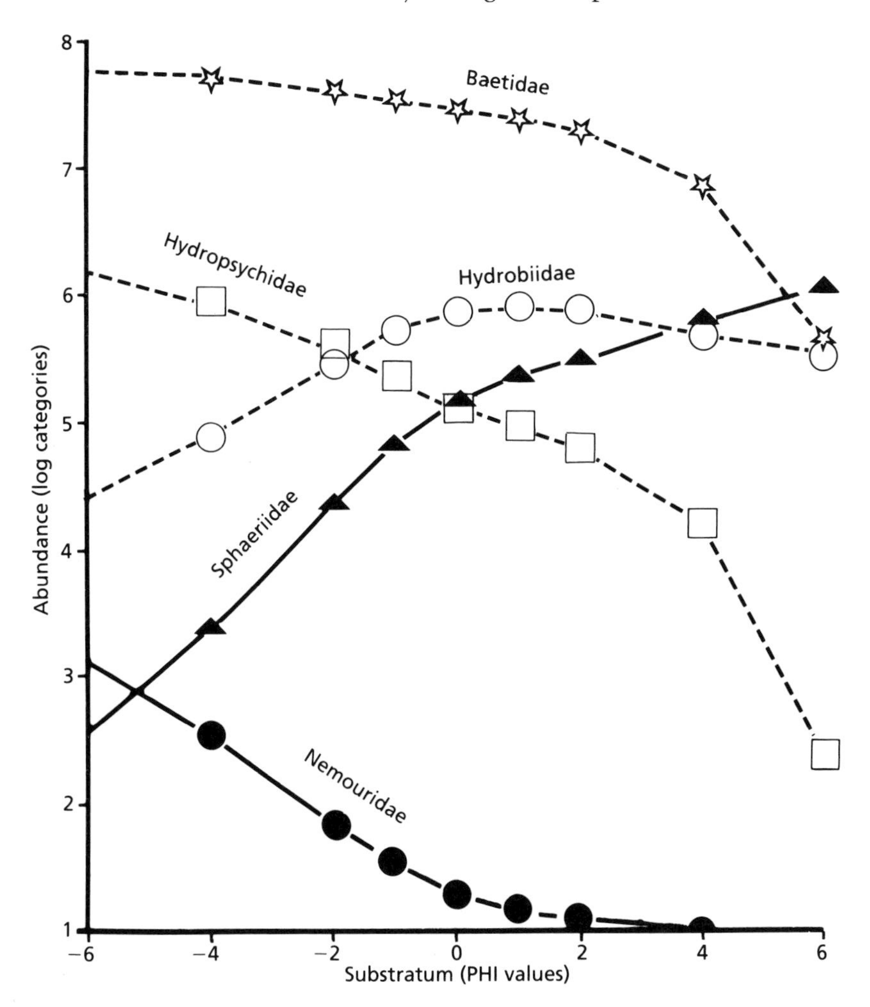

Fig. 12.3 The effect of simulated change in mean substratum particle size on predictions of abundance of five families (from Armitage 1989, by permission of CRC Press).

target values as do the two methods described.

The only general system of assessment which features bacteria, protozoans and algae is the 'saprobien' (Kolkwitz & Marsson 1909) although this system can also use all other river biota. The use of bacteria and protozoans for general assessments (Evison 1979; Sladacek 1981) is rare but they have been widely used to examine the effects of contaminants in field and laboratory systems (Mathews *et al* 1982). Algae alone have been used to assess environmental quality in rivers (Patrick 1954; Descy 1979; Rumeau & Coste 1988). However, none of these systems offers a predictive methodology.

Macrophytes have some advantages as indicators of pollution in that they are stationary, are visible to the naked eye, and manpower demands for macrophyte surveys are low provided experienced operators are used. Harding and Whitton (1981) and Haslam (1982) have developed schemes to assess disturbance to the environment but these do not allow prediction. However, Best

(1990), in reviewing 12 macrophyte models, indicates how some of them, based on oxygen flow, can be used to predict the probable effects of waterway management measures (Westlake 1966), and farming, urbanization, effluent treatment and thermal stress (Rutherford 1975).

The use of fish in routine environmental monitoring has the advantage that the public at large and other interested parties such as anglers can readily identify major disturbance to the river system by the observation of fish kills. Nevertheless fish community surveys require extensive manpower and are difficult to implement in fast-flowing deep rivers. Perhaps as a result of this fish have been used rarely in comprehensive monitoring programmes, although they are used commonly to detect contaminants (Cairns & van der Schalie 1980). There is regular monitoring of fish populations in the UK but not usually so frequently as for invertebrates. The monitoring is for fisheries management purposes rather than pollution assessment except where fish are used for bioaccumulation work. The mobility of fish away from a pollution source detracts from their value as pollution monitors. In the USA, Karr *et al* (1986) have developed an 'index of biotic integrity' based predominantly on fish, which is being adopted for widespread use in the USA (Plafkin *et al* 1989). While not a predictive methodology it does provide a means of disturbance. The index is based on data for species richness and composition, trophic composition, abundance and condition.

Acidification

The recent acidification of freshwaters in parts of Europe and eastern North America has had severe effects on aquatic life (Overrein *et al* 1981; Battarbee & Charles 1986; Schindler 1988). Many studies have been devoted to examining lake sediments since they contain decipherable biological and chemical information about changes in lake and watershed conditions through recent geological time (Warwick 1980; Holmes *et al* 1989). Such retrospective analyses can be used to predict or infer probable biological responses to changing environmental variables.

It is important to be able to detect and simulate change due to acidification and the advantages of model simulations are becoming apparent. Hydrochemical models such as MAGIC and ILWAS focus exclusively on deposition and acidification elements (Reuss *et al* 1986). Marmorek *et al* (1990) present a hydrochemical model that estimates the pre-acidification ('original') lake chemistry and then predicts the eventual chemistry to be expected given a specified level of acid sulphate deposition. Such models are an integral part of those models which predict biological responses. MAGIC has been used in a whole series of studies on the biological responses of stream animals to acidification in Wales (Ormerod *et al* 1988; Ormerod & Tyler 1989; Ormerod *et al* 1990a). The hydrochemical model is coupled with biological models, employing similar basic multivariate methods to those used in the construction of RIVPACS (Moss *et al* 1987), but developed for catchments in mid-Wales (Weatherley & Ormerod 1987). The technique is used to predict the response of invertebrate assemblages to alterations in stream chemistry due to acidification as a result of changing ionic deposition or land-use modifications (Rutt *et al* 1990).

In fish studies, models have been used to determine egg and fry mortalities of Norwegian trout at low pH (Sadler 1983), predict changes in fish distribution under different acid loadings by simulating pH and alkalinity changes in lakes (Small & Sutton 1986), examine the loss of Atlantic salmon from acidic brown waters in Canada (Lacroix 1987), and determine fish response to acidification in lakes (Reckhow *et al* 1987). Minns *et al* (1986) have simulated changing alkalinity and total dissolved solids to assess temporal trends in the fish populations of lakes of the Canadian Shield. Kelso *et al* (1990) have reviewed the effects of atmospheric acidic deposition on fish and the fishery resource of Canada. They investigated integration of cause and effect through model development by examining 11 models which incorporate a biotic damage element. They concluded that of the models available those that attempt to predict biomass, production or yield receive limited support from available evidence, whereas those that incorporate critical thresholds of occurrence and to some extent community structure receive more support.

Ecotoxicology

The prediction of biological responses is an integral part of ecotoxicological tests and bioassays, whose aim is to assess the ecological risks of polluting chemicals. The response of organisms to specific substances together with data on the properties of those substances in field and laboratory can be used to estimate threats to the environment (see Munawar *et al* 1989 for a comprehensive presentation of these techniques).

Maltby and Calow (1989) have reviewed tests used for resolving environmental problems. They distinguished between bioassays concerned with prediction (anticipating environmental impacts) and assessment (monitoring actual impacts). An environmental impact can be considered to be a response of the biotic community as a whole to some stressor. However, over 90% of the bioassays classified as predictive by Maltby and Calow (1989) were single-species tests mainly using death as a criterion of toxicity and were carried out over short periods of time. Chronic and sublethal bioassays represented less than 20% of all single-species tests and involved a variety of response criteria including growth, reproduction, physiology/biochemistry, behaviour and morphology. Multi-species tests have been in the minority and have ranged from small mixed-flask cultures (Larsen *et al* 1986) through *in situ* trough-type mesocosms and large-scale channel macrocosms (Giesy & Allred 1985), to the manipulation of natural systems (Hall *et al* 1980; Ormerod *et al.* 1987). Criteria used in these studies have involved population density, species composition and diversity (structural properties), and primary productivity, P/R ratios, bioaccumulation and nutrient cycling (functional properties).

The relative simplicity of single-species acute tests and their replicability has meant that they have dominated ecotoxicological work. However, in these and also in multi-species tests there are difficulties in extrapolating from laboratory response to the field situation. Interpretation can be complex, sensitivity will differ between taxonomic units, and variability and replicability may cause problems (Cairns 1985; Slooff 1985). Giesy and Allred (1985) note that when copper was added to limnocorrals in an experimental heavy metal study, the population densities of phytoplankton increased. This effect was attributed to depressed grazing by zooplankton and would not have been predicted from single-species tests used to evaluate the effects of metals on phytoplankton.

Frequently, insufficient data are available to determine the long-term chronic effects of toxicants on the survival, growth or reproduction of aquatic biota. As a result the chronic effects must generally be inferred or estimated from observations recorded during short-term acute studies which may be conducted at higher levels of toxicity. These observations are then related to the chronic effects by some statistical relationship. Giesy and Graney (1989) note that if all the information about a chemical and the ecosystem with which it interacts were known it would still be impossible to predict the effect of the chemical on the ecosystem because of the vast amount of information and the ecological uncertainty principle. With these limitations the authors ask what is possible; what use can be made of short-term toxicity tests, and can such tests be used to predict long-term effects? It is clearly impossible to predict the effects of many toxicants from simple single-species bioassays (Cairns 1988). Conversely more complex tests are not necessarily more predictive (Giesy 1985). In view of these limitations it is likely for the foreseeable future that reliance will have to be placed on traditional assays of acute toxicity testing.

Giesy and Graney (1989), in their detailed review, discuss the importance of different levels of organization in ecotoxicological studies. Biochemical measures of toxic effects such as protein content, RNA/DNA ratio, energy content or enzyme concentrations have been used as functional measures of effects which can be used to predict the effects of chronic exposures to pollutants on the survival and fecundity of organisms. However, Woltering (1985) suggests that before biochemical and physiological parameters can be employed to predict chronic effects it is necessary to understand the relationship between short-term biochemical effects and individual or population effects.

The situation is further complicated at community and ecosystem level (Giesy & Allred 1985) by the increase in the number of biotic interactions and habitat diversity. Some interactions which occur within and among individuals make it difficult to predict effects at one level of organization from observations at another level of organization. The example provided by Giesy and Graney (1989) concerns the partitioning behaviour and metabolism of compounds in whole animals which may be such that the dose to a particular biochemical system is much different under *in vivo* field conditions than in laboratory-based studies of tissues, cells or subcellular preparations. Similarly the prediction of survival, growth and reproduction of organisms based on results from exposure to toxicants under controlled laboratory experiments may bear little resemblance to the responses of the organisms in the dynamic conditions of their environment, population and community.

Despite the difficulties, bioassays are used routinely to assess the impact of pollution loads in water courses (Cairns 1980), and the range of studies incorporating bioassays and early-warning systems is enormous throughout the world. A review of the literature is not appropriate to this chapter and the reader is referred to Cairns (1985) and Munawar *et al* (1989) for more information.

General management

The prediction of environmental responses to management activities other than those directly concerned with toxic pollution is an important part of water resource development. Studies are particularly well advanced with respect to the effects of nutrient enrichment in lake ecosystems (see Vollenweider 1975; Aldenberg & Peters 1990). In lotic systems predictions may take the form of inferred effects from previous studies such as the prediction of the environmental impact of a major reservoir development (Edwards 1984) or the prediction of regulation effects on natural biological rhythms in south-central African freshwater fish (Jackson 1989). Alternatively historical analyses may be used to predict possible impacts.

Bravard *et al* (1986) used a historical approach to investigate the impact of civil engineering works on the successions of communities in a fluvial system. By comparing present ecosystem dynamics and ecological succession with past data from abandoned channels it was possible to predict or forecast possible ecological succession resulting from changes in river management. The value of these studies lies in their recognition of the fact that many physical and biological events occur too slowly to be perceived directly within the period available for study. The historical approach provides data on rates of change in response to agricultural development, other land-use impacts, and river use (Petts *et al* 1989), and is particularly valuable in assessing the long-term effects of climate change (Holmes *et al* 1989).

Restoration and biomanipulation

Recently there has been considerable interest in the restoration (Gore 1985a) and biomanipulation (Gulati *et al* 1990) of water bodies. Stream channel modifications can be used to enhance fish and invertebrate habitat (Wesche 1985; Gore 1985b). Case histories can be used to develop guidelines for improving techniques which will be based on past observations of the responses of fish and invertebrates to restoration measures. Gore (1985b) discusses the prediction of rates and trends in colonization processes and examines the establishment of stable communities in relation to habitat improvement. Sheldon (1984) has used data on emigration and immigration of aquatic insects to propose a model describing the colonization of stream ecosystems. The model incorporated elements of competition between species together with information on periphyton production and detritus accumulation, and was able to simulate colonization processes. However, the author concluded that environmental variability precludes the generalized application of this model.

Despite this inability to produce mathematical models with a general application Gore (1985b) notes that basic trends have been reported which will assist in the assessment and monitoring of reclamation works. A useful endpoint is the demonstration of a permanent and stable (macroinvertebrate) community; but it must be stressed that the whole question of environmental stability and hence biological stability frequently depends on the time-scale of observations (Petts 1987). For recent studies on the recovery of lotic communities and ecosystems following disturbance see Yount and Niemi (1990). Gore (1985b) suggests that the best monitoring regime for the determination of stable communities and the success of restoration works involves comparison between the restored communities and communities in unstressed source areas by means of similarity indices (Gore 1982) or biotic indices (Winget 1985).

Biomanipulation is another field of management in which predictive models can have application. Scheffer (1990), working on the restoration of eutrophied lakes, uses simple models to analyse specific ecological relationships. The models contain only four functional 'groups' – turbidity, fish (planktivorous and benthivorous), vegetation and piscivores – and the only driving variable considered is nutrient level. The models indicate that several ecological relationships in freshwater systems potentially give rise to the existence of alternative equilibria over a certain range of nutrient values. In shallow aquatic systems there appear to be two alternative stable states: a clear vegetated state with low fish stocks, and a turbid unvegetated state with high densities of planktivorous and benthivorous fish. The recognition of these stable states increases understanding and helps optimize biomanipulation/restoration measures. Biomanipulation, for example by reducing planktivorous and benthivorous fish to certain levels or introducing predatory fish, can tilt the system over a threshold to a new stable equilibrium in conditions where it is logistically impossible to reduce nutrient levels sufficiently (Hosper & Jagtman 1990).

In forested streams alterations to riparian vegetation may have considerable impact on stream ecology. Duncan *et al* (1989) modified existing techniques for studying energy flow dynamics and have attempted to evaluate these effects by the means of energy flow models which reflect negative and positive production responses with respect to estimated production values. The models predict changes in production of benthic and terrestrial invertebrates and salmonid tissue in response to alteration of riparian cover and composition, and stream nutrients.

Conservation and catchments

Managers of the environment are increasingly required to conserve areas of streams, rivers and wetlands (Boon *et al* 1992). In order to do this effectively it is necessary to know the habitat requirements of the species or communities which are to be protected. Numerous models have been developed which describe species/habitat relationships (Pearsall *et al* 1986). These models do not precisely predict biological response but instead can be used to provide an index of the carrying capacity of a particular habitat in a procedure closely akin to the habitat preference curves used in the Instream Flow Incremental Methodology (see Stalnaker 1994). Pearsall *et al* (1986) describe a range of habitat evaluation procedures all of which use models and they conclude that the definition of carrying capacity and the validation of species habitat models by measuring population densities is problematic. Some reasons put forward are: that stochastic and temporal effects are difficult to account for in experimental design, and site tenacity and social interactions which prevent subdominant animals using suitable habitat may result in 'habitat sinks', or areas of poor habitat and high density. This point is illustrated by a study of habitat selection in muskrats (Messier *et al* 1990) where the relative densities in less suitable habitats increased much faster than those of the other more suitable habitats.

Thermal modelling

Management activities such as clear-cut logging, removal of riparian vegetation, release of thermal effluents, hydropower generation and deep-water releases from impoundments can alter the natural thermal regime in rivers. The effects of increased temperature on aquatic dipterans has been studied by Rempel and Carter (1987) in experimental channels. The population dynamics of aquatic insects were simulated using a stage-projection matrix model. The results indicated that changes in development rate together with changes in fecundity must be considered when predicting the response of aquatic populations to altered thermal regimes.

Water management modelling

Predictive models have a role in general water quality management and the planning of water resource development (Cairns 1990). The emphasis is frequently not on predicting biological responses but rather on forecasting of river flows (Reed 1987), sediment transport (MacMurray & Jaeggi 1990), water quality impact (Crockett *et al* 1989), or pollution transport (Trapp *et al* 1990) Part or all of these may be incorporated into catchment models (Virtanen 1989; Bodo & Unny 1990) to aid and guide management decisions. A catchment model obviously increases in value if specific biological responses can be predicted and the inclusion of biological data in the more general models is the next step forward. The development of geographic information or expert systems is the logical evolution of model integration.

Forecasting impacts of pollution

Frequently it is necessary to be able to forecast the effects of changing environmental variables on river biota. The establishment of guidelines for limiting nutrients in running water is one such need. Welch *et al* (1989) have attempted this by modelling predictions of periphyton biomass as a function of soluble reactive phosphate (SRP). They use the SRP concentration that would produce a threshold nuisance biomass (150–200 mg chlorophyll *a* m^{-2}) to propose an approach to control the length of stream in which biomass exceeds nuisance level. Painter and Jackson (1989) have developed a mathematical model which simulates *Cladophora* internal phosphorus levels. The model was used to determine the necessary reduction in SRP to decrease production of *Cladophora* at a specified site. Dauta (1986) has produced a model which simulates the dynamics of a phytoplankton population in the lower course of the River Lot in France in relation to temperature, light and nutrients. The model generated phases of development and decline for blue-green algal blooms that were identical to those in the natural environment. The model, which is based on laboratory estimates of specific growth rates of algae (diatoms, Cyanophyceae, Chlorophyceae), provides a description of development conditions and succession for algae.

Reynolds and Glaister (in press) modified a model of phytoplankton growth based on lake studies and used it to reconstruct the growth of algae moving downstream according to criteria of water temperature, day length and the time of travel. They then applied the technique to investigate the problems of hydraulic retention in river channels.

Svadlenkova *et al* (1989) mathematically modelled the kinetics of radiocaesium uptake and release by the filamentous alga *Cladophora glomerata* in order to trace radioactivity in running waters. Leps *et al* (1990) have used a transition matrix model to predict changes in ephemeropteran communities, arising mainly from increased eutrophication. These models are the simplest method for predicting further development of ecological systems and have had most application in the study of plant succession (Hulst 1979; Usher 1981). The predictions are basically linear extrapolations of observed present trends.

Habitat and hydraulics

Gore (1989) presents a range of models for describing benthic macroinvertebrate habitat suitability. The basis of some of these models is the 'habitat suitability curve', which represents the preference of the species under test for a range of variables. These data are used in conjunction with predictions of habitat change in relation to flow in order to predict the impact of altered discharge regimes on river biota. This in essence is the Instream Flow Incremental Methodology (see Stalnaker 1994). The situation is complicated by the need to define, for some organisms, habitats at different life history stages. Gore and Bryant (1990), for example, note that the crayfish *Orconectes neglectus* has some size-related flow preferences as well as dimorphic velocity preferences during the period when females migrate to riffles carrying their eggs. In addition there are few studies which quantify biotic responses to extreme discharge events. These may take the form of concentration of organisms in refuges or dead zones which may bear little resemblance to the preferred habitat estimated at 'normal' flows.

Shifts in distribution in relation to flow-related variables such as velocity depth and substratum have been the subject of much study. Much of this concerns fish habitat, particularly the Instream Flow Incremental Methodology (see Stalnaker 1994) but there has been increased interest in the influence of hydraulic factors in determining the behaviour and hence the distribution of river biota, especially invertebrates. For a detailed presentation of these approaches see Statzner *et al* (1988).

Pest control and parasite behaviour

Control programmes for pest species have required predictive techniques to facilitate their implementation. Precise and accurate estimates of secondary production are important for the rational management of aquatic resources and the development and testing of ecological theory. Morin *et al* (1988b) described the growth rates of three blackfly species in relation to individual mass, water temperature and food quality. Allometric models of growth rates were used to estimate production. Benke and Parsons (1990) have also modelled blackfly production dynamics. Two predictive models were used together with intensive field sampling to estimate production of blackflies. One model predicts daily growth rate from temperature and hydrograph pattern, the other predicts habitat abundance (in the form of submerged wood) from river height.

Such predictions can be essential for the efficient application of control measures. In addition it is useful to have information on the ingestion rates of blackfly larvae in order to optimize application of particulate larvicides (Morin *et al* 1988a). Insecticide treatments can be further improved if predictions of blackfly susceptibility to particulate control agents are linked to a one-dimensional transport equation (Chalifour *et al* 1990). In this way it is possible to determine an optimal strategy of treatment which minimizes the quantity of insecticide applied and the logistic movement between injection sites along the river.

Another area in which modelling has contributed to 'control' is in parasitological work, and its relevance to rivers is seen in studies of schistosomiasis. A comprehensive review of helminth infections of humans (Anderson & May 1985) addressed the role of mathematical models in investigations of the population and transmission dynamics of human helminths. These authors concluded that such models can be valuable tools in the design of policies for parasite control. The usefulness of these models depends on a thorough and detailed understanding of host/parasite population dynamics and interrelations. Woolhouse and Chandiwana (1990) have developed a model which describes the population dynamics of the freshwater snail *Bulinus globosus*, which is an intermediate host of *Schistosoma haematobium*. The responses of the snail population to low and high flows are modelled and used to simulate changes in abundance over time-scales of 10 or more years.

Janovy *et al* (1990) constructed a BASIC computer simulation model to mimic the dynamic behaviour of parasite species over a range of abiotic conditions. The model data were compared to a field data set consisting of 33 samples of the fish *Fundulus zebrinus* and its parasites taken during a 7-year period from a single location on a fluctuating river. Over the long term field data did not depart significantly from that predicted by the model.

Fisheries

The close association of the human race with fish as a food source means that studies of fish population dynamics have had important applications. In order to ensure a continuing supply of this food resource it has become necessary to develop models which assess and predict the impact of fishing and fishing technologies on the stocks of fish. The historical development of fish population dynamics and models are reviewed in Gulland (1977), and the practical application of some standard methods and models for estimating production in freshwaters is described in Ricker (1958) and Bagenal (1978).

Fish stock and yield

The mathematical analysis of fish stock dynamics has been the subject of an American Fisheries Society Symposium (Edwards & Megrey 1989).

The proceedings include numerous examples of predictive models. Most deal with marine oceanic fisheries but have some application in freshwaters, and the reader is referred to this publication for source references.

A common objective of fishery managers is the prediction of the future yield of a fish population in response to harvest effort (Gros & Prozet 1988; Vøllestad & Jonsson 1988). Stock production models (models of population growth) (Pella & Tomlinson 1969; Schnute 1977) are commonly used to estimate yield based on data on growth, mortality and reproduction of the population. Information on age structure is, however, notably lacking from these models and this has been redressed to a certain extent by the use of Leslie-matrix or Beverton–Holt (Beverton & Holt 1957) models. These procedures in turn, while employing a basic age structure, make use of the approximation that ages of animals occur in discrete groups whereas an animal can have an age represented by any positive real number. These then are the only two classes of models which are widely used for fisheries management but neither take into account the dependence of mortality, growth and reproduction on food consumption (Bledsoe & Megrey 1989).

On a broader scale it is necessary to assess the effect of general environmental impacts on fish populations and fisheries. Hilden and Kettunen (1985) have reviewed a range of models (basic, age-structured, models with a spatial resolution, multi-species, stochastic and ecosystem models) which attempt to do this. Their analysis suggests that an impact model should preserve the age structure of the population and allow studies of the spatial distribution of the impact and the fish populations. They noted that ecosystem models have been used for simulation tests which may provide information on the fish response to stress. However, a problem arises with increase in model complexity because the model itself becomes more and more difficult to analyse. Hilden and Kettunen concluded that it is almost impossible to tell which predictions simply reflect the behaviour of the model and its computer implementation and which accurately illustrate changes in natural populations and actual fisheries. They suggested that models incorporating a stochastic element are likely to be of increased value in predicting impacts because they would take into account the fact that many environmental impacts are stochastic in nature. Environmental variability tends to increase fluctuations in exploited fish populations (Horwood & Shepherd 1981).

Bioenergetics

Preall and Ringler (1989) noted that despite the fact that energy budgets have been developed for many freshwater fish species (Solomon & Brafield 1972; Niimi & Beamish 1974) little is known of the ways in which growth is governed under natural conditions where food supplies, temperature regimes and numbers of competing individuals often fluctuate unpredictably. Preall and Ringler (1989) draw on the extensive work of Elliott (1975a,b,c,d, 1976a,b,c) on feeding and growth of brown trout. They utilize his bioenergetic equations in a computer model (TROUT) which predicts the maximum growth rate of streamdwelling brown trout and compare the predicted values with actual field estimates. TROUT can be used to study interactions of brown trout with stream invertebrates or other prey and assess the impact of foraging by brown trout on invertebrate production.

Temperature is an important factor controlling growth and is an integral variable in bioenergetic studies of adult fish (Wildhaber & Crowder 1990). Temperature also influences the duration of egg development in freshwater fish (Elliott 1981) and Elliott *et al* (1987) have compared eight mathematical models for the relationship between water temperature and hatching time of salmonid eggs. Saltveit and Brabrand (1987) used a laboratory approach to predict the effects of possible temperature increases, resulting from river regulation, on the eggs of whitefish.

Closer scrutiny is now being given to the importance of activity in bioenergetics models. Boisclair and Leggett (1989) used the Kitchell *et al* (1977) and Kerr (1982) models and field data on *Perca flavescens* (yellow perch) to estimate energy allocated to activity. They concluded that this (activity) component of the bioenergetics budget of fishes can help to explain inter-population differences in perch growth and, by extension, the variance in growth of other actively foraging species.

Fish habitat conditions

Where fish habitat is likely to be changed through flow regulation or other disturbance it is useful to be able to predict the response of the fish population or individual to that change. The techniques range from those involved in the Instream Flow Incremental Methodology (Stalnaker 1994) through simpler habitat models such as HABSCORE, a computer model which has been developed from work started by Milner *et al* (1985). The model, currently in use by the Welsh Region of the National Rivers Authority (Milner, personal communication) can predict fish densities (salmonids) from selected habitat variables.

There have also been studies which have attempted to predict: the impact of thermal effluent on a fish population (Shuter *et al* 1985); the effect of climatic warming on the loss of suitable habitat of brook trout (*Salvelinus fontinalis*) (Meisner 1990); the effects of draining releases from reservoirs on brown trout populations in the river below the dam (Garric *et al* 1990); the response of fish to macrophyte control in a lake (Bettoli *et al* 1990); and the growth of loach in cadmium-loaded environments (Douben 1990).

Feeding behaviour

Basic studies of fish behaviour are not necessarily directly linked with environmental impacts, as they are, for example, in pollution early-warning systems (Cairns & Schalie 1980). However, the work of Hughes and Dill (1990) on position choice of grayling (*Thymallus arcticus*) in mountain streams shows how such work can have relevance to environmental change. The position chosen by the drift-feeding fish is dependent on biotic factors, water depth and velocity. The demonstrated ability to predict position could be used to determine the relationship between flow regulation or abstraction and habitat loss. Grant and Kramer (1990) examined the role of territory size as a predictor of the upper limit to population density of juvenile salmonids in streams. Their results indicated that the territory size model has practical value for predicting maximum densities in shallow habitats. Territory size and position choice both relate to food consumption, distribution and availability. Several predictive feeding models have been used to evaluate size-dependent feeding by lentic fish (Eggers 1977; Wright & O'Brien 1984). In lotic environments there has been less effort (Dunbrack & Dill 1984). Newman (1987) has compared encounter model predictions, i.e. models based on the reactive distance of fish to prey of different sizes, with observed size selectivity by stream-dwelling trout. Such models provide insight into mechanisms of prey selection and the predictions may be useful in estimating the effects of fish predation on benthos production and population dynamics. Such studies have less to do with monitoring and more to do with the development of ecological theory.

12.3 ECOLOGICAL THEORY AND MODELLING

The main emphasis of this chapter so far has been on the application of predictive models in water resource management. However, some investigations which predict the response of an individual or population to specific variables have little direct application to management objectives. They can be thought of as providing the raw data for future predictive models and stimulating further thought and the development of ecological theories. It is beyond the scope of this chapter to present an exhaustive account of such studies and here only some recent examples are presented.

McCauley *et al* (1990) studied the physiological ecology of *Daphnia*. They found that existing models did not describe some essential features of individual growth, particularly when food availability was low. This led the investigators to examine how *Daphnia* stores energy and uses reserves. Their findings formed the basis of a new model for the growth of *Daphnia pulex*. Gurney *et al* (1990) then further developed and tested this model. Their success in matching their predictions to a target data set provides strong evidence that measurements of short-term physiological rates can be used successfully as predictors of long-term growth and fecundity. Hallam *et al* (1990), also working with *Daphnia*, developed an energetics model specifically for females. The model mimics the life history of an individual as it progresses from egg through juvenile to adult instars.

Pockl and Timischl (1990) studied the quantitative relationship between water temperature and brood development time of two species of *Gammarus* (Amphipoda), *G. fossarum*, a headwater species, and *G. roeseli*, which is usually found farther downstream. They suggested that models which can be linearized are more useful than non-linear models, and recommended the use of the 'mixed power-exponential' model for comparing intra- and interspecific variations in the relationship between brood development time and water temperature.

Kokkinn (1990) examined the hypothesis that the rate of embryonic development in *Tanytarsus barbitarsis* (Chironomidae: Diptera) is a predictor of overall development rate. The results showed that the power function that closely described embryonic development could not be fitted to the generation time data, which suggests that laboratory egg-hatching experiments could not account for the increased field development time caused by non-optimal environmental conditions.

Interest in the dynamics of ecological processes in space as well as in time (Allen 1975; Taylor 1990) has stimulated the study of competition in 'patchy' environments (Yodzis 1986; Shorrocks & Rosewell 1987). Many of these studies use models to simulate this competition and the results assist in the formulation of new theories of population dynamics. These may then be adapted for application in stream community ecology (Townsend 1989) and act as a focus for new lines of research (Pringle *et al* 1988). The inevitable disagreements in the utility of new concepts (see Frid & Townsend 1989; Downes 1990) only serve to stimulate the scientific community to more detailed examination of the concepts and reasons for disagreement.

Disturbance and stability

The effect of stress on ecosystems has been the subject of much basic scientific research (see Lugo 1978 for a review) and it is a field of considerable relevance to aquatic ecology (Rapport *et al* 1985; Resh *et al* 1988). The search for generality, which is an inherent characteristic of the scientific method and a feature which facilitates the prediction of biological responses, is made harder by the environmental and economic constraints which force researchers to examine only limited data sets and from these make inferences to larger systems (Perry *et al* 1987). Lately, however, the concern over acidification has prompted the implementation of whole ecosystem experiments involving experimental acidification (Hall *et al* 1980) and liming (Elser *et al* 1986; Ormerod *et al* 1990b). Perry *et al* (1987) have examined the effects of stress on a stream channel ecosystem and in a lake. Further disturbance studies are referred to in Perry *et al* (1987), who concluded that generalities in the ways that community and ecosystem structure respond to disturbance do follow hypothesized trends, including increased nutrient export and decreases in efficiency in system energetics, nutrient standing stock, diversity, evenness and total numbers of organisms. The consistency of response among a variety of ecosystems augurs well for the future development of predictive models, but a note of caution (Hilden & Kettunen 1985) must be added with respect to the difficulties of interpretation of complex ecosystem models.

The relationships between environmental predictability (with respect to lotic communities) and disturbance have been studied without particular reference to predictive models by Rader and Ward (1989) and Poff and Ward (1990). Weatherley and Ormerod (1990), however, address specifically the implications of the constancy (persistence) of invertebrate assemblages in softwater streams to the prediction and detection of environmental change. Disturbance is the dominant organizing factor in stream ecology (Resh *et al* 1988). Its relationship to community stability will have a powerful influence on the design and interpretation of predictive models.

On a different scale Wilson and Botkin (1990) examined the effect of ecosystem complexity on stability using models of simple plankton microcosms and addressed questions relating to the reductionist and holistic approaches to research.

They concluded from their studies that the community and ecosystem-level properties of the models were emergent (i.e. unpredictable solely from observations of the components of the system) and therefore could not in general be predicted only from a knowledge of individual species responses to different environmental conditions.

12.4 FUTURE DIRECTIONS

The future of predictive limnology (in lakes) has been discussed by Peters (1986). Six approaches are identified as being central to improvement.

1 The selection of variables which are easier to measure, less uncertain and better in prediction.
2 The use of multiple regression to focus on the effects of more than one independent variable.
3 The definition of the range over which the model is effective.
4 The replacement of simple arithmetic and logarithmic models with more sophisticated models as our mathematical and statistical knowledge increases.
5 The application of predictive methodologies to a wide range of organisms and habitats.
6 The culling of competing theories.

In addition Peters (1986) notes an increasing interest in heterogeneity at finer temporal and spatial scales. The recognition that dramatic differences in biotic responses can occur in successive years or within a single season in single lakes suggests an increasing ability to think probabilistically and the author anticipates a growth in models and analytical tools to express and use the observed levels of variation.

Peters' suggestions generally still hold true, and although there have been improvements there is a plethora of predictive models available for a variety of applications; there is probably a need for Peters' (1986) suggestion that extensive reviews of the current field should be promoted by leading journals to facilitate the evaluation of differences among similar theories.

In addition to these comments I would add a further six points for future development.

1 Despite the unglamorous nature of the work it is vital that large good-quality databases are collected. The acquisition of the database for RIVPACS depended on the infrastructure of the water industry in Great Britain, which meant that data could be collected in a standardized form throughout the country. Similar databases provide a firm foundation for model development.
2 There should be greater emphasis on linking disciplines as well exemplified by the Instream Flow Incremental Methodology (IFIM) work (Stalnaker 1994) and the work of Statzner *et al* (1988) where hydrology is linked with animal distribution. The IFIM is strictly a management tool but nevertheless provides a new way of looking at the lotic environment and the habitat requirements of all the river biota and encourages dialogue between engineer and scientist. Divisions between basic research and management are being broken down by the need for applied research which borrows from basic research. In turn management requirements encourage new directions in both applied and basic research.
3 It is essential that managers and central government do not render research workers intellectually bankrupt by only supporting applied work. It is vital that work which aims to elucidate mechanisms of response is also funded to build a bank of knowledge on which future researchers can draw.
4 In order to predict biological responses usefully it is important to know the natural variation in response over long and short periods of time in different types of river habitat.
5 Emphasis should be placed on the development of models which predict the biological consequences of environmental disturbance. The ultimate aim of predictive models is the ability to alter variables and or groups of variables which may be affected by disturbance and estimate their consequences. This would be a pre-disturbance step and would determine the implementation and operation of any planned perturbation.
6 Integration of a wide range of models (hydrological, hydrochemical, biological) into geographic or knowledge-based systems. These systems represent an amalgamation of data from many sources which may include information on land use, water quality, and urbanization effects, and will incorporate predictive models. Their aim is to facilitate water management by making these diverse databases and models accessible and providing links between them which can be

used in decision-making. However, some caution is required: as Hedgpeth (1977) points out, bureaucracies are notoriously resistant to change and once a concept or principle is accepted by bureaucrats it may persist and affect policy long after it has been passed by in the 'real world'. This puts the onus on scientists to either get it right first time (clearly unlikely) or maintain closer contacts with management in order to continually revise their predictive methods in the light of increased knowledge.

ACKNOWLEDGEMENTS

This work has been supported financially by the Natural Environment Research Council. I am grateful to the Council and to my colleagues at the Institute of Freshwater Ecology and Dr Jim Gore for providing information.

REFERENCES

Aldenberg T, Peters JS. (1990) On relating empirical water quality diagrams and plankton dynamical models: the SAMPLE methodology applied to a drinking water storage reservoir. *Archiv für Hydrobiologie Beiheft. Ergebnisse der Limnologie* **33**: 893–911. [12.2]

Allen JC. (1975) Mathematical models of species interactions in time and space. *American Naturalist* **109**: 319–42. [12.3]

Anderson RM, May RM. (1985) Helminth infections of humans: mathematical models, population dynamics, and control. *Advances in Parasitology* **24**: 1–101. [12.2]

Armitage PD. (1989) The application of a classification and prediction technique based on macroinvertebrates to assess the effects of river regulation. In: Gore JA, Petts GE (eds) *Alternatives in Regulated River Management*, pp. 267–93. CRC Press, Boca Raton, Florida. [12.2]

Armitage PD, Pardo I, Furse MT, Wright JF. (1990) Assessment and prediction of biological quality. A demonstration of a British macroinvertebrate-based method in two Spanish rivers. *Limnetica* **6**: 147–56. [12.2]

Armitage PD, Furse MT, Wright JF. (1992) Anexo: Calidad medioambiental y valoracion Biologica en los rios britanicos. Perspectivas de pasado y futuro. In: Departamento de Economia Planificacion y Medio Ambiente del Gobierno Vasco (ed) *Caracterizacion Hidrobiologica de la red fluvial de Alava y Gipuzcoa*, pp. 477–511. Itxarapena S.A., Zarantz. [12.2]

Bagenal T. (ed) (1978) *IBP Handbook No. 3. Methods for Assessment of Fish Production in Fresh Waters.* Blackwell Scientific Publications, Oxford. [12.2]

Battarbee RW, Charles DF. (1986) Diatom based pH reconstruction studies of acid lakes in Europe and N. America: a synthesis. *Water, Air and Soil Pollution* **30**: 347–54. [12.2]

Benke AC, Parsons KA. (1990) Modelling black fly production dynamics in blackwater streams. *Freshwater Biology* **24**: 167–80. [12.2]

Best EPH. (1990) Models on metabolism of aquatic weeds and their application potential. In: Pieterse AH, Murphy KJ (eds) *Aquatic Weeds. The Ecology and Management of Nuisance Aquatic Vegetation*, pp. 254–73. Oxford University Press, Oxford. [12.2]

Bettoli PW, Springer T, Noble RL. (1990) A deterministic model of the response of threadfin shade to aquatic macrophyte control. *Journal of Freshwater Ecology* **5**: 445–54. [12.2]

Beverton RJH, Holt SJ. (1957) On the dynamics of exploited fish populations. *UK Ministry of Agriculture Fisheries Investigations (London). Series 2*, p. 19. HMSO, London. [12.2]

Bledsoe LJ, Megrey BA. (1989) Chaos and pseudoperiodicity in the dynamics of a bioenergetic food web model. *American Fisheries Society Symposium* **6**: 121–37. [12.2]

Bodo BA, Unny TE. (1990) Linear stochastic conceptual response models for small catchments. *Revue de la Science Eau* **3**(2): 151–82. [12.2]

Boisclair D, Leggett WC. (1989) The importance of activity in bioenergetics models applied to actively foraging fishes. *Canadian Journal of Fisheries and Aquatic Sciences* **46**: 1859–67. [12.2]

Boon PJ, Calow P, Petts GE. (eds) (1992) *River Conservation and Management.* John Wiley, Chichester. [12.2]

Bovee KD. (1982) *A guide to Stream Habitat Analysis using the Instream Flow Incremental Methodology.* Instream Flow Information Paper 12, Report FWS/OBS-82/86, U.S. Fish Wildlife Service, Fort Collins, Colorado. [12.1]

Bravard J-P, Amoros C, Pautou G. (1986) Impact of civil engineering works on the successions of communities in a fluvial system. *Oikos* **47**: 92–111. [12.2]

Cairns J Jr. (1980) Estimating hazard. *Bioscience* **30**: 101–7. [12.2]

Cairns J Jr. (ed) (1985) *Multispecies Toxicity Testing.* Pergamon Press, New York. [12.2]

Cairns J Jr. (1988) Should regulatory criteria and standards be based on multispecies evidence? *Environmental Professional* **10**: 157–65. [12.2]

Cairns J Jr. (1990) Lack of theoretical basis for predicting rate and pathways of recovery. *Environmental Management* **14**: 517–26. [12.2]

Cairns J Jr, van der Schalie WH. (1980) Biological monitoring. Part I – Early warning systems. *Water Research* **14**: 1179–96. [12.2]

Chalifour A, Boisvert J, Back C. (1990) Optimization of insecticide treatments in rivers: an application of graph theory for planning a blackfly larvae control program. *Canadian Journal of Fisheries and Aquatic Sciences* **47**: 2049–56. [12.2]

Crockett CP, Crabtree RW, Markland HR. (1989) SPRAT, a simple river quality impact model for intermittent discharges. *Water Science and Technology* **21**: 1793–6. [12.2]

Dauta A. (1986) Modélisation du développement du phytoplankton dans une rivière canalisée eutrophe le Lot (France). *Annales de Limnologie* **22**: 119–32. [12.2]

Descy JP. (1979) A new approach to water quality estimation using diatoms. *Nova Hedwigia* **64**: 305–23. [12.2]

Douben PET. (1990) A mathematical model for cadmium in the stone loach, *Noemcheilus barbatulus* L. from the river Ecclesbourne, Derbyshire, England, UK. *Ecotoxicology and Environmental Safety* **19**: 160–83. [12.2]

Downes BJ. (1990) Patch dynamics and mobility of fauna in streams and other habitats. *Oikos* **59**: 411–13. [12.3]

Dunbrack RL, Dill LM. (1984) Three-dimensional prey reaction field of the juvenile coho salmon (*Oncorhynchus kisutch*). *Canadian Journal of Fisheries and Aquatic Sciences* **41**: 1176–82. [12.2]

Duncan WFA, Brusven MA, Bjornn TC. (1989) Energy flow response models for evaluation of altered riparian vegetation in three southeast Alaskan streams. *Water Research* **23**: 965–74. [12.2]

Edwards EF, Megrey BA. (1989) Mathematical analysis of fish stock dynamics. *American Fisheries Society Symposium* No. 6, pp. 1–214. [12.2]

Edwards RW. (1984) Predicting the environmental impact of a major reservoir development. In: Roberts RD, Roberts TM (eds) *Planning and Ecology*, pp. 55–79. Chapman & Hall, London. [12.2]

Eggers DM. (1977) The nature of food selection by planktivorous fish. *Ecology* **58**: 46–59. [12.2]

Elliott JM. (1975a) Weight of food and time required to satiate brown trout, *Salmo trutta* L. *Freshwater Biology* **5**: 51–64. [12.2]

Elliott JM. (1975b) Number of meals in a day, maximum weight of food consumed in a day and maximum rate of feeding for brown trout, *Salmo trutta* L. *Freshwater Biology* **5**: 287–308. [12.2]

Elliott JM. (1975c) The growth rate of brown trout (*Salmo trutta* L.) fed on maximum rations. *Journal of Animal Ecology* **44**: 805–21. [12.2]

Elliott JM. (1975d) The growth rate of brown trout (*Salmo trutta* L.) fed on reduced rations. *Journal of Animal Ecology* **44**: 823–42. [12.2]

Elliott JM. (1976a) Body composition of brown trout (*Salmo trutta* L.) in relation to temperature and ration size. *Journal of Animal Ecology* **45**: 273–89. [12.2]

Elliott JM. (1976b) Energy losses in the waste products of brown trout (*Salmo trutta* L.). *Journal of Animal Ecology* **45**: 561–80. [12.2]

Elliott JM. (1976c) The energetics of feeding metabolism and growth of brown trout (*Salmo trutta* L.) in relation to body weight, water temperature and ration size. *Journal of Animal Ecology* **45**: 923–48. [12.2]

Elliott JM. (1981) Some aspects of thermal stress on freshwater teleosts. In: Pickering AD (ed) *Stress and Fish*, pp. 209–45. Academic Press, London. [12.2]

Elliott JM, Humpesch UH, Hurley MA. (1987) A comparative study of eight mathematical models for the relationship between water temperature and the hatching time of eggs of freshwater fish. *Archiv für Hydrobiologie* **109**: 257–77. [12.2]

Elser MM, Elser JG, Carpenter SR. (1986) Peter and Paul lakes. A liming experiment revisited. *American Midland Naturalist* **116**: 282–95. [12.3]

Evison LM. (1979) Microbial parameters of raw water quality. In: *Biological Indicators of Water Quality. Proceedings of a Symposium at the University of Newcastle-upon-Tyne, UK, September 1978*. Wiley, Chichester. [12.2]

Frid CLJ, Townsend CR. (1989) An appraisal of the patch dynamics concept in stream and marine benthic communities whose members are highly mobile. *Oikos* **56**: 137–41. [12.3]

Furse MT, Moss D, Wright JF, Armitage PD. (1987) Freshwater site assessment using multi-variate techniques. In: Luff ML (ed) *The Use of Invertebrates in Site Assessment for Conservation. Proceedings of a Meeting Held at the University of Newcastle upon Tyne, 7 January 1987*. Agricultural Environment Research Group, University of Newcastle-upon-Tyne. [12.2]

Garric J, Migeon B, Vindimian E. (1990) Lethal effects of draining on brown trout. A predictive model based on field and laboratory studies. *Water Research* **24**(1): 59–65. [12.2]

Giesy JP. (1985) Multispecies tests: Research needs to assess the effects of chemicals on aquatic life. In: Bahner RC, Hansen DJ (eds) *Aquatic Toxicology and Hazard Assessment: Eighth Symposium, ASTM STP 891*, pp. 67–77. American Society for Testing and Materials, Philadelphia, PA. [12.2]

Giesy JP, Allred PM. (1985) Replicability of aquatic multispecies test systems. In: Cairns J (ed) *Multispecies Toxicity Testing*, pp. 187–247. Pergamon Press, New York. [12.2]

Giesy JP, Graney RL. (1989) Recent developments in and intercomparisons of acute and chronic bioassays and bioindicators. *Hydrobiologia* **188/189**: 21–60. [12.2]

Gore JA. (1982) Benthic invertebrate colonization: source distance effects on community composition. *Hydrobiologia* **94**: 183–93. [12.2]

Gore JA. (ed) (1985a) *The Restoration of Rivers and Streams – Theories and Experience.* Butterworth Publishers, Boston. [12.2]

Gore JA. (1985b) Mechanisms of colonization and habitat enhancement for benthic macroinvertebrates in restored river channels. In: Gore JA (ed) *The Restoration of Rivers and Streams – Theories and Experience*, pp. 81–102. Butterworth Publishers, Boston. [12.2]

Gore JA. (1989) Models for predicting benthic macroinvertebrate habitat suitability under regulated flows. In: Gore JA, Petts GE (eds) *Alternatives in Regulated River Management*, pp. 253–66. CRC Press, Boca Raton, Florida. [12.2]

Gore JA, Bryant RM Jr. (1990) Temporal shifts in physical habitat of the crayfish, *Orconectes neglectus* (Faxon). *Hydrobiologia* **199**: 131–42. [12.2]

Grant JWA, Kramer DL. (1990) Territory size as a predictor of the upper limit to population density of juvenile salmonids in streams. *Canadian Journal of Fisheries and Aquatic Sciences* **47**: 1724–37. [12.2]

Gros P, Prozet P. (1988) Forecasting stochastic model of spring salmon catches (*Salmo salar*) on Alune River, Brittany, France. Elements of fishery management. *Acta Oecologica Applicata* **9**: 3–24. [12.2]

Gulati RD, Lammens EHRR, Meijer M-L, Donk van E. (eds) (1990) Biomanipulation – tools for water management. *Hydrobiologia* **200/201**: 628. [12.2]

Gulland JA. (ed) (1977) *Fish Population Dynamics*. John Wiley, London. [12.2]

Gurney WSC, McCauley E, Nisbet RM, Murdoch WW. (1990) The physiological ecology of *Daphnia*, a dynamic model of growth and reproduction. *Ecology* **71**: 716–32. [12.3]

Guyomard R. (1989) Principaux concepts et methodes relatifs à la description de la diversité génétique d'une espêce. *Bulletin Francais de la Pêche et de la Pisculture* **314**: 98–108. [12.1]

Hall RJ, Likens GE, Fiance SB, Hendrey GR. (1980) Experimental acidification of a stream in the Hubbard Brook Experimental Forest, New Hampshire. *Ecology* **61**: 976–89. [12.2, 12.3]

Hallam TG, Lassiter RR, Li J, Suarez LA. (1990) Modelling individuals employing an integrated energy response: application to *Daphnia*. *Ecology* **71**(3): 938–54. [12.3]

Harding JPC, Whitton BA. (1981) Accumulation of zinc cadmium and lead by field populations of *Lemanea*. *Water Research* **15**: 301–19. [12.2]

Haslam SM. (1982) A proposed method for monitoring river pollution using macrophytes. *Environmental Technology Letters* **3**: 19–34. [12.2]

Hedgpeth JW. (1977) Models and muddles – some philosophical observations. *Helgoländer Wissenschaftliche Meeresuntersuchungen Helgoland* **30**: 91–104. [12.4]

Hellawell JM. (1989) *Biological Indicators of Freshwater Pollution and Environmental Management.* Elsevier Applied Science, London. [12.2]

Hilden M, Kettunen J. (1985) Models for the assessment of environmental impacts on fish populations and fisheries. *Finnish Fish News* **6**: 35–42. [12.2, 12.3]

Holmes RW, Whiting MC, Stodard JL. (1989) Changes in diatom-inferred pH and acid neutralizing capacity in a dilute, high elevation, Sierra Nevada lake since A.D. 1825. *Freshwater Biology* **21**: 295–310. [12.2]

Horwood JW, Shepherd JG. (1981) The sensitivity of age structured populations to environmental variability. *Mathematical Biosciences* **57**: 59–82. [12.2]

Hosper SH, Jagtman E. (1990) Biomanipulation additional to nutrient control for restoration of shallow lakes in the Netherlands. *Hydrobiologia* **200/201**: 523–34. [12.2]

Hughes NF, Dill LM. (1990) Position choice by drift feeding salmonids: model and test for Arctic grayling (*Thymallus arcticus*) in subarctic mountain streams, interior Alaska. *Canadian Journal of Fisheries and Aquatic Sciences* **47**: 2039–48. [12.2]

Hulst R van. (1979) On the dynamics of vegetation: Markov chains as models of succession. *Vegetatio* **40**: 3–14. [12.2]

Jackson PBN. (1989) Prediction of regulation effects on natural biological rhythms in South-Central African freshwater fish. *Regulated Rivers: Research and Management* **3**: 205–20. [12.2]

Janovy J Jr, Ferdig MT, McDowell MA. (1990) A model of dynamic behaviour of a parasite species assemblage. *Journal of Theoretical Biology* **142**: 517–30. [12.2]

Jeffers JNR. (1982) *Modelling*. Chapman & Hall, London. [12.1]

Karr JR, Fausch KD, Angermeier PL, Yant PR, Schlosser IJ. (1986) *Assessing Biological Integrity in Running Waters. A Method and its Rationale.* Special Publication No. 5, Illinois Natural History Survey, Champaign, Illinois. [12.2]

Kelso JRM, Shaw MA, Minns CK, Mills KH. (1990) An evaluation of the effects of atmospheric acidic deposition on fish and the fishery resource of Canada. *Canadian Journal of Fisheries and Aquatic Sciences* **47**: 644–55. [12.2]

Kerr SR. (1982) Estimating the energy budgets of actively predatory fishes. *Canadian Journal of Fisheries and Aquatic Sciences* **39**: 371–9. [12.2]

Kitchell JF, Stewart DJ, Weininger D. (1977) Applications of a bioenergetics model to yellow perch (*Perca flavescens*) and walleye (*Stizostedion vitreum vitreum*). *Journal of the Fisheries Research Board of Canada* **34**: 1922–35. [12.2]

Kokkinn MJ. (1990) Is the rate of embryonic development a predictor of overall development rate in

Tanytarsus barbitarsis Freeman (Diptera: Chironomidae)? *Australian Journal of Marine and Freshwater Research* **41**(5): 575–9. [12.3]

Kolkwitz R, Marsson M. (1909) Oekologie der tierischen Saprobien. *Internationale Revue der gesamten Hydrobiologie und Hydrographie* **2**: 126–52. [12.2]

Lacroix GL. (1987) Model for loss of atlantic salmon from acidic brown waters of Canada. In: Perry R, Harrison RM, Bell JNB, Lester JN (eds) *Acid Rain: Scientific and Chemical Advances*, pp. 516–22. Selper, London. [12.2]

Larsen DP, de Noyelles F, Stay F, Shiroyama T. (1986) Comparison of single species microcosm and experimental pond responses to atrazine exposure. *Environmental Toxicology and Chemistry* **5**: 179–90. [12.2]

Leps J, Soldan T, Landa V. (1990) Prediction of changes in ephemeropteran communities – a transition matrix approach. In: Campbell IC (ed) *Mayflies & Stoneflies: Life Histories and Biology*, pp. 281–7. Kluwer, Dordrecht. [12.2]

Lugo AE. (1978) Stress and ecosystems. In: Thorp JH, Gibbons JW (eds) *Energy and Environmental Stress in Aquatic Systems*, pp. 62–101. Department of Energy Symposium Series 78, National Technical Information Service, Springfield, Virginia. [12.3]

McCauley E, Murdoch WW, Nisbet RM, Gurney WSC. (1990) The physiological ecology of *Daphnia*: development of a model of growth and reproduction. *Ecology* **71**: 703–15. [12.3]

MacMurray HL, Jaeggi MNR. (1990) Modeling erosion of sand and silt bed river. *Journal of Hydraulic Engineering* **116**(9): 1080–9. [12.2]

Maltby L, Calow P. (1989) The application of bioassays in the resolution of environmental problems: past, present and future. *Hydrobiologia* **188/189**: 65–6. [12.2]

Marmorek DR, Jones ML, Minns CK, Elder FC. (1990) Assessing the potential extent of damage to inland lakes in eastern Canada due to acidic deposition. I. Development and evaluation of a simple 'site' model. *Canadian Journal of Fisheries and Aquatic Sciences* **47**: 55–66. [12.2]

Matthews RA, Buikema LA Jr, Cairns J Jr, Rodgers JH Jr. (1982) Biological monitoring. Part IIA – Receiving system functional methods, relationships and indices. *Water Research* **16**: 129–39. [12.2]

Meisner JD. (1990) Effect of climatic warming on the southern margins of the native range of brook trout, *Salvelinus fontinalis*. *Canadian Journal of Fisheries and Aquatic Sciences* **47**: 1065–70. [12.2]

Messier F, Virgl JA, Marinelli L. (1990) Density-dependent habitat selection in muskrats: a test of the ideal free distribution model. *Oecologia* **84**(3): 380–5. [12.2]

Metcalfe-Smith JL. (1994) Biological water-quality assessment of rivers: use of macroinvertebrate communities. In: Calow P, Petts GE (eds) *The Rivers Handbook*, vol. 2, pp. 144–70. Blackwell Scientific Publications, Oxford. [12.2]

Milner NJ, Hemsworth RJ, Jones BE. (1985) Habitat evaluation as a fisheries management tool. *Journal of Fish Biology, Suppl. A* **27**: 85–108. [12.2]

Minns CK, Kelso JRM, Johnson MG. (1986) Large-scale risk assessment of acid rain impacts on fisheries: models and lessons. *Canadian Journal of Fisheries and Aquatic Sciences* **43**: 900–21. [12.2]

Morin A, Back C, Chalifour A, Boisvert J, Peters RH. (1988a) Empirical models predicting ingestion rates of black fly larvae. *Canadian Journal of Fisheries and Aquatic Sciences* **45**: 1711–19. [12.2]

Morin A, Constantin M, Peters RH. (1988b) Allometric models of simuliid growth rates and their use for estimation of production. *Canadian Journal of Fisheries and Aquatic Sciences* **45**: 315–24. [12.2]

Moss D, Furse MT, Wright JF, Armitage PD. (1987) The prediction of the macroinvertebrate fauna of unpolluted running-water sites in Great Britain using environmental data. *Freshwater Biology* **17**(1): 41–52. [12.2]

Munawar M, Dixon G, Mayfield CI, Reynoldson T, Sadar MH. (eds) (1989) Environmental bioassay techniques and their application. *Hydrobiologia* **188/189**: 1–680. [12.2]

Newman RM. (1987) Comparison of encounter model predictions with observed size-selectivity by stream trout. *Journal of the North American Benthological Society* **6**: 56–64. [12.2]

Niimi AJ, Beamish FWH. (1974) Bioenergetics and growth of largemouth bass (*Micropterus salmoides*) in relation to body weight and temperature. *Canadian Journal of Zoology* **52**: 447–56. [12.2]

Ormerod SJ, Tyler SJ. (1989) Long-term change in the suitability of Welsh streams for Dippers *Cinclus cinclus* as a result of acidification and recovery: a modelling study. *Environmental Pollution* **62**: 171–82. [12.2]

Ormerod SJ, Boole P, McCahon CP, Weatherley NS, Pascoe D, Edwards RW. (1987) Short-term experimental acidification of a Welsh stream: comparing the biological effects of hydrogen ions and aluminium. *Freshwater Biology* **17**: 341–56. [12.2]

Ormerod SJ, Weatherley NS, Varallo PV, Whitehead PG. (1988) Preliminary empirical models of the historical and future impact of acidification on the ecology of Welsh streams. *Freshwater Biology* **20**: 127–40. [12.2]

Ormerod SJ, Weatherley NS, Gee AS. (1990a) Modelling the ecological impact of changing acidity in Welsh streams. In: Edwards RW, Gee AS, Stoner JH (eds) *Acid Waters in Wales*, pp. 279–97. Kluwer, Dordrecht. [12.2]

Ormerod SJ, Weatherley NS, Merrett WJ, Gee AS, Whitehead PG. (1990b) Restoring acidified streams in upland Wales: a modelling comparison of the chemical and biological effects of liming and reduced sulphate deposition. *Environmental Pollution* **64**: 67–85. [12.2, 12.3]

Overrein LN, Seip HM, Tollan A. (1981) *Acid Precipitation – Effects on Forest and Fish.* Final report of the SNSF-project 1972–1980. SNSF, Oslo. [12.2]

Painter DS, Jackson MB. (1990) Cladophora internal phosphorus modeling: verification. *Journal of Great Lakes Research* **15**(4): 700–8. [12.2]

Patrick R. (1954) Diatoms as an indication of river change. In: *Proceedings of the 9th Industrial Waste Conference, Purdue University.* Engineering Extension Series No. 87, pp. 325–30. [12.2]

Pearsall SH, Durham D, Eagar DC. (1986) Evaluation methods in the United States. In: Usher MB (ed) *Wildlife Conservation Evaluation*, pp. 111–33. Chapman & Hall, London. [12.2]

Pella JJ, Tomlinson PK. (1969) A generalised stock production model. *Bulletin of the Inter-American Tropical Tuna Commission* **13**: 421–96. [12.2]

Perry JA, Troelstrup NH Jr, Newsom M, Shelley B. (1987) Whole ecosystem manipulation experiments: the search for generality. *Water Science Technology* **19**: 55–71. [12.3]

Peters RH. (1986) The role of prediction in limnology. *Limnology and Oceanography* **31**: 1143–59. [12.1, 12.4]

Petts GE. (1987) Time scales for ecological change in regulated rivers. In: Craig JF, Kemper JB (eds) *Regulated Streams – Advances in Ecology*, pp. 257–66. Plenum Press, New York. [12.2]

Petts GE, Müller H, Roux AL. (eds) (1989) *Historical Change of Large Alluvial Rivers: Western Europe.* John Wiley & Sons, Chichester. [12.2]

Plafkin JL, Barbour MT, Porter KD, Gross SK, Hughes RM. (1989) *Rapid Bioassessment Protocols for Use in Streams and Rivers: Benthic Macroinvertebrates and Fish.* Report EPA/444/4-89-001. USEPA, Washington, DC. [12.2]

Pockl M, Timischl W. (1990) Comparative study of mathematical models for the relationship between water temperature and brood development time of *Gammarus fossarum* and *G. roeseli* (Crustacea: Amphipoda). *Freshwater Biology* **23**(3): 433–40. [12.3]

Poff NL, Ward JV. (1990) Physical habitat template of lotic systems: recovery in the context of historical pattern of spatiotemporal heterogeneity. *Environmental Management* **14**: 629–46. [12.3]

Preall RJ, Ringler NH. (1989) Comparison of actual and potential growth rates of brown trout (*Salmo trutta*) in natural streams based on bioenergetic models. *Canadian Journal of Fisheries and Aquatic Sciences* **46**: 1067–76. [12.2]

Pringle CM, Naiman RJ, Bretschko G, Karr JR, Oswood MW, Webster JR, Welcomme RL, Winterbourn MJ. (1988) Patch dynamics in lotic systems: the stream as a mosaic. *Journal of the North American Benthological Society* **7**: 503–24. [12.3]

Rader RB, Ward JV. (1989) The influence of environmental predictability/disturbance characteristics on the structure of a guild of mountain stream insects. *Oikos* **54**: 107–16. [12.3]

Radford PJ. (1988) Model-monitoring relationships. In: Salmons W, Bayne B, Dursma E, Foestrer V (eds) *Pollution of the North Sea: An Assessment*, pp. 666–75. Springer Publishing, New York. [12.1]

Rapport DJ, Regier HA, Hutchinson TC. (1985) Ecosystem behavior under stress. *The American Naturalist* **125**: 617–40. [12.3]

Reckhow KH, Black RW, Stockton TB, Vogt JD, Wood JG. (1987) Empirical models of fish response to lake acidification. *Canadian Journal of Fisheries and Aquatic Sciences* **44**: 1432–42. [12.2]

Reed DW. (1987) UK flood forecasting in the 1980's. In: Collinge VK, Kirby C (eds) *Weather Radar and Flood Forecasting*, pp. 129–42. John Wiley, Chichester. [12.1, 12.2]

Rempel RS, Carter JCH. (1987) Modelling the effects of accelerated development and reduced fecundity on the population dynamics of aquatic diptera. *Canadian Journal of Fisheries and Aquatic Sciences* **44**: 1737–42. [12.2]

Resh VH, Brown AV, Covich AP, Gurtz ME, Li HW, Minchal GW, Reice, SR, Sheldon AL, Wallace JB, Wissmar RC. (1988) The role of disturbance in stream ecology. *Journal of the North American Benthological Society* **7**: 433–55. [12.3]

Reuss JO, Christophersen N, Seip HM. (1986) A critique of models for freshwater and soil acidification. *Water, Air and Soil Pollution* **30**: 909–30. [12.2]

Reynolds CS, Glaister MS. (1993) Spatial and temporal changes in phytoplankton abundance in the upper and middle reaches of the River Severn. *Archiv für Hydrobiologie.* [12.2]

Ricker WE. (1958) Handbook of computations for biological statistics of fish populations. *Bulletin of the Fisheries Research Board of Canada* **119**: 1–300. [12.2]

Rumeau A, Coste M. (1988) Initiation à la systémique des diatomées d'eau douce pour l'utilisation pratique d'un indice diatomique générique. *Bulletin Francais de la Peche et de la Pisciculture* **309**: 1–69. [12.2]

Rutherford JC. (1975) *Simulation of Water Quality in the Waikato and Tarawera Rivers.* Report 119, School of Engineering, University of Auckland, Auckland, New Zealand. [12.2]

Rutt GP, Weatherley NS, Ormerod SJ. (1990) Relationships between the physicochemistry and macroinvertebrates of British upland streams: the development of modelling and indicator systems for predicting fauna and detecting acidity. *Freshwater Biology* **24**: 463–80. [12.2]

Sadler K. (1983) A model relating the results of low pH bioassay experiments to fishery status in Norwegian lakes. *Freshwater Biology* **13**: 453–63. [12.2]

Saltveit SJ, Brabrand A. (1987) Predicting the effects of a possible temperature increase due to stream regulation on the eggs of whitefish (*Coregonus lavaretus*). In: Craig JF, Kemper JB (eds) *Regulated Streams – Advances in Ecology*, pp. 219–28. Plenum Press, New York. [12.2]

Scheffer M. (1990) Multiplicity of stable states in freshwater systems. *Hydrobiologia* **200/201**: 475–86. [12.2]

Schindler DW. (1988) Effects of acid rain on freshwater ecosystem. *Science* **239**: 149–57. [12.2]

Schnute J. (1977) Improved estimates from the Schaefer production model; theoretical considerations. *Journal of the Fisheries Research Board of Canada* **34**: 583–603. [12.2]

Sheldon AL. (1984) Colonization dynamics of aquatic insects. In: Resh VH, Rosenberg DM (eds) *The Ecology of Aquatic Insects*, pp. 401–29. Praeger, New York. [12.2]

Shorrocks B, Rosewell J. (1987) Spatial pathiness and community structure: coexistence and guild size of drosophilids on ephemeral resources. In: Gee JHR, Giller PS (eds) *Organisation of Communities: Past and Present*, pp. 29–51. Blackwell Scientific Publications, Oxford. [12.3]

Shuter BJ, Wismer DA, Regier HA, Matuszek JE. (1985) An application of ecological modelling: impact of thermal effluent on a smallmouth bass population. *Transactions of the American Fisheries Society* **114**: 631–51. [12.2]

Sladecek V. (1981) Indicator value of the genus *Opercularia* (Ciliata). *Hydrobiologia* **79**: 229–32. [12.2]

Slooff W. (1985) The role of multispecies testing in aquatic toxicology. In Cairns Jr J (ed) *Multispecies Toxicity Testing*, pp. 45–60. Pergamon Press, New York. [12.2]

Small MJ, Sutton MC. (1986) A direct distribution model for regional aquatic acidification. *Water Resources Research* **22**: 1749–58. [12.2]

Solomon DJ, Brafield AE. (1972) The energetics of feeding, metabolism and growth of perch (*Perca fluviatilis* L.). *Journal of Animal Ecology* **41**: 699–718. [12.2]

Stalnaker CB. (1994) Evolution of instream flow habitat modelling. In: Calow P, Petts GE (eds) *The Rivers Handbook*, vol. 2, pp. 276–86. Blackwell Scientific Publications, Oxford. [12.1, 12.2, 12.4]

Statzner B, Gore JA, Resh VH. (1988) Hydraulic stream ecology, observed patterns and potential applications. *Journal of the North American Benthological Society* **7**: 307–60. [12.2, 12.4]

Svadlenkova M, Dvorak Z, Slavik O. (1989) A mathematical model for Cs uptake and release by filamentous algae. *Internationale Revue der gesamten Hydrobiologie* **74**(5): 461–9. [12.2]

Taylor AD. (1990) Metapopulations, dispersal and predator–prey dynamics: an overview. *Ecology* **71**: 429–33. [12.3]

Tipping E. (1989) Acid-sensitive waters of the English Lake District; a steady-state model of stream water chemistry in the Upper Duddon catchment. *Environmental Pollution* **60**: 181–208. [12.1]

Townsend CR. (1989) The patch dynamics concept of stream community ecology. *Journal of the North American Benthological Society* **8**: 36–50. [12.3]

Trapp S, Brueggeman R, Muenzer B. (1990) Estimation of releases into rivers with the steady state surface water model EXWAT using dichloromethane. *Ecotoxicology and Environmental Safety* **19**: 72–80. [12.1, 12.2]

Usher MB. (1981) Modelling ecological succession, with particular reference to Markovian models. *Vegetatio* **46**: 11–18. [12.2]

Virtanen M. (1989) Mathematical models within river basin management – an overview. *Aqua Fennica* **19**(2): 145–52. [12.2]

Vollenweider RA. (1975) Input–output models with special reference to the phosphorus loading concept in limnology. *Schweizerische Zeitschrift für Hydrobiologie* **37**: 53–84. [12.2]

Vøllestad LA, Jonsson B. (1988) A 13 year study of the population dynamics and growth of the European eel *Anguilla anguilla* in a Norwegian river: evidence for density dependent mortality and development of a model for predicting yield. *Journal of Animal Ecology* **57**: 983–97. [12.2]

Warwick WF. (1980) Palaeolimnology of the Bay of Quinte, Lake Ontario: 2800 years of cultural influence. *Canadian Bulletin of Fisheries and Aquatic Sciences* **206**: 1–117. [12.2]

Weatherley NS, Ormerod SJ. (1987) The impact of acidification on macroinvertebrate assemblages in Welsh streams: towards an empirical model. *Environmental Pollution* **46**: 223–40. [12.2]

Weatherley NS, Ormerod SJ. (1990) The constancy of invertebrate assemblages in soft-water streams: implications for the prediction and detection of environmental change. *Journal of Applied Ecology* **27**: 952–64. [12.2, 12.3]

Welch EB, Horner RR, Patmont CR. (1989) Prediction of nuisance periphytic biomass: a management approach. *Water Research* **23**: 401–5. [12.2]

Wesche TA. (1985) Stream channel modifications and reclamation structures to enhance fish habitat. In: Gore JA (ed) *The Restoration of Rivers and Streams – Theories and Experience*, pp. 103–64. Butterworth Publishers, Boston. [12.2]

Westlake DF. (1966) A model for quantitative studies of

photosynthesis by higher plants in streams. *Air and Water Pollution International Journal* **10**: 883–96. [12.2]

Wildhaber ML, Crowder LB. (1990) Testing a bioenergetics-based habitat choice model: bluegill (*Lepomis macrochirus*) responses to food availability and temperature. *Canadian Journal of Fisheries and Aquatic Sciences* **47**: 1664–71. [12.2]

Wilson MW, Botkin DB. (1990) Models of sample microcosms: emergent properties and the effect of complexity on stability. *The American Naturalist* **135**: 414–34. [12.3]

Winget RN. (1985) Methods for determining successful reclamation of stream ecosystems. In: Gore JA (ed) *The Restoration of Rivers and Streams – Theories and Experience*, pp. 165–92. Butterworth Publishers, Boston. [12.2]

Woltering DM. (1985) Population responses to chemical exposure in aquatic multispecies systems. In: Cairns Jr J (ed) *Multispecies Toxicity Testing*, pp. 61–75. Pergamon Press, New York. [12.2]

Woolhouse MEJ, Chandiwana SK. (1990) Population dynamics model for *Bulinus globosus* intermediate host for *Schistosoma haematobium* in river habitats. *Acta Tropica* **47**: 151–60. [12.2]

Wright DI, O'Brien WJ. (1984) The development and field test of a tactical model of the planktivorous feeding of white crappie (*Pomoxis annularis*). *Ecological Monographs* **54**: 65–98. [12.2]

Wright JF, Armitage PD, Furse MT, Moss D. (1989) Prediction of invertebrate communities using stream measurements. *Regulated Rivers: Research and Management* **4**: 147–55. [12.2]

Yodzis P. (1986) Competition, mortality and community structure. In: Diamond J, Case TJ (eds) *Community Ecology*, pp. 480–91. Harper & Row, New York. [12.3]

Yount JD, Niemi GJ. (eds) (1990) Recovery of lotic communities and ecosystems following disturbance: theory and application. *Environmental Management* **14**: 515–762. [12.2]

Index

Note: Page numbers in *italics* refer to pages with figures and/or tables.